A Practical Guide to Welding Solutions

A Practical Guide to Welding Solutions

Overcoming Technical and Material-Specific Issues

Robert W. Messler, Jr.

WILEY-VCH

Author

Robert W. Messler, Jr.
Emeritus Professor
Materials Science & Engineering
Rensselaer Polytechnic Institute
Ballston Lake, NY
United States

Cover
© Marine Construction Photos/Alamy Stock Photo

Library of Congress Card No.:
applied for

British Library Cataloguing-in-Publication Data
A catalogue record for this book is available from the British Library.

Bibliographic information published by the Deutsche Nationalbibliothek
The Deutsche Nationalbibliothek lists this publication in the Deutsche Nationalbibliografie; detailed bibliographic data are available on the Internet at <http://dnb.d-nb.de>.

Print ISBN: 978-3-527-34543-4
ePDF ISBN: 978-3-527-81878-5
ePub ISBN: 978-3-527-81880-8
oBook ISBN: 978-3-527-81881-5

Typesetting SPi Global, Chennai, India
Printing and Binding

Printed on acid-free paper

10 9 8 7 6 5 4 3 2 1

Contents

Preface

A convincing argument can be made that joining by mechanical fastening or design feature interlocking, adhesive bonding, or welding, including brazing and soldering, is the most important process in manufacturing and most construction because it usually occurs after a considerable value has already been added to produce the near-net-shaped detail parts for assembly. As welding is typically the most technically elaborate, requiring the most skilled practitioners, and accounts for about half of all joining, it almost certainly accounts for much more than half of all joining by value.[1] Hence, any problem associated with welding that cannot be avoided or resolved is serious.

Major industrial sectors dependent on welding worldwide, as of 2017, included, in a descending order, the following: energy (~23%), construction (~20.5%), transportation (~19.5%), process and others (~15%), heavy machinery (~11%), ship building (~8%), and aerospace and defense (~3%). Major applications include, in alphabetical order, the following: agricultural equipment; aircraft; airport support equipment; automobiles; bridges; buildings; chemical-processing equipment; earthmoving equipment; food- and beverage-processing equipment; gas, oil, and water pipelines; heavy machines; locomotives and railcars; marine power plants; mining equipment; oil and gas drilling and recovery equipment; petroleum-processing equipment; pharmaceutical-processing equipment; power generation equipment; railroad cars and track equipment; ships; and trucks, buses, and RVs. Although digital electronics has transformed *how* we live, welding has enabled us *to* live in our modern world.

Understanding how welding works in terms of various processes and the metallurgy is well treated in books, colleges, and continuing education. Far less well treated is why welding sometimes does not work as expected or needed, what problems can occur, and, most importantly, how to solve problems that do appear. What every practitioner of welding needs to know – and often most aggressively seeks – is how to avoid problems if possible and how to resolve problems once they have occurred.

A Practical Guide to Welding Solutions: Overcoming Technical and Material-Specific Issues is a first! It focuses precisely on what everyone involved with

1 The total global market for welding consumables alone (e.g. stick electrodes and solid and flux-cored wires) exceeded $15B in 2017 and has been forecast to grow by 1.7× over the next decade. As the cost of labor far exceeds the cost of consumables in welding, with consumables typically accounting for only around 5% of the total cost, the total annual value of welding is staggering.

welding wants to know, i.e. how to identify the root cause of a weld-induced or welding-related problem. To be most practical, problems are treated in three ways one may encounter them: (i) by how they manifest themselves (e.g. as distortion or defects); (ii) by where they are located (e.g. in the weld metal or in the surrounding heat-affected zone); and (iii) by the specific material in which certain problems are most apt to occur or exclusively occur (e.g. brittle martensite formation in hardenable steels or sensitization of stainless steels).

A Practical Guide to Welding Solutions is the latest, if not the last, of the eight technical books I have written over the past 25 years. It has taken me that long to fully appreciate that engineers solve problems, so they need a book that cuts-to-the-chase by focusing on solving problems with welding, welds, and weldments.

My deepest thanks to my assistant editor, Lesley Jebaraj, for all the hard work and patience with a finicky author, and to my production editor, Vishnu Priya, for her talents. Thanks too to my editor, Martin Preuss, for his enthusiasm for my idea for this book.

Sorry so late. However, as they say: Better late than never!

July 2018

Robert W. Messler, Jr., Ph.D.
FASM, FAWS
Emeritus Professor
Materials Science & Engineering
Rensselaer Polytechnic Institute

1
Introduction

Welding as a process for joining materials, in general, and metals and alloys, in particular, is a double-edged sword.[1] On the one hand, welding offers one of the best methods for obtaining joints with strength comparable to (or even superior to) the physical elements being joined, with a lesser weight penalty than mechanical fastening (e.g. bolting or riveting) and a greater environmental durability than adhesive bonding (whether using organic adhesives, such as epoxies, or inorganic adhesives, such as cement). It also offers one of the assured ways of achieving leak tightness against fluids (i.e. gases and liquids), can be performed indoors or outdoors, manually or automatically (using mechanization or robots) using a wide variety of process embodiments, and, for better or worse, produces joints that are permanent.[2] On the other hand, the use of welding always demands thoughtful structures and joint designs, proper equipment and consumables (e.g. shielding gases or fluxes and fillers), skilled operators, appropriate quality assurance for joint performance demands, and, most importantly, an understanding of what it takes to produce a sound weld. The latter requirement typically leads to most problems encountered with welding.

Problems with welding normally relate to unacceptable welds, i.e. welds that fail to pass nondestructive evaluation immediately following their production *or* welds that fail to provide intended functions in service. Some examples of the former include welded assemblies that fail to meet the geometric and dimensional criteria (i.e. do not provide needed fit and/or function), welds that contain surface or internal flaws or defects that fail to meet the required quality specifications (e.g. freedom from cracks and freedom from porosity), or welds that degraded the base material components (e.g. because of cracking, severe oxidation, hardness loss,

1 From the notion that if two sides of the same blade are sharp, it cuts both ways. The metaphor may have originated in Arabic, in the expression حـدين ذو ســيف (sayf zou hadayn, "**double-edged sword**"), but it is first attested in English in the fifteenth century.

2 The permanency of a joint is desirable only if a structure is never intended to be disassembled, particularly without destroying the components of the assembly or having the process employed be simple. The inability to disassemble the welded components of the #4 light water graphite-moderated nuclear reactor at the Chernobyl Nuclear Power Plant near Pripyat in Ukraine, which suffered catastrophic failure on 26 April 1986 (Figure 1.1), is a prime example, as highly radioactive remains had to be entombed in a massive concrete "Object Shelter" (Figure 1.2), or sarcophagus, for the next 1000 years, as the massive welded containment vessel could not be disassembled!

A Practical Guide to Welding Solutions: Overcoming Technical and Material-Specific Issues,
First Edition. Robert W. Messler, Jr.

Figure 1.1 Remains of the #4 light water graphite-moderated reactor unit at the Chernobyl Nuclear Power Plant near Pripyat in Ukraine shortly after it catastrophically failed on 26 April 1986, because of a series of errors by Soviet operators during safety check tests. Source: Photograph by an unknown source posted by Garvey STS on en.wikibooks.org. Freely used under Creative Commons ShareAlike CC BY-SA 4.0.

or, contrarily, embrittlement). Also not to be ignored are welds that do not look good, as, in welding, internal "beauty" (i.e. quality) is often related to external "beauty" (i.e. appearance) the reason being a lack of care in welding, in particular, suggest a lack of care in manufacturing, in general, and, ultimately, a lack of care in design, marketing, senior management, etc. Very typically, the quality of an organization, starts at the top, with leadership by example meaning more than rules and regulations.

Figure 1.2 After several iterations to safely contain the highly radioactive remnants of the #4 reactor unit at the Chernobyl Nuclear Power Plant, the current New Safe Confinement or NSC was in position as of October 2017. Entombment was necessary because much of the steel structure used in the reactor containment vessel could not be disassembled as it was welded to be permanent and because it is highly radioactive. Besides safe containment of radiation, the €1.5B structure prevents damage by weather and runoff of lingering radioactive contamination. Source: Wikipedia.com "Chernobyl new safe confinement". Freely used under CC BY-SA 4.0; posted by Tim Porter on 13 October 2017.

Figure 1.3a,b shows a couple of examples of extremely well-executed welds made in stainless steel and an Al alloy using the gas tungsten arc process with a filler wire, whereas Figure 1.3c shows a very badly executed repair weld on a steel automobile part, and Figure 1.3d shows a badly factory-made gas–metal arc repair weld on an Al alloy boat.

To date, books that deal with the welding of metals and alloys, at least, have been found to deal with one or the other of (i) the processes employed to make welds *or* (ii) the metallurgy that underlies welding (i.e. welding metallurgy).[3] The former seldom, if ever, mention problems with welding or welds, as welding, not welds, is their purpose. The latter typically spend the first 80–90% of the book presenting the underlying physical metallurgy that allow welds to be made in metals and alloys in the first place and that can produce sound structure in the weld (i.e. fusion zone and surrounding heat-affected zone using fusion-welding processes), if everything is done properly. The remaining 10–20% on what can go wrong, how to detect such short-comings, and, finally, how to resolve any short-coming(s). It is almost as if the author is telling a story and carefully avoiding the outcome, as in a mystery. Not surprisingly, producing high-quality welds through welding *is* a mystery for many users.

3 A notable exception is *Principles of Welding: Processes, Physics, Chemistry and Metallurgy* by R.W. Messler, Jr., 1st ed., 24 March 1999, Wiley VCH, ISBN-13: 978-0471253761 or ISBN-10: 0471253766.

Figure 1.3 Two examples of superbly made fusion arc welds using manual gas tungsten arc (TIG) welding in (a) steel fittings and (b) Al alloy bicycle frames. In addition, two badly made welds: using a gas–metal arc to repair (c) a steel automobile part and (d) an Al alloy boat structure at a factory repair shop, with neither source being identified – fortunately for them – as images are in the public domain! Source: The former by Scott Raabe at his Clean Cut Metal Works, Houston, TX, USA and used with his kind permission; and the latter on the website www.cycling.zanconato.com by Mike Zanconato at Zanconato Custom Cycles, Sutton, MA, USA and used with his kind permission.

This book will approach the problems with welding and the welds produced in a reverse order: starting with the problem(s) and working backward to the cause(s) and resolution(s).[4] As such, not to underestimate the ultimate importance of understanding the process (i.e. physics and chemistry) that is used to make a weld and, even more importantly, the physical metallurgy that underlies and enables the production of welds of sound quality and properties, but to simply deal with the nature of real-world engineering in which pragmatism often prevails over detailed understanding of principles, the reason for the rising

4 See Part IV, chapter 34, pp. 237–244 of *Engineering Problem-Solving 101: Time-Tested and Timeless Techniques* by R.W. Messler, Jr., 1st ed., 5 October 2012, McGraw-Hill Education, ISBN-13: 007199966 or ISBN-10: 007199966.

of various problems encountered will be covered briefly. Details will be left to the reader to seek information on welding metallurgy from other references. The rationale behind the approach of this book is as follows: engineers seek answers to problems and often achieve their goal(s) without having to delve into every detail. Every young engineer soon learns upon entering practice from college: the solution to a problem often only needs to be good enough, not perfect. A minimalist approach to engineering is often just as good as the minimalist approach used by a jockey to get a thoroughbred to win a race. Encourage the horse by clicks and chortles, tugs on the mane, and the light snap of a riding crop to increase the length of its stride at full gallop, without needing to know and understand all the details of equine physiology, like a veterinarian. After all, few veterinarians could ever ride a horse to victory in any race, no less in the Kentucky Derby!

This approach will work because those electing to use welding to create a structural assembly employ a backward problem-solving technique anyway. Knowing the end goal of a challenge (e.g. to get a man onto the Moon and back to the Earth safely), they work backward from the desired goal to identify the steps, methods, and procedures needed at each step to incrementally reach that goal from some given starting point. Regrettably, this enlightenment only dawns on young engineers once they leave engineering school, where most of what they are taught is the step-by-step process for reaching a goal by starting from first principles and seeing where the steps lead.

The reason a backward problem-solving approach often works, and often suffices, is that the first step will be to recognize the shortcoming (e.g. a severely distorted structure following cooling after welding; cracks in the fusion zone of a weld made in an austenitic stainless steel, such as type 304, using a recommended filler metal; cracks in the heat-affected zone of an arc weld made in a low-alloy steel that has been successfully welded before using the same process, same operators, and same parameters and procedures; and cracking in the base metal in some component after service). With this as a start, one could – and often does – begin to "troubleshoot" by checking each and every step for some potential cause–effect relationship. However, in this book, an organized collection of problems, categorized by the way in which they manifest themselves (i.e. distortion, cracking, porosity, and inclusions) and/or, in addition or alternatively, by where they are located (e.g. in the fusion zone with or without the use of a filler metal, in the high-temperature portion of the heat-affected zone or in the low-temperature portion of the heat-affected zone, and in the unaffected base metal), and/or a few problems predominantly, if not uniquely, associated with certain types of alloys (e.g. as-quenched martensite in hardenable steels; reheat cracking in some age-hardenable alloys) or, occasionally, pure metals (e.g. abnormal grain-growth or germination in cold-worked pure copper, as well as in some brasses), will guide the user toward the means of either avoiding such a problem in the future or, in some cases, attempting to resolve the already-present problem. Some of these latter-type problems may, in fact, be covered by manifestation or location but are covered here as well for easier searching by readers.

To facilitate problem solving, each chapter ends with a "Troubleshooting Guide." Each guide tabulates the problems covered therein, the most likely cause, and a suggested approach to correct the problem. As described in Chapter 2, the chapters are divided into groups (i.e. parts) in the following manner: (i) how the most commonly encountered problems manifest themselves (Part I), (ii) where the problems with welds arise in or around a weld by location (Part II), and (iii) what problems tend to arise only in (or most often in) specific materials (Part III). This arrangement is also intended to guide users during troubleshooting.

Once again, so as not to underestimate the importance of – and, hopefully, natural technical curiosity of – an engineer from seeking to understand *why* something happens, each problem addressed will include a brief explanation of the cause, with readers desiring more details being encouraged to refer to other references (such as those listed at the end of this chapter).

The goal of this book is simple: help practicing engineers practice their profession and achieve their desired and needed outcomes.

In conclusion, it should not go unnoticed that all professionals – medical doctors, surgeons, attorneys at law, dentists, veterinarians, etc. – refer to what they do as a "practice." Rather than any intent to downplay the rigor with which each obtains their formal education or to suggest any sense that what one does is take a stab in the dark to achieve the goal, what is meant is simply that only gets better and better at what they do as they do it – repeating and building upon successes and learning from and voiding any repeat of failures.

Further Reading

Althouse, A.D., Turnquist, C.H., Bowditch, W.A. et al. (2004). *Modern Welding*, 10e. London: Goodheart-Willcox.

American Welding Society (AWS) (2001–2015). *Welding Handbook*, 9, in five volumes. Miami, FL: AWS Volume 1 – Welding Science and Technology, 2001; Volume 2 – Welding Processes, Part 1, 2001; Volume 3 – Welding Processes, Part 2, 2007; Volume 4 – Materials and Applications, Part 1, 2011; Volume 5 – Materials and Applications, Part 2, 2015.

Cary, H.B. and Helzer, S. (2004). *Modern Welding Technology*, 6e. Hoboken, NJ: Pearson Education.

Easterling, K. (1992). *Introduction to the Physical Metallurgy of Welding*, 2e. Oxford: Butterworth-Heinemann.

Geary, D. and Miller, R. (2011). *Welding*, 2e. New York, NY: McGraw-Hill Education.

Granjon, H. (1991). *Fundamentals of Welding Metallurgy*. Cambridge: Abington Publishing/Woodhead Publishing.

Jeffus, L. (2016). *Welding Principles and Applications*, 8e. Boston, MA: Cengage Learning.

Kou, S. (2003). *Welding Metallurgy*, 2e. Hoboken, NJ: Wiley Interscience, Wiley.

Lancaster, J.F. (1993). *Metallurgy of Welding*, 6e. Cambridge: Woodhead Publishing.

Lippold, J.C. (2014). *Welding Metallurgy and Weldability*, 1e. Hoboken, NJ: Wiley.

Messler, R.W. Jr. (2004/1999). *Principles of Welding: Processes, Physics, Chemistry, and Metallurgy*. Weinheim/New York, NY: Wiley-VCH Verlag/Wiley.

Messler, R.W. Jr. (2004). *Joining of Materials and Structures: From Pragmatic Process to Enabling Technology*. Oxford: Elsevier Butterworth-Heinemann.

2

Categorization of Welding and Weld Problems

2.1 What Is Welding?

Common published and online dictionaries tend to define welding in a utilitarian sense, appropriately intended for the general public rather than for engineers, in general, and materials and welding engineers, in particular. A good and typical example found in *The Merriam-Webster Dictionary New Edition*,[1] advertised as "America's Best Selling Dictionary," defines *welding* as: "1.a. To unite (metallic parts) by heating and allowing the metals to flow together or by hammering or compressing with or without previous heat. b. To unite (plastics) in a similar manner by heating." Not bad, overall! In all cases, welding is a joining process intended to unite parts in order to create an assembly, usually (but not only) for structural applications. Beyond this, a specialist – like the author – could quibble over details. So, let us quibble for a moment.

Metals can, indeed, be welded – to one another! In fact, virtually, every solid metallic element, which only excludes mercury at room temperature,[2] can be welded to itself and to most other metals by one means or another, with widely varying degrees of difficulty. One means for forming a joint is to heat to melt all the metals to be joined, allowing the melts to intermix and, then, solidify to form the joint. The only pressure needed is that to hold the parts to be joined in contact throughout welding. Performing welding this way is known as *fusion welding*. Alternatively, a joint can be created by forcing metals together using enough pressure to cause plastic deformation in one or all joint elements, either at the macroscopic (i.e. gross) level or at the microscopic level. Performing welding this way is known as *non-fusion welding*.

As stated by the second part of the above definition (i.e. 1.b), plastics (more correctly, *polymers* and, most precisely, *thermoplastic polymers*) can also be joined by welding, quite correctly using some heat to cause softening, as opposed to

1 *The Merriam-Webster Dictionary New Edition*, 2016, Merriam-Webster, Springfield, MA. The parentheses in the definition shown are those of the publisher.
2 Mercury, atomic number 80 with symbol Hg, is a metallic element formerly named *hydrargyrum* (hence, the symbol Hg) and commonly called *quicksilver*, for its silvery luster and fluidity at room temperature. With a melting point of −37.89 °F/−38.83 °C, mercury is the only metallic element that is liquid at room temperature, joining only the nonmetallic element bromine, atomic number 35 with symbol Br, to have that physical state at room temperature. Bromine is a fuming red–brown liquid above 19.04 °F/−7.2 °C, which rapidly evaporates to form a like-colored gas.

A Practical Guide to Welding Solutions: Overcoming Technical and Material-Specific Issues, First Edition. Robert W. Messler, Jr.

melting. Once softened, the thermoplastics being joined move by a viscous flow under some applied pressure (at either a macroscopic or microscopic level) to unite by intimate contact and some slight intermixing in their plastic state. Welding of thermoplastic polymers is commonly referred to as *plastic welding* or, by specialists in polymers, *thermal bonding*.

Like thermoplastic polymers that progressively soften with increasing temperature, rather than melt to form a liquid from a solid at some point, glass materials (i.e. *glasses*) can also be welded, by what is commonly referred to as *fusing*. Many people have seen it performed by glassblowers attaching a handle or handles to a blown pitcher or vase or sticking a rod of glass to a figurine being made at a kiosk in a mall.

Finally, it is also possible to weld some, albeit not all, ceramic materials (i.e. *ceramics*). For those ceramics that melt (as opposed to sublime, directly transforming from a solid to a vapor, or that decompose upon heating at some point), certain fusion welding processes such as laser-beam welding and electron-beam welding can be employed. For others too high-melting (i.e. refractory) to melt or that sublime or decompose rather than melt, welding can sometimes be accomplished in the solid state using friction, which involves microscopic viscous flow in areas heated by friction. It would suffice to say that welding of crystalline ceramics is never easy and is sometimes impossible!

Figure 2.1 shows the (a) welding of metals (here, using the gas tungsten arc process), (b) welding or thermal bonding of a thermoplastic polymer (here, using a resistance-heated, metal-tipped tool, commonly referred to as "an iron"), and (c) welding or fusing of glass (here, two tubes made from lead (Pb) glass using a high-temperature oxy-fuel torch).

However, it is not possible to weld a material from one fundamental class (e.g. metal, ceramic, or polymer) to a material from another fundamental class. Let us find out why.

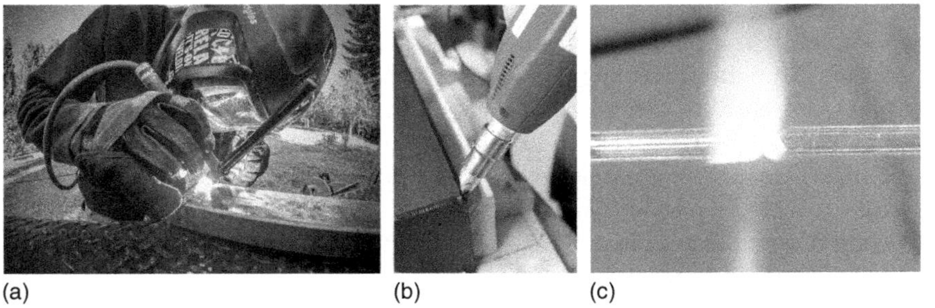

(a) (b) (c)

Figure 2.1 Photographs showing (a) the welding of a metal (here, using the gas tungsten arc process). Source: Photograph in "a" is from the Wikipedia article on "Welding," attributed to Prowelder87 and used under Creative Commons CC BY-SA 4.0. (b) The welding or thermal bonding of plastic (here, a thermoplastic polymer using a resistance-heated metal-tipped tool). Source: Photograph in "b" is used with permission of ECA Enterprises, Inc. (c) The welding or fusion of glass (here, using a high-temperature oxy-fuel torch to join two tubes made from leaded glass). Source: Photograph in "c" is from section on "Glass and plastic welding" under Wikipedia article on "Welding," attributed to Zaereth and used under Creative Commons CC BY-SA 4.0.

In a most general and unambiguous sense, welding is *a process in which materials of the same fundamental class or type are brought together and caused to coalesce through the formation of primary (and, occasionally, secondary) chemical bonds under the combined action of heat and pressure, without or with any filler* (Messler, 1993). Most important in this definition, and in the process, is the need for the materials being joined to *coalesce*, by which the materials are meant to "come together to form one continuous body," i.e. to obtain *material continuity*. This is made clear in the definition of welding found in the International Organization of Standardization (ISO) Standard R 857 (1958), which states that "Welding is an operation in which [material] continuity is obtained between parts for assembly, by various means," although, by this definition, the related processes of brazing and soldering would be considered to be accomplishing welding. Some would find this perfectly acceptable, whereas others would not. For the former, all three processes – welding, brazing, and soldering – accomplish joining metals through the formation of primary metallic bonds.[3] For the latter, bond formation (whether primary in nature, as is the case in metals, glasses, and ceramics,

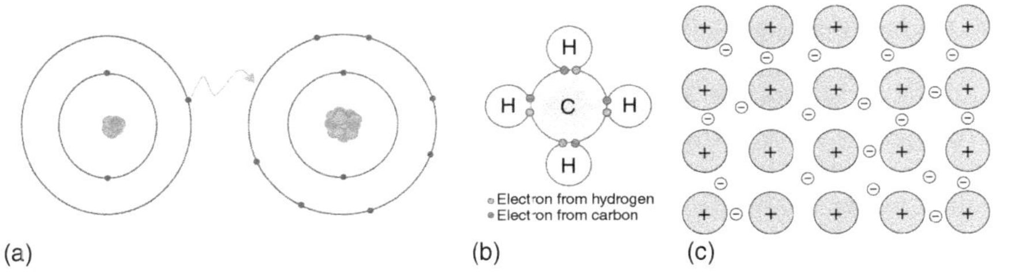

Figure 2.2 Schematic illustrations showing the three strong primary types of atomic bonding, i.e. ionic (a), covalent (b), and metallic (c). Ionic and covalent bonding (or a mix of the two) hold solid ceramics together. Metallic bonding (sometimes with a degree of covalent bonding) holds solid metallic materials (i.e. metals and alloys) together. Although strong covalent bonding holds the atoms in the long-chain C-based and Si-based polymers together, the chains are held together in the solid material by weaker secondary van der Waals bonds that involve either permanent, induced, or fluctuating dipoles. Source: All images are from Wikipedia, and are freely used under Creative Commons ShareAlike, CC BY SA 4.0, as follows: Ionic bonding by EliseEtc on 5 February 2012; covalent bonding by DynaBlast on 28 January 2006; and metallic bonding by Muskid on 2 May 2012.

3 There are three types of primary chemical bonding: (i) *ionic bonding*, which involves the formation of positive ions through the loss of one or more electrons by an atom of a metallic element and the formation of negative ions by the gain of one or more electrons by an atom of a nonmetallic element and, then, the electrostatic attraction of oppositely charged ions to form a compound known as ceramic; (ii) *covalent bonding*, which involves the intimate sharing of one, two, three, or four outermost (valence) electrons of like or dislike atoms of nonmetallic elements to form either molecules (such as diatomic gases such as oxygen, water, carbon dioxide, and methane) or solid materials (e.g. carbon diamond, some ceramics, and polymer macromolecules or chains); and (iii) *metallic bonding*, which involves extended sharing of outermost (conduction) electrons among atoms of metals to form solid, crystalline aggregates. Figure 2.2 shows these three types of primary atomic bonding (see footnote 5). Within ceramic materials, it is often the case that the bonding is mixed ionic–covalent, with the degree of one or the other varying across a wide spectrum.

or secondary,[4] as is the case in thermoplastic polymers) is considered critical for obtaining material continuity. The major difference is that the subprocesses of brazing and soldering are restricted in their use only to metals, whereas welding, as stated before, can be applied to glass, thermoplastic, and some ceramic materials as well.

The reason welding is restricted to joining materials from the same fundamental class is that the type of bonding inherent to each class is different: metallic bonding in metals, ionic or mixed ionic–covalent bonding in ceramics and glasses, and secondary molecular bonding in polymers.

In this book, the process of welding will be considered only when it is applied to metals and alloys. For this reason, before moving on, it is worth considering the achievement of material continuity in metals, i.e. metallic continuity, for the purpose of understanding some of the problems that can arise during welding and/or in welds.

It is fundamental to producing a weld between two or more pieces of material is to obtain *material continuity*. It suits the intent of this book to focus on obtaining continuity between crystalline (versus amorphous or semicrystalline) materials, in general, and between crystalline metals or alloys, in particular.[5]

As shown schematically in Figure 2.3, there are three distinct mechanisms for obtaining metallic continuity, as articulated by Granjon (1991), that are employed in welding by one process embodiment or another: (i) solid-phase plastic deformation, without or with dynamic recrystallization (as with cold welding and ultrasonic welding and with various forms of hot pressure or forge welding, respectively), (ii) diffusion (as with diffusion welding, sometimes called diffusion bonding), and (iii) melting and solidification (as with all forms of fusion welding using exothermic chemical or electric arc, plasma, beam, or resistance heat sources).

Each of these three mechanisms involves bringing the atoms of one substrate into intimate contact with the atoms of another substrate in order to allow the formation of metallic bonds in sufficiently large numbers to produce a strong joint in the final solid state. Atoms from one component can be forced to come into contact with the atoms of another component using pressure, with the pressure required being made lower by heating, as is the case of the so-called *pressure welding* in the broader class of *non-fusion welding*. Atom movement, and attendant bond formation, can be facilitated by diffusion during the process of dynamic recrystallization in hot pressure processes, including friction welding. An intermediary filler metal is seldom required in non-fusion welding, but may be useful when the substrate metals are difficult to plastically deform.

As an alternative to relying on plastic deformation to move atoms of mating substrates into intimate contact, atoms can be moved from one substrate to mix with atoms from another substrate by melting a portion (i.e. some small volume) of each substrate or, if necessary, a supplemental metal can be added in the molten

4 Secondary bonding arises from dipole attraction, whether dipoles are permanent (as in polar molecules), induced, or fluctuate. In all cases, secondary bonds are much weaker, by one or more orders of magnitude, than primary bonds. Secondary bonding, along with physical entangling, holds long-chain molecules together in a polymer.

5 The discussion actually pertains generally to all materials, but with some subtleties for noncrystalline types, such as thermoplastic polymers and glasses.

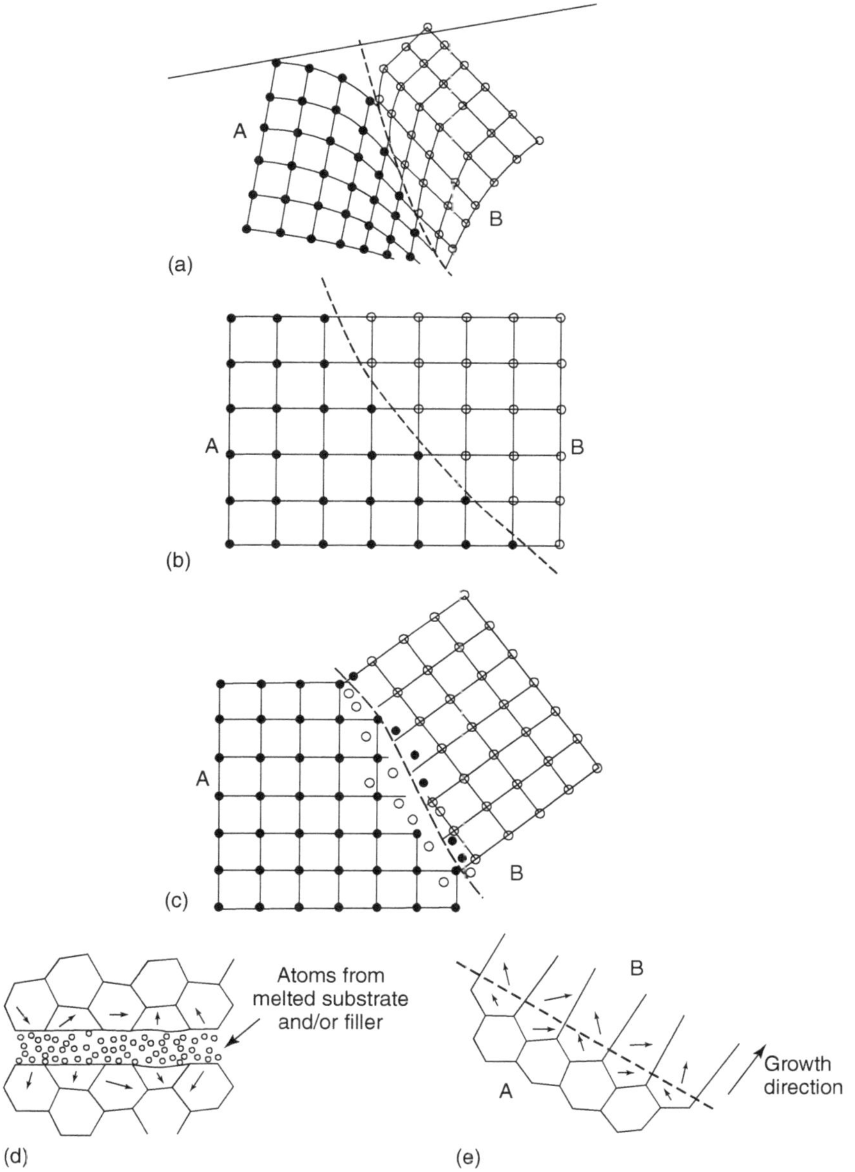

Figure 2.3 Schematic illustrations showing the three distinct ways in which metallic continuity can be achieved during welding by bringing atoms together using, first, (a) cold plastic deformation and lattice strain or (b) hot deformation and dynamic recrystallization; second, (c) solid-phase diffusion across the original interface (shown by a dashed line); and, third, (d) liquid provided by melting the parent materials (or substrates), without or with an added filler, and (e) establishing bond upon epitaxial solidification of this liquid. Source: Messler (1999), figure 2.1, p. 19. Used with permission of John Wiley & Sons.

form to fill any gap and help melt some portion of each substrate via contained superheat. In either case, the process is known as *fusion welding*; the former, without a filler, is said to be autogenous.

Figure 2.4 shows one classification or taxonomy of the major welding processes, including the related processes of brazing and soldering, although there are other essential similar schemes.

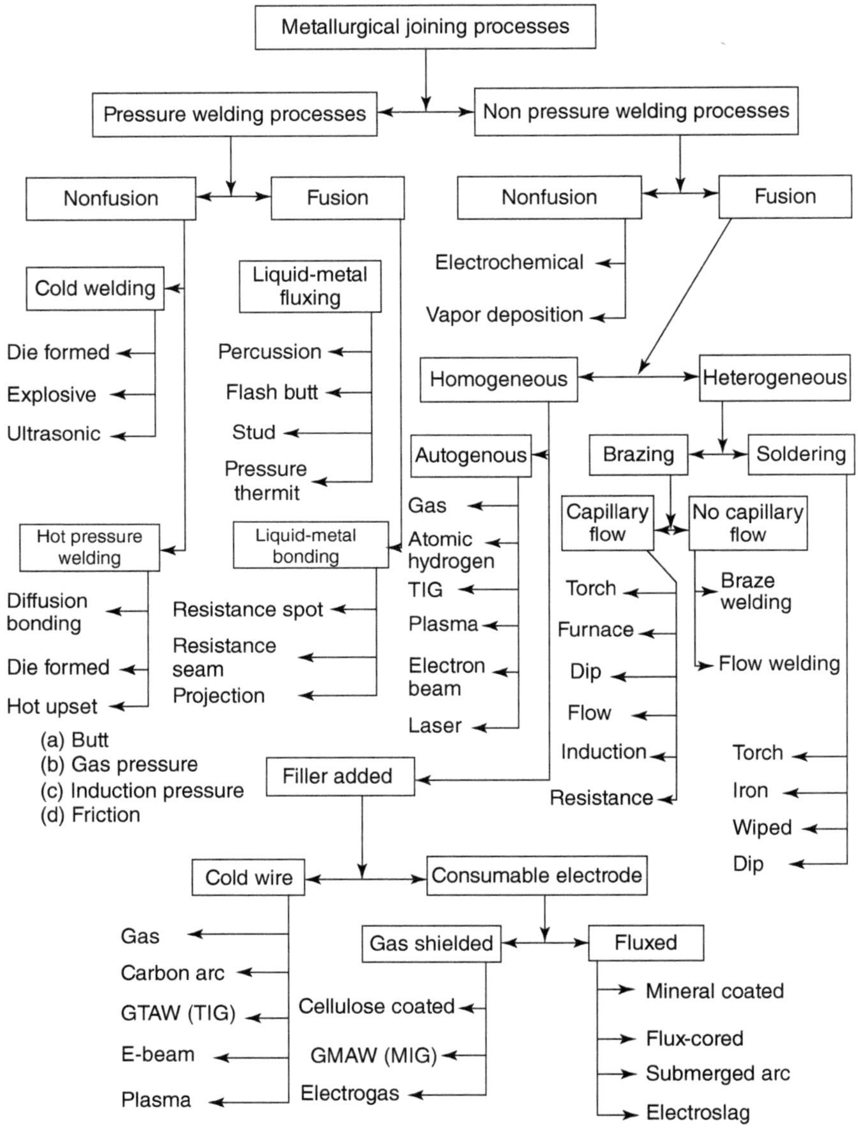

Figure 2.4 A typical comprehensive scheme for classifying welding processes. This scheme shows how the related processes of brazing and soldering fit into the taxonomy. Source: Messler (1999), figure 2.3, p. 28. Used with permission of John Wiley & Sons.

With welding defined, let us look at a weld, which is, after all, the intent of welding.

2.2 Microstructural Zones of Welds

Welding is used to fabricate new equipment or structures by building up detail parts (e.g. mill products, such as U-channels or I-beams, near-net-shaped forgings of net-shaped castings, or machined components) into a unitized assembly or weldment *or* to repair equipment or structures that have suffered loss of metal by mechanical wear or chemical corrosion or have cracked or fractured in service, regardless of whether the original was assembled by welding or not. The former is known as *OEM welding* (original equipment manufacturer), and the latter is known as *repair welding*. Although OEM welding might be accomplished using either a fusion or a non-fusion welding process, repair welding is almost always performed using a fusion welding process.

The fundamental types of welds used in both OEM and repair welding to accomplish joining include those that are used to make *butt joints, corner joints, tee* or *T-joints, lap joints,* and *edge joints.* These are shown schematically in Figure 2.5. In repair welding, fusion welding may additionally be used to fill a machined out or gouged out volume formerly containing a defect (e.g. crack). The other major application of welding than for joining components is for either restoring metal lost by wear or corrosion or, alternatively, for providing protection against wear or corrosion. Both are known as surface welding. For corrosion protection, the process is typically called *weld cladding*, while for wear protection, the process is typically called *hard-facing*.

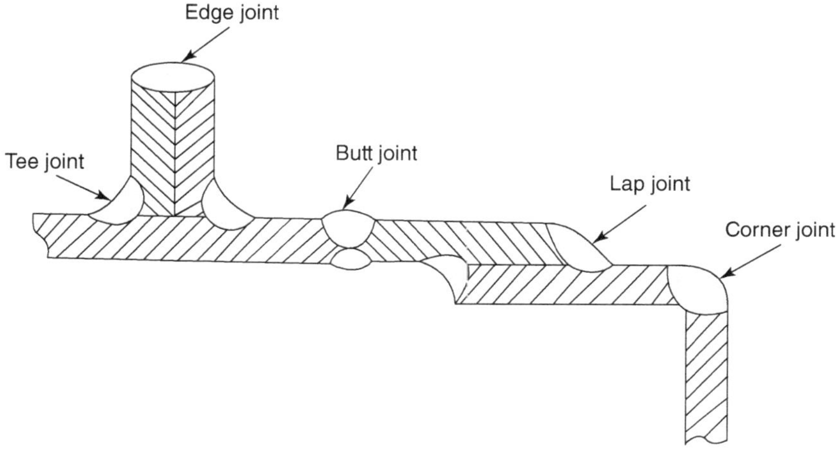

Figure 2.5 Schematic of the cross section of a hypothetical weldment showing the five fundamental types of joints found in welding: butt joint, corner joint, tee joint, lap joint, and edge joint. Although fusion welds are shown, similar joints can be made using non-fusion welding, albeit not by every non-fusion welding method and often with minor adjustments.

An underlying relationship in materials engineering is known as the *structure–property–processing relationship* or, more correctly, interrelationship, as each of the three affects the other in all directions. Whenever welding is used to make a joint, one needs to consider how the joint may differ in structure at the microscopic level (i.e. microstructure) and, thus, exhibit different properties as *structure determines properties in materials.*

The third key factor in this interrelationship is *processing*, which, in the broadest sense, could involve net-shape creation by *casting* from a melt to form a cast product; *deformation processing* using rolling, forging, or extrusion to produce a wrought product; or *machining* (including abrasive grinding) to remove materials to create a finished shape. From details produced by one or more of these processes, larger and/or geometrically more complex assemblies can be made by mechanical joining (e.g. using fasteners or interlocking integral design features), adhesive bonding, or welding. As it involves either considerable heat to cause fusion or considerable pressure to force the material into intimate contact, welding inevitably affects the structure of a material and, thereby, the properties of that material.

Depending on the category of the welding process used, i.e. fusion welding or non-fusion welding, and the type of the base metal(s) or alloy(s) involved, a weld actually consists of several zones of microstructures (i.e. *microstructural zones*). Because the problems that can arise during or after welding often manifest themselves as either physical defects (e.g. cracks) or degraded properties (e.g. lost strength or increased brittleness) in one of these zones, it is worth considering which zones are found in welds made with or without fusion in pure metals or in alloys.

A central paradigm underlying materials in all of engineering is *Structure as influenced by processing determines properties, which critically influence performance*, as shown schematically in Figure 2.6.

For our purpose, as we explore practical ways to either avoid problems that can arise with the process of welding or, alternatively, overcome problems that have already been encountered during welding or in the welds produced (either of which would detract from the performance of a welded assembly or structure), we need to focus on the *structure–property–process interrelationship*. Here, interactions can occur in any direction between any of the three factors. As an example, the process of welding (i.e. embodiment and procedure) influences the structure of the materials being joined (which influences joint properties), but, contrarily, the structure of base materials to be welded influences the choice of welding process (i.e. embodiment and procedure). Thus, it is essential to observe what happens to the structure in and around a weld.

Because the properties of a welded joint are affected to the greatest degree by the changes in the microstructure of the base metal(s) as altered by either the melting and surrounding heat of fusion welding or strain, and perhaps associated heat, of non-fusion welding, it is essential to first know how the properties of the base metal(s) were obtained. This means knowing how the base metal(s) obtained their strength before welding, i.e. to know the *strengthening mechanism(s)* employed.

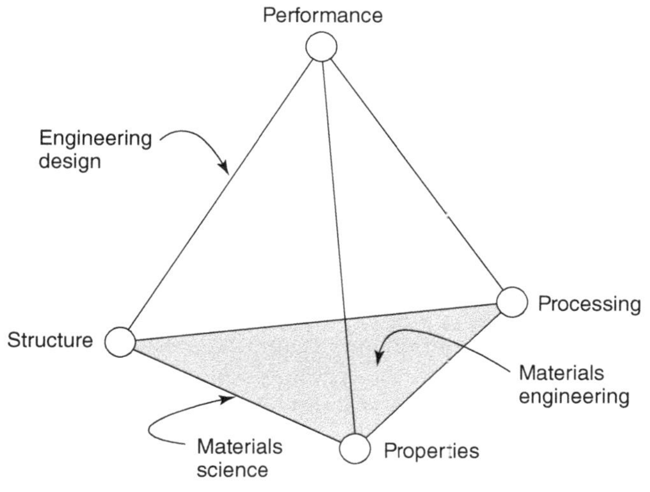

Figure 2.6 Schematic illustration showing the inextricable interrelationship among the structure, processing, and properties of a material as these three influence performance in a design. Source: Messler (2011). Used with permission of Jones & Bartlett Learning.

Pure metals can be strengthened by one or more of the seven basic strengthening mechanisms, some of which leave the metal pure, whereas others turn it into an *alloy*, defined as a mixture of two or more elements, at least the one in greatest concentration being metallic, to purposely increase the strength or resistance to corrosion. When multiple mechanisms are used, the effect on strength is additive, whereas the effect on corrosion may be positive or negative and depends on the particular mechanism. These seven mechanisms of strengthening are as follows:

1. Refining grain size
2. Cold-working to strain (or work) harden
3. Alloying to solid solution strengthen
4. Adding enough solute to create a second solid solution phase that gives rise to a phase boundary
5. Causing solute above a solubility limit to precipitate out of the solid solution matrix as a stoichiometric ceramic or intermetallic compound to form a coherent interface
6. Inducing massive shear transformation of a parent phase, as in deformation-induced or quench-induced martensite
7. Adding a new phase that is not inherent in the base metal or alloy system to create a metal–matrix composite (MMC).

Quite simply, what gives rise to strengthening in each case is the creation of an obstacle to the movement of dislocations[6] in the crystalline metal or alloy matrix phase to hinder slip and plastic deformation leading to shape change or,

6 A *dislocation* is a linear imperfection found in virtually all crystalline materials (e.g. metals and ceramics), having length but essentially no width. Like all imperfections found in crystals (e.g. vacancies, impurity atoms, free surfaces, and grain and phase boundaries), dislocations are a normal and inevitable consequence of entropy, which seek to lower the free energy of a system by

ultimately, fracture. Briefly, with details being left to the reader to seek information either in Part III of this book or in other sources, here is what occurs with each mechanism:

1. In refining the grain size (either during metal or alloy production at the mill or subsequently by cold-work and recrystallization), increased numbers of grain boundaries block moving dislocations, create pileups and back-pressure on trailing dislocations, and raise yield strength by the Hall–Petch effect.
2. Cold-work plastic deformation generates dislocations on intersecting slip planes, which, in turn, lead to tangles and pileups that raise yield strength.
3. Adding any foreign atom (as a solute) to a pure host metal (as the solvent) creates either a tensile or a compressive strain field around each solute atom that interacts with either the compressive or the tensile region of strain that surrounds every dislocation to repel and block that dislocation to prevent further movement and raise the yield strength.
4. If adding a solute above some level results in a second solid solution phase (i.e. in a two-phase region of the system's phase diagram), the boundaries (known as phase boundaries) between these two different phases block moving dislocations as the strain field associated with the boundaries interacts with the strain field of the dislocations, thereby raising the yield strength.
5. For an alloy system in which there is limited solid solubility of a solute in a solvent, with a second-phase stoichiometric ceramic or intermetallic compound (e.g. A_xB_y) resulting under equilibrium if that limit is exceeded, the second phase can be precipitated out following a solution heat treatment, followed by rapid quenching, room temperature (natural) aging, or an aging heat treatment to accelerate the rate at which the second phase develops to a size where coherency at phase interfaces block dislocations and raises the yield strength, in the process of precipitation–strengthening–hardening.
6. In some small number of alloy systems, the most common and notable being iron with interstitial carbon to form steel, an elevated temperature solid solution phase that would decompose on cooling, relying on diffusion, into two phases under equilibrium, can be made to undergo a massive shear transformation during rapid enough quenching to prevent diffusion to form a metastable martensitic phase. The high degree of strain found in the martensitic phase (because of carbon atoms being trapped in unfavorable sites) blocks dislocation movement and dramatically raises the yield strength. In steel, the elevated temperature phase is austenite, whereas the metastable phase formed upon quenching above some critical cooling rate is ferrous martensite.

---

introducing disorder. There are two basic types of dislocations: (i) *edge dislocations*, which are found at the edge of an extra half-plane of atoms in a crystal, and (ii) *screw dislocations*, which are found at the core of a spiral ramp of atoms in a crystal. In reality, most real dislocations in crystalline materials, such as metals, are *mixed dislocations*, consisting of short components of both edge and screw dislocations. In the simplest sense, edge dislocations lessen the energy needed to move one plane of atoms over another via shear to cause slip and plastic deformation shape change, allowing slip to occur in small steps rather than all at once. By disrupting the perfect arrangement of atoms in a crystal, dislocations are surrounded by both compressive and tensile strain fields.

7. It is possible to add a new solid-phase constituent to a continuous matrix, whether a pure metal, a single- of multi-phase alloy, or even a ceramic or a polymer, such that the new phase acts to hinder dislocation movement in the matrix, creating the so-called MMC. The new solid phase can be in the form of fine particles (known as *dispersoids*), as short single-crystal fibers (known as *whiskers*), as chopped/discontinuous or continuous fibers (known as *fibers*), as woven fabrics or meshes, or as elastically continuous laminations (known a *lay-ups* or *laminates*). Regardless of the form of the reinforcing phase, it should be of a composition not found within the system of the matrix, if an alloy.

With the above discussion as a suitable background, the various microstructural zones found around a weld are schematically shown in Figure 2.7, for a fusion weld made between a pure metal (at the right) and a generic alloy (at the left) in panel (a) and for a non-fusion weld involving some heat made between a pure metal (at the right) and a generic alloy (at the left) in panel (b).

The weld zone (WZ) of a fusion weld always includes, from the centerline of the joint outward, a fusion zone (FZ), a heat-affected zone (HAZ), and an unaffected base metal (UBM). For pure metals, in which melting occurs at a discrete temperature, known as the melting point, the fusion zone would consist of a cast, dendritic microstructure of uniform (i.e. homogeneous) composition throughout. The effect of heating below the melting point would always lead to some grain

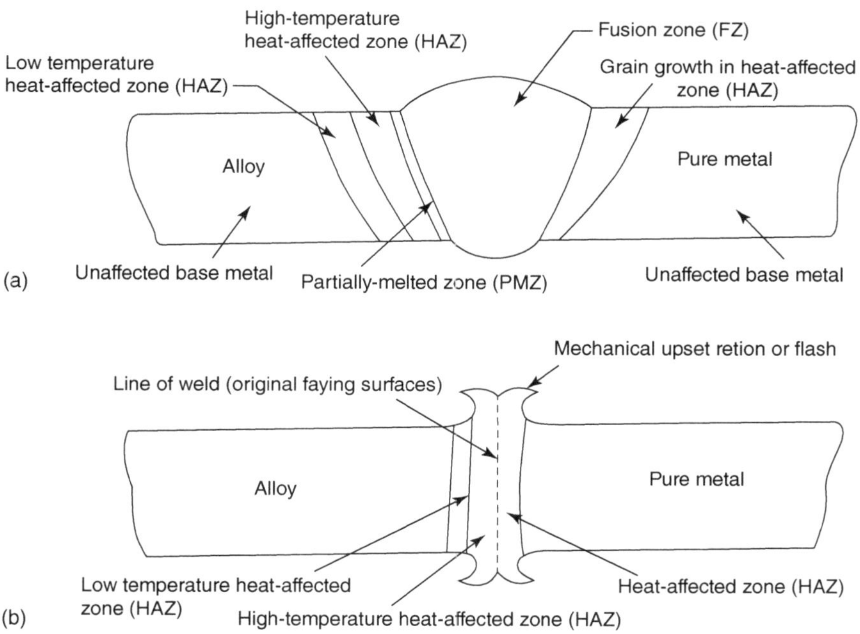

Figure 2.7 Schematic illustrations showing various microstructural zones that are typically found in welds made between a generic pure metal (on the right) and a generic alloy (on the left) using a fusion welding process (a) or a non-fusion welding process (b). Source: Messler (1993), figure 6.5, p. 193. Used under reciprocal agreement between Wiley and Elsevier for fair use.

growth in the highest temperature region of the HAZ (nearest to the FZ), with the extent of growth usually being less because of the short time-at-temperature to allow rearrangement of atoms in the solid state. If, and only if, the pure base metal had been cold-worked before welding, creating elongated deformed and strain-hardened grains, the HAZ would be more extensive, consisting, from an unaffected base metal inward, of recovery (only revealed by a decrease in electrical resistivity), recrystallization (evident from the appearance of new, strain-free, small equiaxed grains, decreased hardness and yield strength, and increased ductility), and grain growth (with little change in properties).[7]

For an alloy welded by a fusion process, the weld zone, moving from the highest temperature weld centerline outward, always includes (i) a FZ with cast, dendritic structure, with some degree of solute segregation (i.e. inhomogeneity) always present at the microscopic level (i.e. microsegregation) and, possibly, more or less segregation at the macroscopic level because of incomplete mixing of any filler used with the melted base metal(s) (i.e. dilution), and (ii) a partially melted zone (PMZ) consisting of a mixed unmelted and melted and re-solidified base metal structure where the temperature was below the solidus but above the liquidus temperature of the alloy. Outside the PMZ, there is a HAZ whose details depend on the mechanism(s) used to develop the properties of base metal alloys. Typically, the HAZ is divided into a high-temperature and low-temperature region; again, with details depending on pertinent strengthening mechanism(s). As two examples: First, a carbon steel would exhibit 100% brittle as-quenched martensite in the high-temperature HAZ (as well as in the PMZ) and lesser percentages of as-quenched martensite (with either ferrite or bainite) in the low-temperature HAZ, wherever austenite had existed at-temperature and, second, reversion of precipitates in the high-temperature HAZ and overaged precipitates in the low-temperature HAZ of an age-hardened alloy; all being detectable using microhardness traverses and/or metallographic examination.

When a non-fusion welding process is used to make a weld, the weld zone most notably differs in the complete absence of any FZ.[8] Whether any HAZ occurs with non-fusion welding depends on whether heat is either used to facilitate the formation of the joint (usually by making the regions of the parts to be welded easier to deform to the immediate vicinity of the abutting surfaces) or whether the physical process of bringing metal into intimate contact to form a proper joint generates heat as a by-product of mechanical work. In most cases, however, there is little or no heat effect on the microstructure (and, thus, properties) of the metals or alloys being joined; this being the principal reason a non-fusion process was selected in the first place.

The lower part of Figure 2.7 shows the possible regions of a non-fusion weld made between a pure metal and a generic alloy. Because the level of heating in

---

7 The exception is when *abnormal grain growth*, sometimes referred to as "germination," occurs in some metals (e.g. pure copper) or alloys (e.g. some brasses) following cold-work to a level of around 4–6% thickness or cross-sectional area change. Such an excessive grain growth can degrade impact toughness and fatigue resistance.

8 It is possible that there is some evidence of melting at the very point where parts are being welded by hot pressure non-fusion processes or even friction. This, however, is accidental and often indicates a lack of proper process control.

non-fusion welding is often modest and is never high enough to cause melting, there is seldom any effect even in the highest temperature region of a pure metal. Depending on the mechanism(s) of strengthening in an alloy being welded by a non-fusion process, there may be some heat effect, but rarely much more than some low-temperature effect.

It should be noted, however, that in lieu of any fusion zone, there is often (depending on the specific process) a region of mechanical upsetting of the microstructure. This may be manifested in deformed grains, by so-called deformation flow lines or by refined grain size because of dynamic recrystallization during hot upsetting.

In short, real welds consist of zones of microstructure that have been altered and caused to differ in structure from the original base metal(s) from the effect of extreme heat of fusion welding or deformation and, possibly, modest heat of non-fusion welding. Understanding the origin and nature of these zones helps determine the cause of some problems encountered in welding and in welds.

2.3 Origin of Problems in Welding and Welds

It goes – or should go – without saying: *The problems one has in welding are in producing what one considers an acceptable weld or weldment.* "Weld" means any individual joint or deposit (in the case of surface welding as the build-up of lost metal, the overlay of a more corrosive-resistant metal or alloy, or the application of hard-facing with a more wear-resistant alloy) produced by a welding process. "Weldment" means the entire assembly of components comprising a part (e.g. the casing of a catalytic converter, tank for a liquid propane transporter truck, and blade assembly for a bulldozer), a piece of equipment (e.g. mountable snowplow for a pickup truck), a product (e.g. automobile), or a structural element (e.g. built-up beam), a substructure (e.g. truss), or a structure (e.g. bridge and ship). Weldments, as used here, involve the use of many welds to join components. By "acceptable" means meeting the applicable design specification, which may be that of the manufacturer, the ultimate customer, or that imposed by a code (e.g. ASME Boiler and Pressure Vessel Code). Specifications typically dictate the required geometry, dimensions, finishes, and quality, which, in the case of welded joints, primarily means freedom from unacceptable defects. Codes applicable to welds typically dictate quality and minimally acceptable properties (e.g. yield strength, tensile strength, hardness, fracture toughness, and impact toughness).

What makes a weld or weldment unacceptable is anything deemed by prior experience or engineering judgment or determined by prior analysis or testing to render it less-than-adequate for providing needed performance. By this definition, anything about a weld or a weldment that prevents proper fit and/or function is a problem.

"Failure to fit" usually means some problem that makes subsequent assembly during sequential assembly difficult or impossible. Although minor problems can often be dealt with, more serious ones cannot. Worse, serious problems with fit can lead to compromised performance (e.g. reduced fatigue life from residual stresses introduced by forced fit) and may compromise the safety and/or longevity of a structure. "Failure to function" is potentially much more serious, as it is always experienced by the end user, which, obviously, can lead to an unhappy (or lost) customer or, worse, a lawsuit for injuries or death.

With these as the consequences of welding problems, some of which are quite severe, one needs to consider their origins in order to hope to avoid the occurrence of problems in the future or resolve problems that have already occurred. Having said this, there are several logical general origins, namely, in approximate order based on how one proceeds to make a weld and weldment: (i) the preparation of joint elements before welding, (ii) the setup of joint elements before welding, (iii) the chosen welding process embodiment, (iv) the choice and control of key welding parameters, (v) the choice and control of welding procedures, (vi) ignored or improper post-weld treatment, and (vii) ignored or inadequate post-weld inspection for both internal quality and external appearance. Not to be forgotten is the possibility of problems that can arise from the materials being welded, singly or in certain dissimilar combinations. These problems typically fall into one of two categories: problems arising from the inherent nature of the particular material (i.e. its weldability) or some problem associated with particular batch or heat of material (e.g. excessive impurities).

With this as a starting point for potential problems, let us consider how else the potential problems that can arise be logically categorized?

2.4 How Problems Can Be Logically Categorized?

Troubleshooting a problem associated with welding begins with determining the likely origin or source of the problem. Only then is it even possible to either avoid such a problem in the future or resolve problems that have already occurred and could logically continue to occur in the production of more welds or weldments of the same type (i.e. base materials; joint configurations; welding processes, parameters, and procedures; structural arrangements; equipment; and operators).

The process of determining the source of a particular problem is aided by dividing possible problems that can arise in welds or weldments into logical categories that serve as clues as to: (i) how the problem manifests itself in a weld or weldment; (ii) where a problem in a weld is located; and (iii) in what material the problem is encountered. Without going into great detail here, as details are the purpose of the rest of this book, suffice to say:

The *manifestation of problems* begins with whether these are associated with an individual weld or a weldment, as defined earlier, and include the following:

- Problems with joint setup or joints
- Geometric distortion, dimensional shrinkage, or instability of weldments
- Porosity in welds
- Nonmetallic or metallic inclusions in welds
- Cracks in welds or weldments
- Weld appearance.

When associated with welds, a key factor for cause is the *location of problems* relative to the microstructural zones (shown in Figure 2.7), as follows:

- Fusion zone of fusion welds
- Partially melted zone of fusion welds in alloys
- Heat-affected zone of fusion welds
- Non-fusion welds.

Finally, there are *material-specific problems* that include the following:

- Embrittlement in carbon-, low-, and medium-alloy steels
- Sensitization weld decay and knife-line attack in stainless steels
- Reheat-cracking in age-hardenable alloys
- Long-term embrittlement with high-chromium contents
- Weld dilution and segregation in welds
- Dissimilar metal welds.

The rest of this book addresses each category in three parts: Part I – Manifestation of Problems; Part II – Location of Problems; and Part III – Material-Specific Problems. A chapter is devoted to each bulleted item.

Hopefully, this format provides a practical guide to users of welding who encounter problems.

As a further aid to users, each chapter ends with a "Troubleshooting Guide," which includes a tabulation of the problems covered in that chapter, the most likely cause of each, and a suggested approach for resolution.

References

Messler, R.W. Jr. (1993). *Joining of Advanced Materials*. Stoneham, MA: Butterworth-Heinemann (now part of Elsevier, Ltd.).

Messler, R.W. Jr. (1999). *Principles of Welding: Processes, Physics, Chemistry, and Metallurgy*. New York, NY: Wiley.

Messler, R.W. Jr. (2011). *The Essence of Materials for Engineers*. Sudbury, MA: Jones & Bartlett Learning.

Granjon, H. (1991). *Fundamentals of Welding Metallurgy*. Cambridge, UK: Abington Publishing, Woodhead Publishing.

Further Reading

Ashby, M.F. (2017). *Materials Selection in Mechanical Design*, 5e. Oxford, UK: Butterworth-Heinemann/Elsevier Ltd.

Callister, W.D. Jr. and Rethwisch, D.G. (2013). *Materials Science and Engineering: An Introduction*, 9e. Hoboken, NJ: Wiley.

Hibbeler, R.C. (2016). *Mechanics of Materials*, 10e. Upper Saddle River, NJ: Pearson Publishing.

Messler, R.W. Jr. (2004a). *Joining of Materials and Structures: From Pragmatic Process to Enabling Technology*, 1e. Burlington, MA: Butterworth-Heinemann, Elsevier.

Messler, R.W. Jr. (2004b). *Principles of Welding: Processes, Physics, Chemistry, and Metallurgy*. Weinheim, Germany/New York, NY: Wiley-VCH/Wiley.

Messler, R.W. Jr. (2010). *The Essence of Materials for Engineers*, 1e. Sudbury, MA: Jones & Bartlett Learning.

Part I

Manifestation of Problems with Welds and Weldments

The problems that can be encountered with welding, whether they arise from some aspect of the design, the base material(s), the environment, the choice of process, the choice of parameters, the welding procedure or technique, or pre- or post-weld treatment manifest themselves in either the welds, the weldment, or both. *How* problems manifest themselves is the first clue to their root cause and, thus, solution.

3

Problems with Joint Setup and Weld Joints

3.1 Joint Efficiency

One of the principal reasons welding is often selected as the preferred process for joining components into an assembly is the high joint efficiency obtained; *joint efficiency* being the percentage of the strength in the joint compared to the strength of the adjacent joint elements. High *joint efficiency* by welding is the combination of two key factors: (i) the ability of the weld (i.e. the entire weld zone) itself to provide strength comparable to the base metals[1] and (ii) the nearly negligible added weight of a weld to a structure. A third factor may be that welds can be made continuous, either to meet structural demands and/or to provide leak tightness of fluids, whether into or out of a weldment.

The strength that can be obtained in a properly made weld depends on whether there is a fusion zone and, if there is, the choice of a filler metal. As a fusion weld consists of a cast microstructure, it is inevitable that there is some loss of strength (and ductility) compared to base metals having a wrought microstructure.[2] Under best conditions, a cast structure results in a yield and an ultimate strength about 70–75% that of a similar wrought structure and ductility that is typically reduced by 50% or more. This said, one can select a filler metal (usually richer in alloy content) that has higher strength than the base metal(s), which is known as an *overmatched filler*. Another factor that tends to offset the knockdown of strength associated with fusion welding is that a properly made butt weld is supposed to exhibit reinforcement in the form of a convex crown bead and root bead (if the weld is a full penetration type). For certain applications, the portion of the joint elements where a fusion weld is to be made is made thicker using a "raised weld land," further offsetting any knockdown of properties from fusion welding. As will be seen later (in Chapter 10), the heat-affected zone (HAZ) of a weld must

1 Throughout this book, the term *base metal* or *base metals* refers to the metallic material or materials being welded in a particular joint and may be referring to either or both of a pure metal or an alloy. If it is important to a particular discussion whether the base metal(s) is(are) pure or alloyed, it will be indicated by the specific use of the term pure metal(s) or alloy(s).

2 Wrought microstructures result from rolling, forging, or extruding (i.e. working or deformation processing) a previously cast billet produced at the mill, directly from the melt. The process of working the metal breaks up the dendritic structure always found with casting, along with inevitable inhomogeneity due to microsegregation under nonequilibrium solidification. It also tends to "heal" any porosity or shrinkage cracks formed during casting.

A Practical Guide to Welding Solutions: Overcoming Technical and Material-Specific Issues,
First Edition. Robert W. Messler, Jr.

also be considered; certainly to assure there has not been an unacceptable loss of some key property of the chosen base metal(s).

As implied at the start of the second paragraph of this chapter, one of the principal advantages of using a non-fusion welding process to make welds is that virtually every embodiment, e.g. hot pressure welding (HPW), forge welding (FOW), roll welding (ROW), explosion welding (EXW), various forms of friction welding (FRW), and diffusion welding (DFW) or diffusion bonding (DB), provides high joint efficiency, as there is no fusion zone, and there is usually minimal heat effect to the base metal(s).

If one considers *adhesive bonding* using organic/polymeric adhesives for joining, achievable joint efficiencies are inevitably low, albeit depending on the materials (i.e. *adherends*) being joined. The reason is that such adhesives typically offer strength under pure shear (i.e. shear strength) of 1.7–23.8 MPa (0.25–3.5 ksi) and strength under a peeling force (i.e. peel strength) of only 10–20% of this (typically, 500–5000 N/m [~3–30 lb./in.]), with out-of-plane peeling almost being inevitable. Obviously, if metallic adherends are being joined, joint efficiencies are extremely low (typically 5–10%), but if paper or wood adherends are being joined, joint efficiencies can be much higher, often exceeding 100%.[3] Because adhesives operate best in pure shear (versus tension or peel), they tend to be applied in thin layers over large surfaces (i.e. surface areas). Even so, as the density of organic/polymeric adhesives is low (at 0.8–1.6 g/cm^3) compared to most potential adherends (e.g. wood at 0.35–0.8, Al alloys at 2.8, and steel at 8 g/cm^3), there is a very small weight penalty associated with adhesive-bonded joints and structures.

If one considers joining using mechanical fasteners, such as nuts and bolts with steel or steel nails with wood, even though fasteners are placed at discrete, spaced locations, there can be a significant weight penalty due to the added weight of bolt heads, washers, nuts, and the threaded portion of the bolt that extends through the nut. On the other hand, the tensile yield and the ultimate strength of steel bolts is quite high and can easily be, and often is chosen to be, higher than the metal (e.g. steel) joint elements being fastened. Likewise, the tensile strength of steel nails is much greater than the tensile or compressive strength of any wood. In most cases, the joint efficiency with mechanical fastening is high, at seldom less than 25%, often near 100%, and possibly over 100%.

Thus, welding potentially offers high joint efficiency, even when fusion welding processes are employed. However, obtaining high efficiency requires welding to be performed properly and welds to be free of problems. So, let us consider what can go wrong – and why.

3.2 Weld Joint Types and Weld Configurations

The problems that can arise during welding and in welds can, not surprisingly, depend on the joints themselves, two key factors being (i) the type of joint and (ii) the configuration of the weld.

---

3 A joint efficiency of over 100% means failure will occur in the joint elements, not in the joint proper. In adhesive bonding, this means one of the adherends will fail, not the adhesive.

The loads in a welded structure are transferred from one structural element to another through welds placed at joints. The type of joint, or the joint geometry, is largely determined by the geometric requirements (or, contrarily, restrictions) of the ultimate structure and the type of loading it must safely endure (e.g. tension, shear, torsion, pressure, or combinations of these). Additional factors affecting the choice of joint type, however, are accessibility for proper welding (i.e. required depth of penetration and width, addition of filler, etc.), available or chosen welding process embodiment (i.e. welding process), accessibility for effective inspection, and cost constraints (e.g. setup time, amount of filler required, and time to weld).

Figure 3.1 schematically illustrates the five basic weld joint type designs used with fusion welding processes,[4] namely, butt joint, corner joint, edge joint, lap joint, and tee joint.[5] Figure 2.5 shows how welds appear for each of these joint types.

For the joint types shown, a fusion weld could penetrate entirely through the butt joint (i.e. be *a full penetration weld*) or only penetrate partially through the joint thickness (i.e. be *a partial penetration weld*). Whether a weld should be a full- or a partial-penetration type depends on the intended function but is clearly influenced by whether welding is to be accomplished using a single pass or multiple passes, the choice of which largely depends on the thickness of the joint elements. For the edge joint, a fusion weld would, obviously, only be a partial penetration type, more appropriately called a "blind weld." For the other joint types, *fillet welds* made at the right angle opening are used. In the case of the tee joint, a fillet weld is normally placed at each side of the vertical leg, whereas for the lap joint, whether a second fillet weld is added (say to the underside of the joint shown) absolutely depends on having accessibility and may depend on the expected loading.

A potential problem associated with the blind weld used for an edge joint, the use of a single fillet weld with a lap joint (at either location), a single fillet weld (as opposed to two opposing fillet welds) for a tee joint, and, to a lesser extent, for opposing fillet welds used to make a tee joint without overlapping is the possibility of extension of the unfused portion of abutting or faying surfaces to cause cracking. Such cracking could occur because of the shrinkage of the sole weld upon solidification wedging the open portion of the joint into the newly completed weld, especially before that weld has fully cooled. Alternatively, it could occur under cyclic (fatigue) loading as the unfused portion of the joint opens and closes with each cycle, although this is usually not a significant problem. Best practice is to use opposing fillet welds on lap joints and on tee joints.

4 When a non-fusion welding process is to be employed, only the butt joint and tee joint types are commonly used, although an edge joint could be welded using the friction stir welding (FSW) process.

5 To be complete, two additional types of welding need to be mentioned: *plug welding*, to join one planar part to another through a hole by making a fillet weld around the inside perimeter of the hole, and *surface welding*, to deposit metal to the surface of a part in order to replace the metal lost by wear, for example, or to enhance resistance to corrosion or wear by applying a more corrosion-resistant or more wear-resistant alloy, respectively.

Related to the above, within the joint types shown in Figure 3.1 are several variations for welding. These are based on specific geometry (i.e. shape and dimensions) and preparation of joint before welding (i.e. weld configurations). Examples are shown schematically in Figure 3.2 and include square butts (sometimes called straight butts), single and double grooves (whether V, U, J-grooves or bevels),[6] single and double fillets, and plugs and slots (not shown). The straight butt is the easiest and cheapest to prepare and provides good joint strength, but is limited

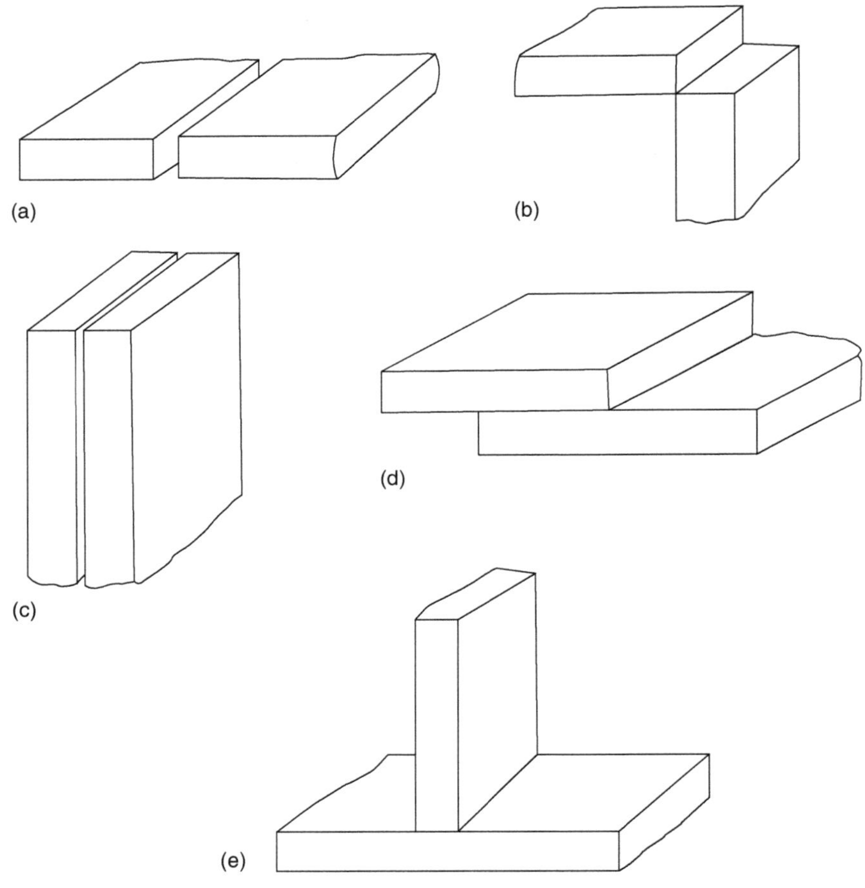

Figure 3.1 Schematic illustrations of the five basic designs for weld joints: (a) butt joint, (b) corner joint, (c) edge joint, (d) lap joint, and (e) tee joint. Not shown are plug welds and surface welds, as there are no actual joint designs for these types of welds, the former being performed through drilled holes to attach an upper to a lower plate and the latter on part surfaces to restore the metal lost by wear or to apply an overlay metal or alloy for improved wear or corrosion resistance. Source: Messler (1993), figure 6.31, p. 229. Used under reciprocal agreement between Wiley and Elsevier for fair use.

6 J and U groove configurations minimize weld metal requirements compared to V grooves, but add to joint preparation costs. Single-bevel and J-groove configurations are more difficult to weld than V or U groove configurations because one edge of the preparation is vertical.

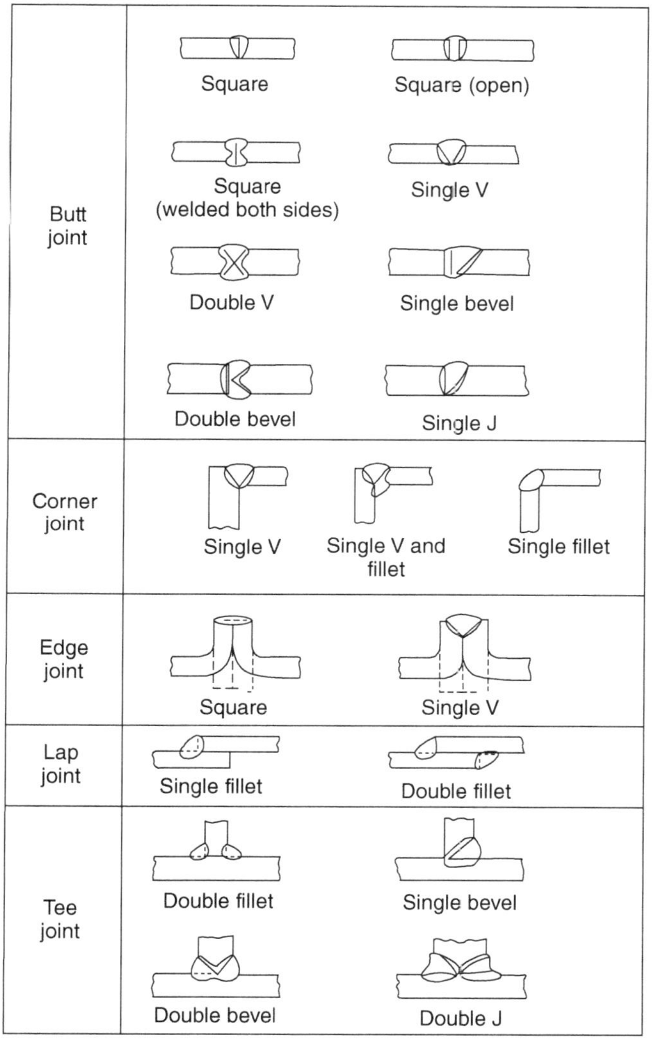

Figure 3.2 Schematic illustrations of some typical welds for various preparations of the five basic joint types. Source: Messler (1993), figure 6.32, p. 230. Used under reciprocal agreement between Wiley and Elsevier for fair use.

by joint thickness (except for narrow-beam welding by electron-beam welding [EBW] or laser-beam welding [LBW]). Single-sided joint preparations are usually made on thinner materials or when access from both sides is not possible. Thick joints require the edge of each joint element to be prepared to a geometry that will provide accessibility for welding (e.g. allow a welding electrode and/or torch to reach the root of the joint preparation), whether accomplished with a single pass or multiple passes, ensure the desired strength and soundness in the weld, and (as will be discussed in Chapter 13) minimize the unwanted dilution of the filler metal by base metal(s). Different weld joint preparations (i.e. weld configurations)

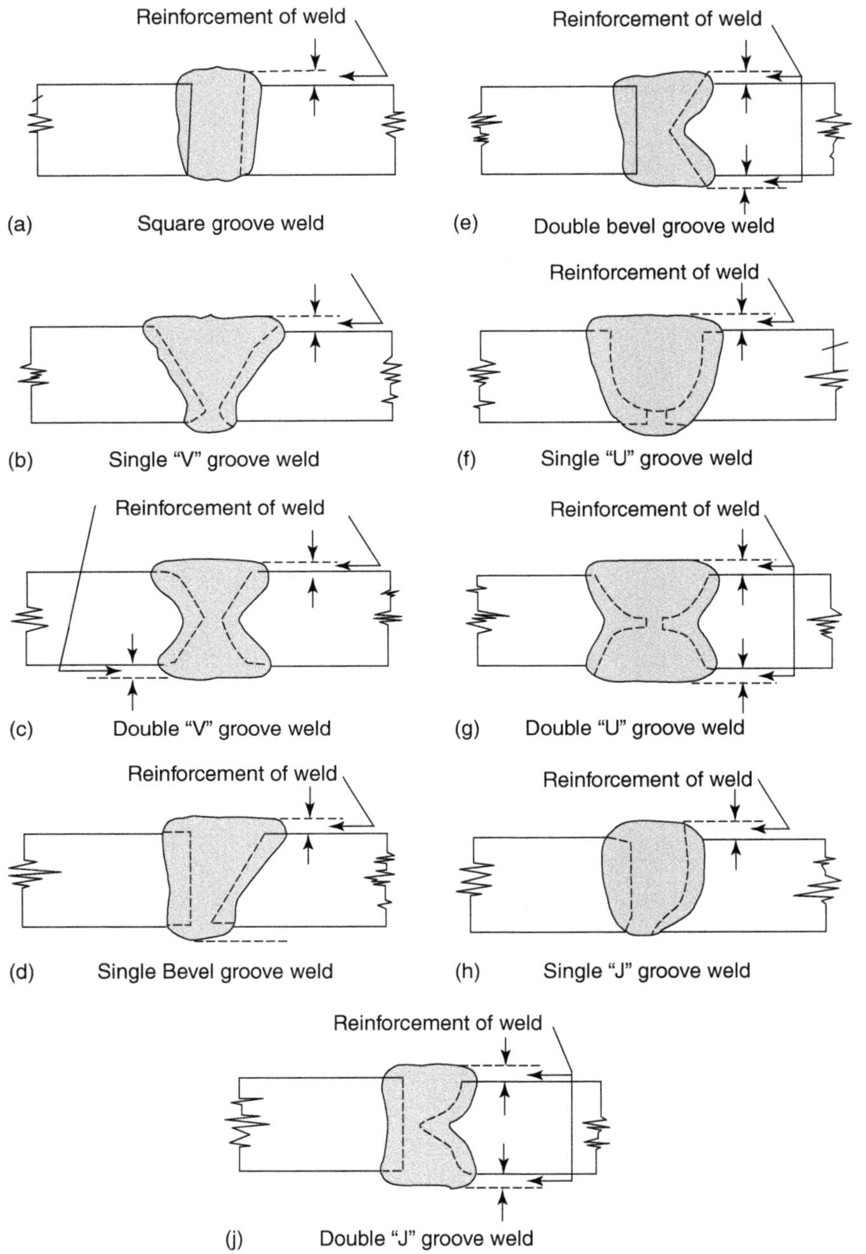

Figure 3.3 Schematic illustrations showing properly made fusion welds (shaded) in various preparations of butt weld joints. Source: Free Education Network. Freely used under Creative Commons ShareAlike, CC BY SA, 4.0.

may exacerbate the entrapment of slag when fusion welding processes using slag for shielding are used. The gap at the bottom or root of a joint (i.e. root gap) should be selected to minimize the volume of the weld filler metal needed to make a sound weld, while also aiding accessibility during welding, yet still meeting joint strength requirements.

Figure 3.3 schematically illustrates how properly executed fusion welds appear (as shaded profiles) for various common joint preparations, i.e. square grooves, and single and double V-grooves, U-grooves, J-grooves, and bevels.

3.3 Joint Setup Problems

The first place (or instance) problems can arise when making a fusion weld during the setup of the components comprising the joint or the joint elements. Three factors are important during joint setups: (i) linear alignment, (ii) angular alignment, and (iii) root gap.

Figure 3.4 schematically illustrates a typical (and representative) single V-groove joint preparation for fusion welding, with key terminology annotated and typical dimensions listed within parenthesis in the figure.[7] Three of the labeled dimensions represent geometric features important to the creation of a sound fusion weld. The *bevel angle* and the *included angle*, as well as the *root gap*, facilitate access of a filler wire (e.g. shielded metal-arc welding [SMAW] electrode, gas tungsten-arc welding [GTAW] wire, flux-cored arc welding [FCAW], or gas metal-arc welding [GMAW] wire electrode) and a torch body (e.g. especially for GTAW, but also for FCAW and GMAW) to allow proper molten metal flow and fill as well as to help prevent the entrapment of any evolved gases as porosity or molten slag as inclusions. The *root face* and *root gap* are important for helping create the desired weld shape (or profile) and produce a structurally sound finished weld. More will be discussed about the role of the root gap later in this section. However, first, let us consider the alignment of joint elements.

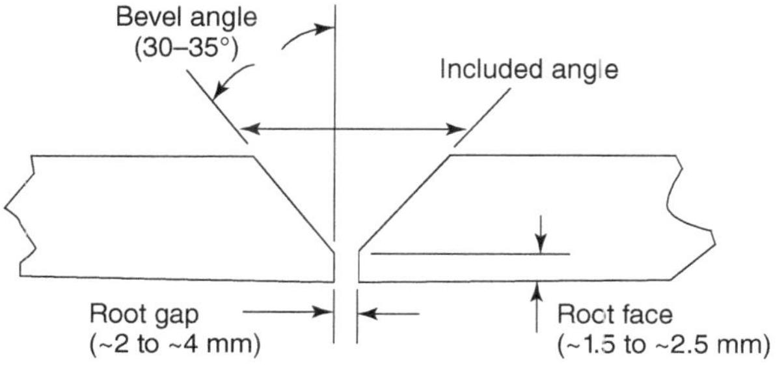

Figure 3.4 Schematic illustration showing a single V-joint weld preparation with key features labeled and typical dimensions provided at the bottom of the figure.

7 Comparable terminology and typical dimensions can be found for the other joint preparations in various references on welding, as well as on the Internet.

If the finished weldment is to have the specified/required geometry, the elements for each joint comprising the weldment must be properly aligned before welding, i.e. during joint setup. Two forms of misalignments can occur during joint setup: (i) *linear misalignment* or (ii) *angular misalignment*. These are shown schematically in Figure 3.5.

The linear misalignment, also known as *mismatch*, as shown schematically for a square groove (or butt) and a single U-groove joint preparation in Figure 3.6, manifests itself in a misshapen, asymmetrical weld profile, as shown schematically in Figure 3.7. More significant than the aesthetic issue, the mismatch compromises the structural integrity of a joint by introducing unintended bending moments across the joint and can reduce the effective load-carrying cross section in some cases. Figure 3.7 also shows the X-radiograph of a weld made in a mismatched joint, revealing the difference in effective cross section.

Angular misalignment, measured in degrees (as shown in the lower schematic illustration in Figure 3.5), often results in nonuniform shrinkage as a fusion weld

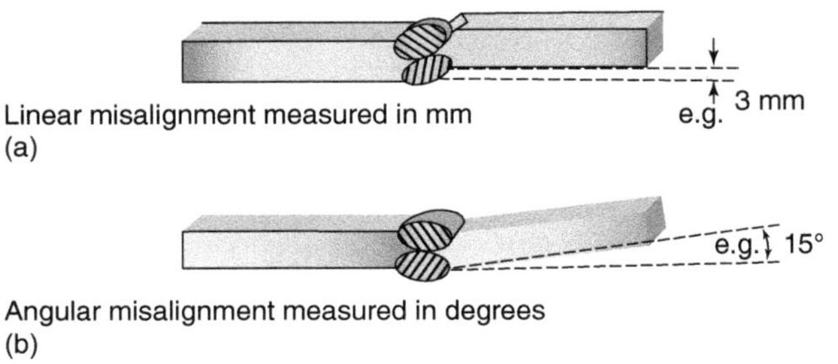

Figure 3.5 Schematic illustration showing the two types of misalignments that can occur in joint setup, i.e. (a) linear misalignment or mismatch and (b) angular misalignment.

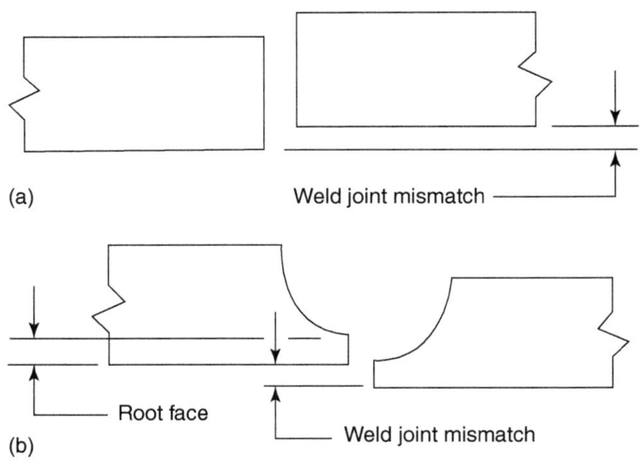

Figure 3.6 Schematic illustration showing weld joint mismatch (or linear misalignment) in two different joint preparations, i.e. a straight- or square-groove butt (a) and a single U-groove butt (b).

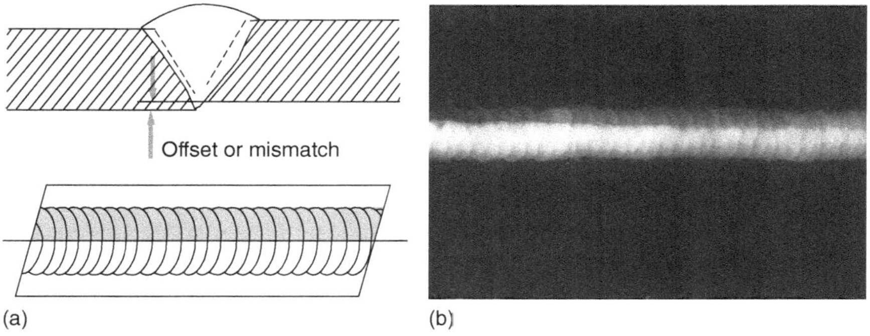

Figure 3.7 (a) Schematic illustration showing offset or mismatch of a single V-groove joint (top), along with the schematic illustration of an X-ray of such a weld (bottom) and (b) an X-radiograph of such a weld. The noticeable difference in the X-ray density between the two pieces is caused by the difference in material thickness, while the dark, straight line is caused by the failure of the weld to fuse with the land area. Source: From the website www.nde-ed.org of the NDT Resource Center of Iowa State University; used with permission of the Center for Nondestructive Evaluation, Iowa State University, Ames, IA, USA.

solidifies and then cools, which, in turn, may result in distortion of the weldment. However, such an angular misalignment often leads to even more insidious and potentially serious consequences when the misshapen joint is forced back into alignment before the next weld is made in a complex weldment of which post-weld stress relief was not scheduled. Such realignment (i.e. straightening), being done "cold," introduces residual stresses that become locked in the weldment as each weld in sequence is made. The result can seriously compromise the life of a weldment expected to perform in fatigue (i.e. cyclic loading) or in a corrosive environment. In the former case, locked-in stresses are added algebraically to applied stresses in unexpected ways, degrading fatigue performance. In the latter case, locked-in stresses, whether tensile or compressive, accelerate corrosion in service. This is an example of a situation made worse by the effort of a worker to correct a problem.

Avoiding misalignment is the simplest and best tactic, as resolving problems once they occur is usually fraught with its own problems, or impossible to achieve (e.g. mismatched joints, once welded, cannot be fixed, only redone). Proper alignment is best done using setup gauges, templates, and/or fixtures. Once angular misalignment presents itself in a weldment, great care should be exercised when attempting straightening. Removing any angular misalignment using cold deformation processes (e.g. hammering, etc.) may break welds and always introduces locked-in or residual stresses. If cold realignment is done, serious consideration should be given to thermal stress relief treatment as per the recommended practice guidelines in references or online. Elimination of any residual stresses is especially important when the service environment for the weldment involves fatigue loading or corrosion.

As a general rule, *joint gaps*, as measured by the root gap for all except straight butt or square groove joints, must be uniform and reasonably tight to avoid severe shrinkage,[8] which can lead to improper finished dimensions and/or distortion (or warpage) in the weldment. Another problem from excessive root gap is *burn-through*, during which a welding arc or a beam blasts through the joint, unimpeded by any metal. This leads to the collapse of the molten weld pool because of excessive penetration and loss of any support from solid, resulting in a hole in the root run or severe slumping (i.e. underfilling and concavity) in a single-pass weld. Burn-through is worse with (i) large root gap, (ii) small root face, (iii) slow welding or travel speed (TS), and (iv) high welding current (i.e. amperage) and/or high welding voltage.

Besides the need for proper root gap and root face during joint setup, precautions should be taken to prevent joints (especially at the root) from opening during welding, as a result of the shrinkage as weld solidification progresses. This can be accomplished by using fixtures or by tacking joints before welding. Tack welds are normally made as small, discrete, spaced spots made by the same process by which the full weld will be made (e.g. shielded metal-arc [SMA], flux-cored arc [FCA], gas tungsten-arc [GTA], and gas metal-arc [GMA]) or, in some cases, using SMAW before beginning full welding by another process. Figure 3.8 shows a typical tack-welded structure.

Another possible solution is to employ a backup bar or strip or specially designed tape, as shown in Figure 3.9. Whether temporary (to be removed after welding is completed) or permanent (to be left in place as part of the joint design), these act to hold molten weld metal in place during welding, thereby preventing underfill, slumping, and even burn-through.

Again, as is the case of all problems that can arise from misassembly, avoidance is best, as resolution of a problem once it occurs is difficult to impossible.

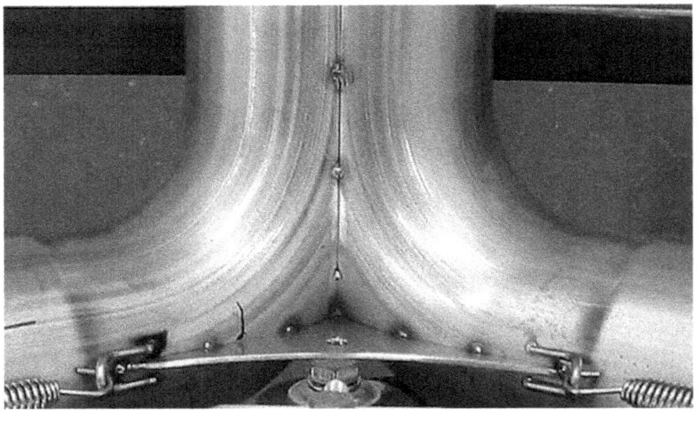

Figure 3.8 Photograph showing properly made tack welds to maintain tight fit-up of details of a stainless steel exhaust system assembly, thereby preventing burn-through. Source: From the website of Burns Stainless LLC, 1013 W. 18th St., Costa Mesa, CA 92627; figure 4 by CPR Fabrication, used with permission of both Burns Stainless LLC and CPR Fabrication.

8 Dimensional shrinkage results from two sources during fusion welding. This first is the fact that all commonly welded metals and alloys undergo volumetric shrinkage when they solidify from a liquid to a solid. For pure metals, which solidify at a discrete temperature, this shrinkage occurs rapidly, whereas for alloys, which solidify over a range between their liquidus and solidus temperatures, shrinkage occurs more slowly. The second source of shrinkage is due to the thermal

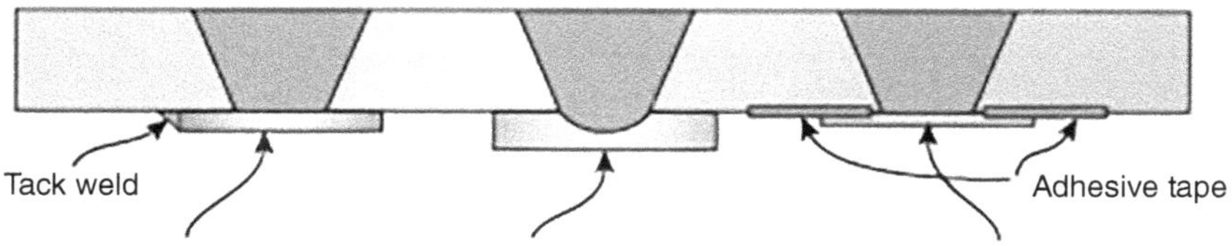

Figure 3.9 Schematic illustrations showing various possible backup bars, strips, and tape, some of which (e.g. strips) are intended to be a permanent part of the finished joint and others of which (e.g. bars and tape) are intended to be removed following welding. Source: Mathers. Used with permission of TWI.

Table 3.1 Joint setup problems: cause and effect.

o Poor fit-up of butts, tees, laps, corners, or edges > Dimensional or shape distortion
o Mismatch of top and/or bottom surfaces of abutted elements > Distortion, unplanned bending moments
o Joint gaps that are excessive, not uniform, or open during welding > Burn-through, weld slumping
o Excessive root gaps > Burn-through, weld slumping, weld underfill
o Insufficient root faces > Burn-through
o Angular misalignment > Dimensional or shape distortion, compromised performance

Table 3.1 summarizes the problems that can arise during joint setup.

Problems from linear misalignment, angular misalignment, and excessive root gap can obviously occur with weld joints of other types, such as corners, tees, and laps, but what has been presented here for butt joints is generally applicable elsewhere.

3.4 Problems with Weld Profile

Many problems can arise during the making of fusion-welded joints. These are generally referred to as *weld joint discontinuities*, with a rather comprehensive list given in Table 3.2. This book will address them all, but in this section, will focus on problems that manifest themselves in the so-called weld profile, which is cross-sectional geometry. Problems associated with internal soundness (e.g. porosity, cracks, and inclusions) and appearance or cosmetics, as well as problems that occur outside the fusion zone, in the HAZ or unaffected base metal, will be discussed in other chapters.

The most common problems that can arise with fusion-welded joints, or simply joints, are listed in the left-hand column of Table 3.2 and include, in the order in which they will be addressed here, (i) incomplete or insufficient penetration, (ii) lack of fusion or incomplete fusion, (iii) underfill, (iv) excessive reinforcement or convexity, (v) weld bead overlap, and (vi) undercut.

---

contraction of solid metals and alloys when they cool because of their coefficient of thermal expansion (CTE) on heating. Shrinkage from this source occurs progressively as cooling progresses.

Table 3.2 Weld joint discontinuities.

* Misalignment – Linear – Angular * Burn-through * Incomplete or insufficient penetration * Incomplete fusion * Underfill * Concavity or slumping * Excessive reinforcement/ convexity * Overlap * Undercut * Arc strikes * Spatter * Bead irregularity – Crown – Root	* Porosity – Uniformly scattered – Clustered – Linear – Wormhole/piping * Inclusions – Nonmetallic – Metallic * Cracks – Hot – Cold or delayed – Longitudinal/centerline – Transverse – Crater – Toe – Underbead – Root	* Base metal discontinuities – Banding/delamination – Lamellar tearing – Laps and seams – Voids * Heat-affected zone alterations – Hardening/embrittlement – Softening/weakening – Sensitization – Excessive grain growth – Reheat cracking * Partially melted zone defects * Dimensional shrinkage * Warpage/distortion * Instability/movement

Incomplete penetration or *insufficient penetration* is a discontinuity that manifests itself in any of three ways: (i) when the weld bead (i.e. fusion zone) does not penetrate the entire thickness of the base plate(s), when it is supposed to, (ii) when two opposing weld beads do not interpenetrate, as they must, and/or (iii) when the weld bead does not penetrate to the point where two joint elements form an inside corner (i.e. to the toe of a fillet weld) but only bridges across it. These different situations are shown in Figure 3.10. For a weld being made with a single pass (i.e. a single-pass weld), the problem arises when the weld bead does not reach the root of the weld groove. For joints requiring multi-pass welding, the problem arises when the first weld bead (or pass) does not start at the root of the weld groove.

The potential causes of incomplete or insufficient penetration in an approximate order of frequency are as follows:

- o Insufficient welding current for arc welding processes (insufficient voltage for EBW)
- o Too rapid welding (i.e. Travel Speed (TS) is too fast)
- o Improper welding technique (for situations (ii) and (iii))
- o Improper joint configuration to allow accessibility (e.g. U-better than J-butt)
- o Root face is too large
- o Root gap is too small (i.e. narrow)
- o Consumable electrode is too large in diameter
- o Joint element misalignment.

Lack of fusion or *incomplete fusion* occurs when molten weld metal fails to wet and adhere to joint faces or, for multi-pass welding, adjoining weld beads. Figure 3.11 schematically illustrates some examples for a single-pass and multi-pass welds.

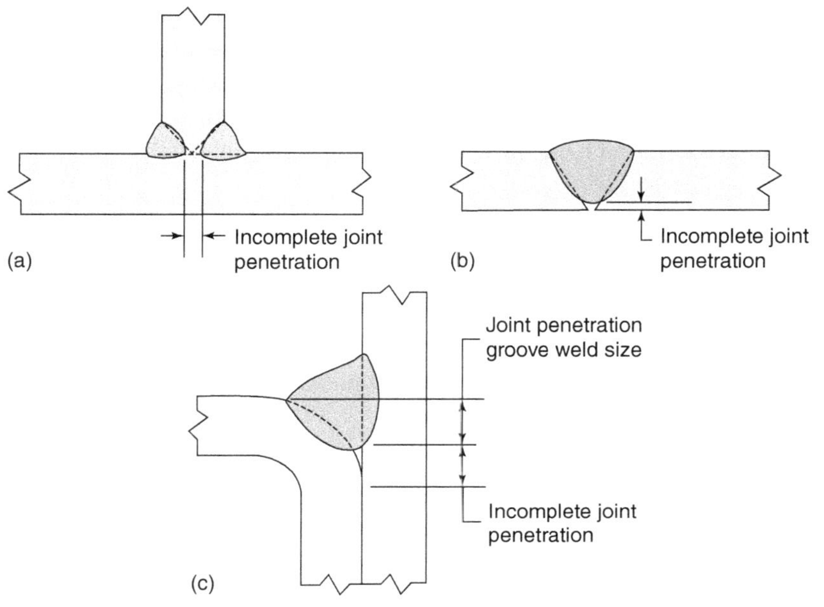

Figure 3.10 Schematic illustrations showing examples of incomplete weld or joint penetration for fusion arc welds made in a tee joint (a), in a single V-groove butt weld (b), and at the junction of two right-angle elements (c). Source: AWS (2012). Reproduced with permission of The American Welding Society.

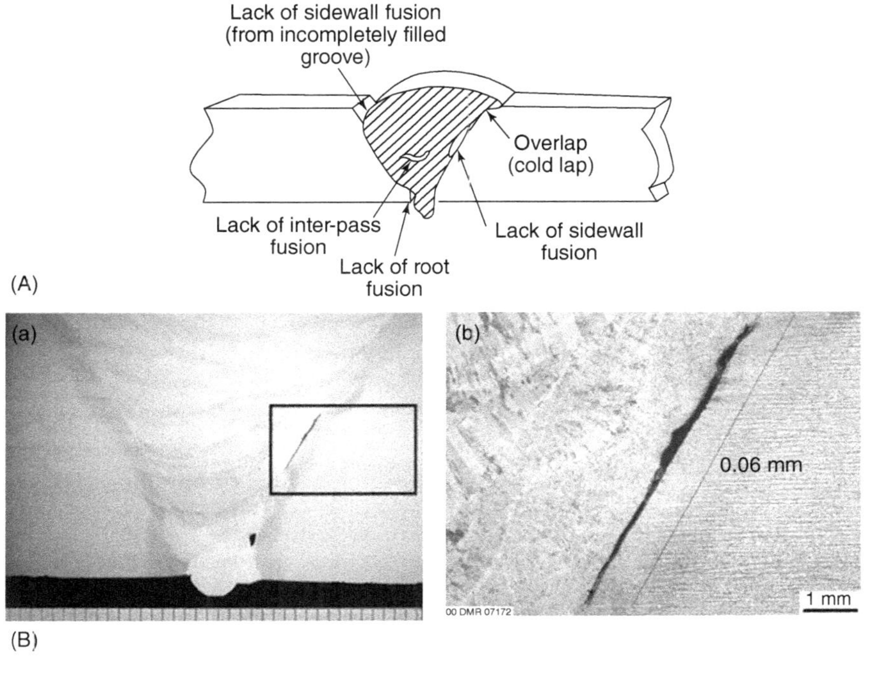

Figure 3.11 (A) Schematic illustration showing the various ways lack of fusion can appear in a fusion weld and (B) photomacrograph showing an actual incident of lack of fusion in a real weld. Source: Consonni et al. (2011). Reproduced with permission from TWI.

The potential causes of lack of fusion or incomplete fusion in an approximate order of frequency are as follows:

- o Improper manipulation of the welding electrode in arc welding
- o Surface oxide from improper cleaning of joint elements before welding (to remove oxides) or from inadequate shielding during welding
- o Improper heat input (i.e. too "cold" from too low current or too high TS)
- o Flooding or smothering the arc with the molten weld pool (from poor manipulation)
- o Arc blow (i.e. deflection by induced electromagnetic field)
- o Incorrect weld joint design to allow proper manipulation.

Freedom from oxide contamination on joint elements is especially important for highly reactive metals and alloys, such as aluminum and magnesium, as well as for stainless steels.

Figure 3.12 shows underfill in the crowns of butt and fillet welds made in a double V-groove and a schematic illustration of root concavity.

Underfill at the crown of a single-pass fusion weld can be caused by any of the following, in an approximate order of frequency:

- o Insufficient filler would definitely cause crown bead concavity
- o Excessive welding or TS may cause crown bead concavity
- o Excessive weld heat input (i.e. heat input per unit length) can cause slumping when root faces are melted too quickly
- o Slumping or even drop-through can occur if the root gap is too great.

For a multi-pass weld, underfill is almost always the result of insufficient weld metal or weld passes.

Underfill at the root of a single-pass fusion weld can be caused by any of the following, in an approximate order of frequency:

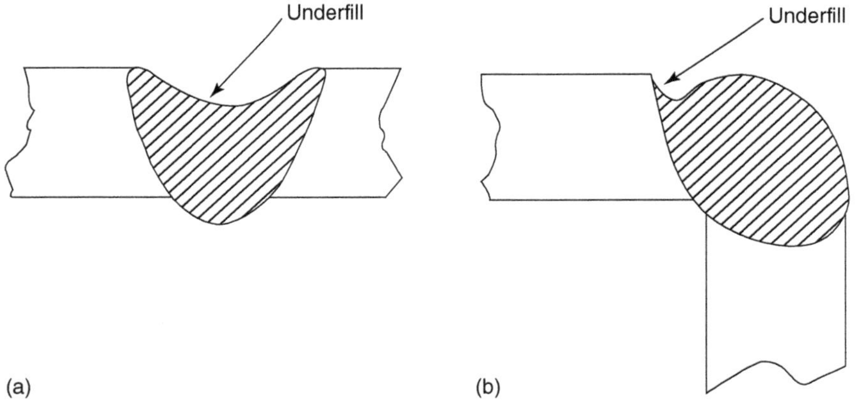

Figure 3.12 Schematic illustrations showing underfill in fusion welds made in a butt (a) and in a corner weld (b).

- Volumetric shrinkage of the weld pool in the through-thickness direction of the weld
- Insufficient arc power to produce positive bead; too low current or too high speed
- Root-bead suck-back can occur with excessive welding current or excessive weaving in the root
- Melting of a root pass by the second filling pass can produce root suck-back
- For slag-protected fusion welding, slag flooding of any backing bar groove[9]
- For GTAW, excessive backing pressure.

Fortunately, underfill from any cause can be fixed by welding to add filler, perhaps after first grinding away some of the prior weld's crown or root.

Excessive reinforcement or *excessive convexity* in a fusion weld is, quite literally, too much of a good thing. A properly made fusion butt weld (as well as fillet weld) should always have some convexity at both the crown or face and at the root (Figure 3.13). The purpose of having more weld metal than is needed to just fill the joint is to increase the cross-sectional area at the weld in order to offset some of the loss of strength in cast versus wrought microstructures. It also has a cosmetic effect in that it adds to a weld's attractiveness. This said, too much convexity at the crown of a butt weld (or fillet weld) can produce a stress riser or notch effect that can compromise the integrity of a structure, especially under fatigue loading, but also under impact.

The possible causes of excessive reinforcement or convexity in a fusion weld are, in an approximate order of frequency, as follows:

- Excessive welding heat from current that is too high or TS that is too slow
- Deposition of too much weld metal, often associated with inadequate weld joint preparation (e.g. shallow edge preparation)
- Faulty electrode manipulation
- Oversize electrode diameter
- Improper electrode angle.

Once encountered, excessive reinforcement can be corrected by removing the excess by machining or grinding to produce a smoother transition from weld bead to base metal or, in the worst cases, removal of the offensive welds and subsequent rewelding.

Figure 3.14 schematically illustrates excessive concavity and excessive convexity for fillet welds, whereas Figure 3.15 schematically illustrates the variety of problems that can affect fillet welds.

Weld bead overlap in fusion welding can occur at the crown or root of butt welds as well as in fillet welds. Overlap is defined as a protrusion of weld metal beyond the toe of the weld at its face or at its root. If not properly fused to the

9 A backing bar or backup bar is sometimes used at the root of a fusion weld to help with shielding and/or to help hold molten weld metal in a prepared groove until it solidifies. In fact, some codes require the use of a backing bar. See Figure 3.9 for examples.

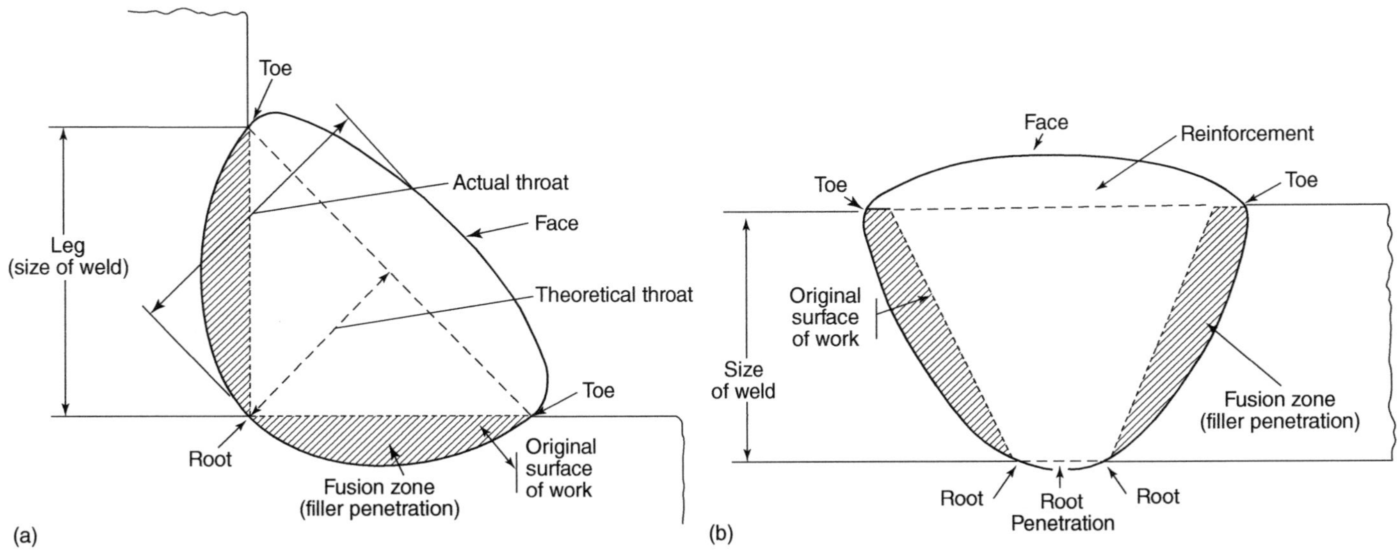

Figure 3.13 Schematic illustrations showing the proper profile for (a) a fillet weld and (b) a butt weld made by a fusion welding process.

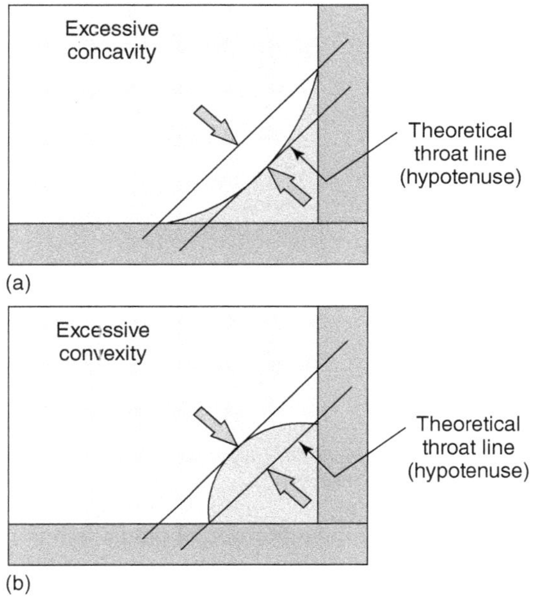

Figure 3.14 Schematic illustrations showing excessive concavity and excessive convexity in a fillet weld made by a fusion welding process. The ideal throat depth to meet design load requirements is shown by the "Miter." (a) With excessive concavity, there is insufficient weld metal to produce needed throat depth. (b) With excessive convexity, there is excessive weld metal and unnecessarily great throat depth.

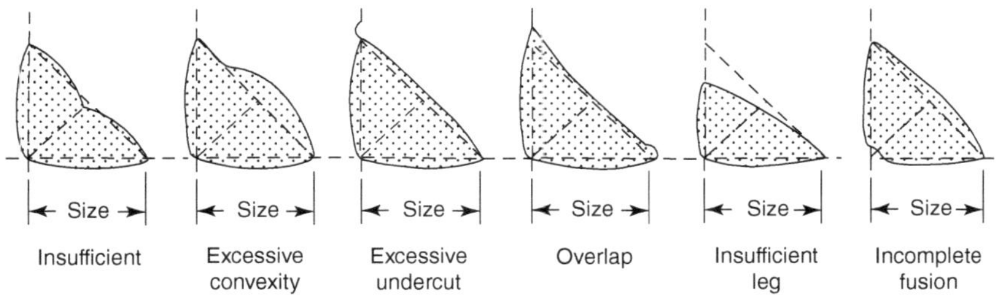

Figure 3.15 Schematic illustrations of the various problems that can appear in fillet welds. Source: AWS (2008). Used with permission of The American Welding Society.

base metal (i.e. forming a "cold seam"), a notch is produced, which gives rise to undesirable stress concentrations under loading, thereby compromising fatigue and impact (i.e. dynamic) properties. The condition is shown schematically in Figure 3.16.

Overlaps in fusion arc welding are caused by, in an order of frequency,

- o Improper welding technique
- o Insufficient welding current.

Some overlaps can be removed by grinding to blend the weld bead smoothly to the base metal, thereby eliminating the offensive notch.

Finally, the term *undercut* refers to either of two specific conditions. The first condition involves the melting away (or gouging or erosion) of the base metal at the side wall of a groove weld or at the edge of a bead, as shown schematically in Figure 3.16. The most obvious concern is the stress concentration that notch causes, with a secondary concern being the reduction of effective load-bearing

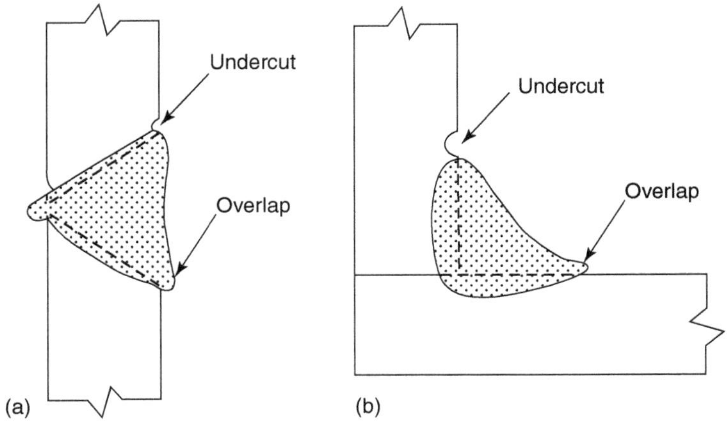

Figure 3.16 Schematic illustrations showing both overlap and undercut discontinuities in (a) butt welds and (b) fillet welds.

cross section at the undercut. Should a follow-up pass be made; it is possible for any slag present during the production of the initial pass to become entrapped as an inclusion. The second condition, also shown schematically in Figure 3.15, is the reduction of the base metal's thickness along a line where the weld bead (on the final layer of weld metal for multi-pass welding) ties to the surface of the base metal, i.e. at the toes of the weld. This condition, like the first, can occur on a fillet weld or a butt weld.

In many cases, the amount of undercut permitted at the surface of a completed weld is usually specified by the standard or code being used.

Both conditions of undercut can be caused in an approximate order of frequency by the following:

- Incorrect welding technique
- Incorrect positioning of a consumable (SMA, FCA, and GMA) electrode
- High welding currents with long arc lengths
- Excessive welding voltage
- Incorrect welding speed (typically, too slow).

As is always the situation in welding: doing things right in the first place is easier than fixing problems.

3.5 Troubleshooting Guide

Table 3.3 lists the various problems covered in this chapter, i.e. (i) problems with joint setup and (ii) problems with weld profile. For each problem listed, the likely consequence(s) is(are) given, followed by a brief proposed solution for the problem.

Table 3.3 Troubleshooting joint and weld profi e problems.

Problems with Joint Set-up:		
o Poor fit-up of elements	Dimensional and/or shape distortion	Gauge, then fixture or tack Change preparation method[a]
o Linear misalignment	Shape distortion	Gauge, then fixture or tack
o Excessive/opening gap	Weld slumping or burn-through	Gauge, then fixture or tack Employ back-up
o Excessive root gap	Weld underfill or burn-through	Redesign, gauge, etc. Employ back-up
o Insufficient root face	Burn-through	Redesign
o Angular misalignment	Shape distortion	Gauge, then fixture or tack
Problems with Weld Profile:		
o Incomplete/insufficient penetration		
	Insufficient current (or V for EBW)	Increase current (V for EBW)
	Too rapid welding	Reduce welding speed
	Improper technique (two-side or fillets)	Modify technique
	Improper prep configuration	Redesign weld groove
	Root face too large	Redesign to reduce root face
	Root gap too narrow	Increase root gap on set up
	Consumable electrode too large	Use smaller diameter
	Joint elements misaligned	Align with gauges, fixture/tack
o Lack of/incomplete fusion		
	Improper electrode manipulation	Adjust manipulation
	Surface oxide not removed	Properly clean of oxide
	Surface oxide from poor shielding	Increase/improve shielding
	Heat input too low	Increase I or decrease TS
	Weld pool flooding arc	Improve manipulation
	Arc blow/deflection	Move ground closer; use alternating current (AC)
	Incorrect joint design to allow proper manipulation	Redesign joint for better access

a) Preparation methods include, in increasing order of precision: oxy-fuel gas cutting, plasma-arc cutting, laser-beam cutting, routering; machining.

(continued)

Table 3.3 (Continued)

o Crown underfill		
	Insufficient fill (concavity)	Increase filler, add passes, apply final cosmetic pass
	Excessive welding speed (concavity)	Reduce welding speed
	Excessive heat input (slumping)	Reduce I or increase TS
	Slumping or drop-through	Reduce root gap
	Underfill of multi-pass weld	Add passes
o Root underfill		
	Excessive volumetric shrinkage	Add filler, redesign for smaller weld
	Insufficient arc penetration	Increase I or reduce TS
	Suck-back with single-pass	Reduce current Reduce/eliminate weaving
	Suck-back after a root-pass	Avoid melting root pass
	For slag-protected welding	Avoid slag flooding of arc
	For GTAW with backing gas-shielding	Reduce backing gas pressure
o Excessive crown or root convexity		
	Excessive welding heat input	Reduce I or increase TS
	Depositing too much filler	Check and reduce filler Check shallow joint edge prep
	Possible shrinkage across joint	Fixture or use strong-backs
	Faulty electrode manipulation	Adjust manipulation
	Oversize consumable electrode	Reduce electrode diameter
	Improper welding angle	Adjust electrode-to-work angle
o Weld bead overlap		
	Improper welding technique	Alter technique to correct
	Insufficient welding current (cold)	Increase I
o Undercut		
	Incorrect welding technique	Alter technique to correct
	Incorrect positioning of electrode	Lower angle 5–15° from 90°
	Excessive current with long arc	Shorten arc length to under 1/8th-in./3-mm
	Incorrect welding speed	Increase welding speed slightly

References

AWS (2008). *Structural Welding Code – Steel*, figure 5.4 A and C, AWS D1.1/D1.1M: 2008. Miami, FL: American Welding Society.
AWS (2012). Figure B.26, H, I and J, AWS A3.0 M/A3.0: 2012. American Welding Society, Miami, FL.
Consonni, M., Wei, C.F., and Schneider, C. (2011). Manufacturing of welded joints with realistic defects. In: *50th Annual Conference of the British Institute of Non-destructive Testing (NDT 2011)*, Telford, UK (13–15 September 2011). Cambridge, UK: TWI.
Free Education Network. *Fundamentals of Professional Welding*, figure 7.17. Free Education Network.
Mathers, G. Design – Part 4, figure 3. In: *Job Knowledge 93*. Cambridge, UK: TWI http://www.twi-global.com.
Messler, R.W. Jr. (1993). *Joining of Advanced Materials*. Stoneham, MA: Butterworth-Heinemann (now part of Elsevier, Ltd.).

Further Reading

American Welding Society (AWS) (2001). *Welding Handbook*, Volume 1: Welding Science and Technology, 9e (ed. C.L. Jenney and A. O'Brien). Miami, FL: American Welding Society.
American Welding Society (AWS) (2015). *Welding Inspection Handbook*, 4e. Miami, FL: American Welding Society.
Mandal, N.R. (2016). *Welding Defects*. SpringerLink, on-line.
Messler, R.W. Jr. (2004/1999). *Principles of Welding: Processes, Physics, Chemistry, and Metallurgy*. Weinheim/New York, NY: Wiley-VCH Verlag/Wiley.
Weber, T. (ed.) (1999). *The Everyday Pocket Handbook for Visual Inspection and Weld Discontinuities – Causes and Remedies*, Part 2. Miami, FL: American Welding Society.

4

Shape Distortion, Dimensional Shrinkage, and Geometric Instability

4.1 Thermal Versus Mechanical Stresses in a Structure

The preponderance of stresses[1] that arise in a material and, therefore, appear in a structure is mechanical in origin. Forces leading to stress may be from a source external to the structure (e.g. wind loads on walls, lift loads on wings, snow loads on roofs, traffic loads on bridges, seismic loads on buildings, and steam pressure on a boiler) or from a source internal to the structure (e.g. the weight of the structure acting on itself).[2] These are collectively known as *mechanical stresses*. However, forces and stresses that are resulted can – and often do – arise from another source, i.e. thermal sources. Unfortunately, these are sometimes forgotten, occasionally with severe consequences.

Thermal stresses or, more accurately, *thermally induced stresses* arise from a material and/or mechanical structure being acted upon by a temperature change or a temperature gradient, and not from temperature itself. There are three principal examples with particular relevance to welding, welds, and weldments: (i) stresses induced by a volumetric change,[3] involving either expansion or contraction associated with some change of phase in the material of construction; (ii) stresses induced by a difference in the coefficient of expansion (CTE)[4] between two materials linked together, known as CTE mismatch; and (iii) stresses induced by a temperature gradient resulting in differential amounts of expansion (on heating) or contraction (on cooling) within the volume of a material or within a structure. For stresses to arise from a phase change, the temperature must change to cause the phase change or transformation; the two prominent examples are the change from liquid to solid ($L \rightarrow S$) during solidification and the change from one solid phase to another ($S1 \leftrightarrow S2$) during allotropic transformation. For stresses to arise from a difference in the CTEs

1 Recall that a stress is the result of a force acting over an area, so that *stress = force/area*.
2 This latter source is sometimes said to involve inertial loads.
3 As used here, "volumetric change" refers to the change in the specific volume of a material, i.e. volume per unit mass, such as cm^3/g or $in.^3/lb$.
4 Relevant to the discussion in this book, metals expand when heated (and contract when cooled) as a result of the net potential energy versus atomic separation distance or spacing curve having an asymmetrical well, such that as temperature increases above absolute zero, the equilibrium interatomic spacing increases, giving rise to a positive *coefficient of thermal expansion* with units of cm/cm/K of °C or in./in./°F.

A Practical Guide to Welding Solutions: Overcoming Technical and Material-Specific Issues, First Edition. Robert W. Messler, Jr.

of abutting materials (i.e. from a CTE mismatch), the temperature may have stabilized or it may be changing. For stresses to arise from differing amounts of expansion or contraction from one region to another in a specific material, the temperature must change and produce a gradient. Whether such a temperature gradient persists or not, the thermally induced stresses from this source persist, provided and until any relaxation process occurs in the material.

To fix each of these three sources of thermally induced stresses in one's mind, respectively, think of the following: (i) how freezing of water can crush a boat as the specific volume of liquid water is greater as ice; (ii) how the bimetal couple in a thermostat bends one way or the other as its temperature is changed (down or up) to close or open an electrical contact to call for or turn off a heating system; and (iii) how a glass baking dish shatters when suddenly placed in cold water because of the different amounts of contraction that occur in the more rapidly cooling surface that slower cooling interior of the glass. However, how does this relate to welding?

4.2 Residual Stresses Versus Distortion

The examples just given exemplify how thermally induced stresses from any of the three possible sources (i.e. phase change involving specific-volume change, CTE mismatch between abutting materials, or temperature gradients within a material) are capable of causing cracking or fracture or geometric shape change or distortion. Whether cracking or distortion results without cracking is determined by one or both of the two obvious factors, in order of importance: (i) the magnitude of the stresses induced and (ii) the ability of the material to tolerate stress (actually, the strain produced by stress) without fracturing. A third, less obvious, factor is the degree to which a material and/or the component it comprises is restrained. If the aforementioned boat cannot escape the expanding ice, it gets crushed. If it can escape, it simply moves away from the growing ice. If the bimetal strip in the thermostat is free to bend, it bends. If it is restrained from bending, however, the stress trying to cause bending becomes locked in the strip, as a residual stress.

Relative to welding, if a material/component is free to move in response to a thermally induced stress during post-weld cooling, it will do so via a dimensional change (i.e. shrinkage) and/or a shape change (i.e. distortion), provided the volume in which stress has been induced is able to move. If it cannot move, the stress that tries to move it (i.e. shrink it or distort it) becomes locked in the material/component as residual stress. Although no one would argue that dimensional and/or shape distortion that occurs in a weldment is a problem that must be prevented or corrected, too many fail to appreciate the potential problems that can arise from disregarded residual stresses.

4.3 Origin and Effect of Volumetric Shrinkage

As the experiential knowledge of most people as to what happens when a liquid freezes to form a solid comes almost exclusively from water, it needs to be recognized that water is a rare – and fortuitous – exception. Liquid water is composed of molecules comprising two atoms of hydrogen covalently bonded to a single, central atom of oxygen with a very specific included angle (i.e. conformation angle[5]) of 104.45° (as shown in Figure 4.1a). These molecules constantly move around in a random and anharmonic manner because of thermal

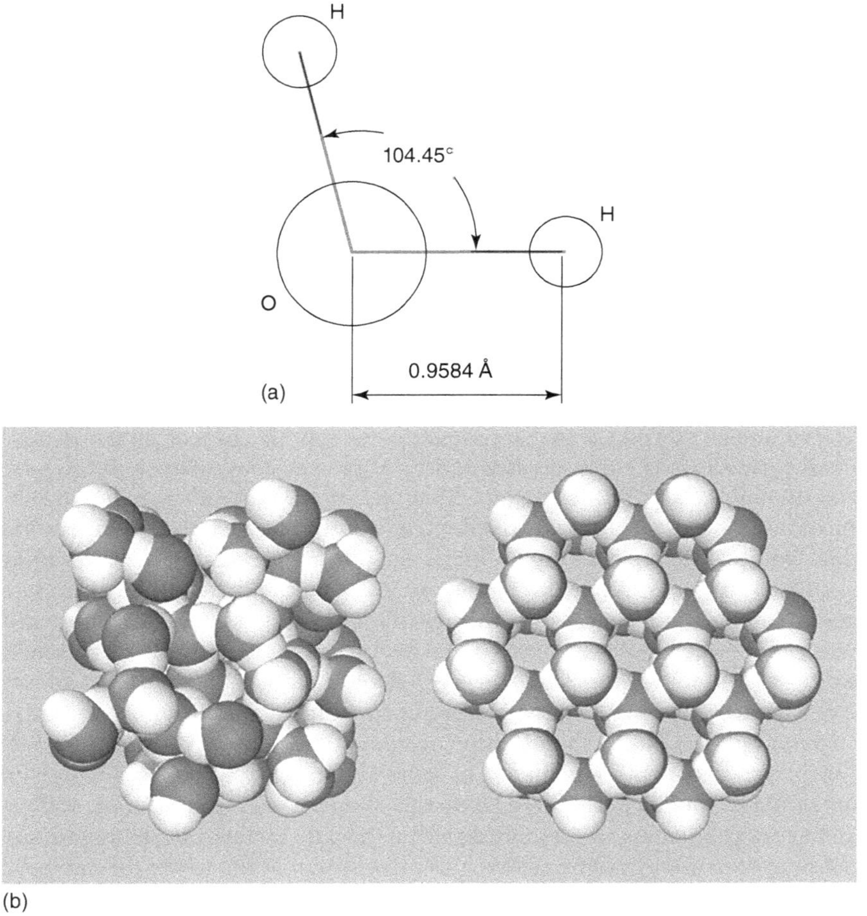

(b)

Figure 4.1 (a) Illustration showing the 104.45° conformation angle between the two hydrogen atoms that covalently bond to an oxygen atom in a molecule of water and (b) the difference in the density of packing of H_2O molecules in liquid water as opposed to in the crystalline form of ice (i.e. Ice III) typically encountered, with the solid ice being less dense. Source: (a) Booyabazooka (2006); (b) From the website www.chemisrty.elmhurst.edu, Elmhurst College, 190 Prospect Avenue, Elmhurst, IL 60126, used with permission of the Department of Chemistry & Biochemistry, Elmhurst College, Elmhurst, IL, USA.

5 The directionality associated with covalent bonding arises from the demands of the *Schrödinger wave equation* that describes the probability of finding an electron around an atom. Specific conformation angles allow the electrons involved in the covalent bond between atoms in a molecule to be shared, i.e. between the two hydrogen atoms and the single, central oxygen atom.

energy, causing liquid water to contract as motion slows down with a decreasing temperature, until the spacing between the moving molecules reaches a minimum at 4 °C, at which the density of liquid water is the greatest. Upon freezing to form a crystalline solid at 0 °C, the spacing between the adjacent molecules actually increases, and, as a consequence, solid ice has a greater specific volume (i.e. cm^3/g) or lower density (g/cm^3) than liquid water (as shown in Figure 4.1b).[6] The approximate 9% volume increase upon solidification causes ice to float in liquid water; a bad thing for moored boats, a good thing for fish!

In almost all metals and alloys, however, this is not the case. Most metals solidify to form close-packed crystal structures (i.e. face-centered cubic, FCC, or hexagonal close-packed, HCP), with 74% of the volume of either crystal structure (or unit cells that represent the crystal structures) being atoms and the rest being empty space (i.e. interstices). For those metals that do not solidify to form close-packed FCC or HCP structure, all but a few solidify to have a body-centered cubic, BCC, crystal structure, with 68% of the volume of the crystal structure being atoms and the rest being empty space. This, it turns out, is still much closer packing than that found for the atoms (in molecules) in solid, crystalline ice. Hence, almost every metal has higher density as a solid than as a liquid, so that the crystals of the solid sink in the liquid, and solidification is accompanied by volumetric shrinkage.[7]

The volumetric shrinkage that accompanies the solidification of almost all metals and their alloys is either greater and/or of greater consequence in the pure metal than in the alloy.[8] The reason is that pure metals (like all pure liquid substances) transform from liquid to solid (i.e. solidify) at a discrete temperature rather than over a range. In a pure metal, solidification takes place at a *melting point* (if one is picky, a freezing point) or MP. In an alloy (as in most impure liquids), solidification occurs over a range of temperatures between the *liquidus temperature* (where solid first can be formed upon cooling under equilibrium conditions) and the *solidus temperature* (below which no solid can exist and/or above which, on heating, the first melting occurs under equilibrium conditions), i.e. a *freezing range*. When solidification occurs over a range on cooling, volumetric shrinkage occurs more gradually as more and more solid replaces more and more liquid as cooling progresses. Upon the completion of solidification, the net shrinkage may be similar (unless altered from the pure metal by the alloy addition or additions), but may not be as much of a disruption to the newly formed solid structure.

Table 4.1 lists the key properties of commonly welded metals and alloys, as these properties affect shrinkage and distortion. Although published values of

---

6 The actual angle found in the most familiar form of ice (i.e. type III) is not quite the tetrahedral angle of 109.47° but, rather, about 107°, with the crystal structure being essentially hexagonal. This hexagonal structure gives snowflakes, and Jack Frost patterns of ice on windows, their beautiful sixfold symmetry, in endless varieties.

7 Should you be asked in a game of Trivial Pursuit™, or wish to end a conversation at a cocktail party, five pure metals (as well as some of their alloys) that expand upon solidification are Bi, Ga, Li, Na, and Sb. The reason is either an unusual non-close-packed crystal structure (e.g. rhombohedral Bi and orthorhombic Ga) or a significant degree of covalency in the mixed covalent–metallic bonding.

8 Most molten metals and alloys also shrink in volume as the liquid cools. However, this contribution to overall volumetric shrinkage, compared to that from solidification, is very small.

Table 4.1 Key properties of commonly welded metals and alloys, as these affect shrinkage and distortion with fusion welding.

Metal/alloy	Melting point/ range (°C)	L–S Δvol. shrinkage (%)	CTE @20 °C (10^{-6}/°C)	Thermal conductivity@ 0–25 °C (W/m-K)	Density L/S @MP/20 °C (g/cm³)	Elastic modulus (GPa)
Aluminum (1100)	660.32	−6.6	23.1	237	2.38/2.72	69
Al–Cu 2xxx	502–638	–	22–24	116–170	2.79S	72–74
Al–Mg + Si 6xxx	582–652	–	23.0–23.5	160–170	2.72S	69–70
Al–Mg–Zn 7xxx	477–635	–	23.2–23.6	125–135	2.83S	71–76
Cobalt	1495	N/A	13.0	100	8.86/8.90	209
Co-alloy	1397–1450	–	11–12	50–70	8.75S	199
Copper	1084.62	−4.9	16.5–16.8	401	8.02/8.96	130
Cu–Zn brasses	900–940	−1.4/−1.6	18–21	120–140	8.5–8.7S	100–125
Cu–Sn bronzes	850–1000	−1.1/−2.2	17.5–19.5	50–106	8.8–8.9S	96–120
Cu–Ni cupro-nickel	1170–1240	–	15.5–17.0	20–65	8.91–8.94S	120–156
Iron	1538	−3.4	11.8	80.4	6.98/7.874	207
Low C steel	1460–1520	−2.5/−3.0	11.5–11.7	60–65	7.85S	211
Medium C steel	1430–1495	−3.5/−4.0	11.0–11.3	50–55	7.85S	211
Low-alloy steel	1425–1500	−2.5/−3.0	12.2–13.8	40–50	7.8–7.9S	211
Cast irons	1127–1204	−2.0/−4.5	10.5–12.0	47–80	7.2–7.4S	165–180
Austenitic SS	1400–1450	–	16.0–17.3	12–40	8.0–8.1S	193
Ferritic SS	1425–1510	–	9.5–11.0	25–35	7.75–7.8S	210
Martensitic SS	1450–1510	–	10.0–10.5	25–30	7.65–7.7S	210
PH SS	1400–1440	–	10.6–10.9	17.5–19.0	7.6–7.66S	204
Lead	327.46	N/A	28.9	35.3	10.66/11.34	16
Pb alloys	240–290	–	20–27	50–60	11–11.3S	12.5–16
Magnesium	650	−4.0/−4.2	24.8	156	1.58/1.74	45
Mg alloys	349–645	–	26–27	80–100	1.8–2.4S	42–47
Molybdenum and alloys	2623	N/A	4.8	138	9.33/10.28	329
Nickel (200 Grade)	1455	N/A	13.4	90.9	7.81/8.91	204
Ni–Cu Monel alloys	1300–1350	–	13.0–14.0	25–35	8.4–8.8S	180–210
Ni–Cr alloys	1400–1480	−0.9/−2.0	9–16	8–18	7.9–8.5S	160–240
Inconel alloys	1390–1425	–	11–12	12–20	8.4–8.5S	205–210
Tin	231.93	−3.2	22.0	66.8	6.99/7.15	50
Sn alloys	170–230	–	21–22	N/A	7.2–7.6S	55–60
Titanium (CP)	1668	N/A	8.6	21.9	4.11/4.51	103
α Alloys	1520–1590	–	9.3–9.5	7.8	4.5–4.6	90–120
Ti-6-4	1878–1933	–	8.8–9.0	7.1–7.3	4.43	113.8
Ti-6-6-2	1588–1716	–	7.6–7.8	7.0–7.2	4.54	110.3
Tungsten and alloys	3422	N/A	4.5	173	17.6/19.25	411
Zinc	419.53	−7.28	30.2	116	6.57/7.14	99.3
Zn alloys	290–350	–	21–22	N/A	7.1–7.2	70–105
Zirconium	1855	N/A	5.8	22.6	5.8/6.52	88
Zr alloys	1750–1950	–	6.0–6.3	21–22	6.5–6.9	88–92

the percentage by which the volume of a liquid shrinks once it has solidified are given (i.e. L–S Δvol. shrinkage), few such values are readily available without thoroughly searching published reports and papers (shown as "N/A" or by "–"). Where a value is not listed, a good approximation can be obtained by using the listed value for the density of the liquid metal at its melting point and the value of density of that same metal at room temperature (i.e. 20 °C). By converting these density values to the values of specific volume by taking the reciprocals, the percentage shrinkage can be calculated by subtracting the specific volume of the solid from that of the liquid, dividing by the value for the liquid, and multiplying by 100. Some error occurs as, to be more precise, the value of specific volumes (perhaps from densities) used in this calculation should be for the liquid just above the MP and for the solid just below the MP. Although values of the density of a solid just below the MP are rarely available, an approximate value can be obtained by adjusting the value correcting for thermal expansion from RT to just below the metal's MP. Readers are cautioned from trying to obtain a precise value, however; adhering to the author's mantra "Never try to measure a manure pile with a micrometer. It's just a pile of poop!"

The result of volumetric shrinkage accompanying solidification is the development of thermally induced stresses in the newly formed solid structure (e.g. weldment). These stresses may cause cracks to form in regions where the last remaining liquid is yet to transform to solid, producing what are known as *hot cracks* (see Chapter 7). Alternatively, these induced stresses may cause dimensional shrinkage in the newly formed structure, if the structure is free to comply (i.e. is not restrained), or give rise to locked-in residual stresses, if the structure is not free to comply (i.e. is restrained). We will consider both of these in Section 4.6.

4.4 Origin and Effect of Thermal Contraction

One of the things that characterizes all metals, without exception, is that: *All solid metals expand on heating and contract on cooling.*[9] The contraction of metals and alloys once they have become entirely solid and are further cooled is frequently called "shrinkage." However, to be correct, and unambiguous, when speaking about cooling materials (herein, metals and alloys), *shrinkage* refers to a lowered specific volume accompanying a phase change,[10] whereas *contraction* refers to the shortening of a linear dimension (albeit perhaps different in different directions in a crystal) whether or not a solid-state phase change is involved.

9 Two other things that characterize all solid metals are as follows: (i) All metals are conductors of both electricity and heat and (ii) the electrical resistivity of all metals increases as temperature increases. These characteristics, like thermal expansion with heating and contraction with cooling, are the consequence of metallic bonding.

10 The change from liquid to solid phase known as solidification, with volumetric shrinkage being the norm, has been discussed above. In the solid state, some metals exhibit different crystal structures at different temperatures, with transformations from one to the other being known as *allotropic transformations.* It is usually the case that solids with different crystal structures, even though they may have the same composition, have different specific volumes, smaller or larger one to the other.

How much a metal or alloy expands on heating or contracts on cooling is given by its coefficient of thermal expansion or CTE, which is positive for heating and negative for cooling, albeit having the same magnitude at the same temperature.

The units of coefficient of thermal expansion are unit length/unit length/degree of temperature change, i.e. mm/mm/°C or K (SI), in./in./°F (English). In both systems, the change per degree of temperature is very small, almost always in microunits, e.g. 10^{-6} mm/mm or 10^{-6} in./in. In order to access the full effect of a temperature change in a solid, one needs to consider not only the CTE (α) for the material but also the elastically continuous length of the structural member (L) and the amount by which temperature changes (i.e. the temperature change ΔT), i.e. $\Delta L = L\Delta T\alpha$. To appreciate the impact of these other factors, consider how much a single steel rail of a railroad track (and, thus, the entire track) that crosses the Great Victoria Desert of Outback of Western Australia changes in length between summer and winter.

As the rails of modern railroads are welded end-to-end, they are "elastically continuous." The Great Victoria Desert is 800 km wide, and the temperature reaches a peak of 50 °C in summer and −10 °C in winter. For rail steel, for which the CTE is 11.7×10^{-6} m/m/°C, the change in length of the track would be [(800 km)(1000 m/km)][50−(−10)°C][11.7×10^{-6} m/m/°C] = 561.6 m or 0.56 km! If this does not seem too bad, look at the buckling that has occurred (Figure 4.2).

As it pertains to welding, and weldments, the thermal contraction as a weld cools following solidification, like solidification shrinkage, gives rise to thermally induced stresses through the equation $\sigma = E\alpha\Delta T$, where E is the modulus of elasticity and α is the coefficient of thermal expansion (here, contraction) of the base metal, whereas ΔT is the temperature change from solidification to room temperature. Unlike shrinkage stresses, thermal contraction stresses that develop after the weld has completely solidified do not act on not-yet-solidified

Figure 4.2 Photograph showing buckling of the steel rails of a railroad track because of thermal expansion under unusually high atmospheric temperatures. Some dramatic examples of the so-called "rail buckling" can be viewed in YouTube. Source: Photograph by the US Department of Transportation and in the public domain for free use.

regions in the fusion weld to produce hot cracks. On the other hand, like shrinkage stresses, they can and do act on the newly formed *and* surrounding ever-solid, heat-affected zone (HAZ) to induce stresses that can distort the structure (i.e. change the structure's shape) if it is free to move or, if it is not free to move, introduce residual stresses.

Figures 4.3–4.6 schematically illustrate how the combination of the volumetric shrinkage that occurs as fusion welds solidify *and* the subsequent thermal contraction of newly formed solid in the fusion zone and ever-solid, HAZ produces residual stresses in and around welds and in some simple but common welded shapes (i.e. weldments).

In Figure 4.3, fundamental dimensional and shape changes that can take place in weldments from shrinkage across the weld (i.e. transverse shrinkage), along the weld (i.e. longitudinal shrinkage), and as angular distortion or bowing are schematically illustrated. *Transverse shrinkage* arises from solidifying the fusion zone. *Longitudinal shrinkage* arises from end effects, as a weld is started on or enters a part from a start tab or is terminated on a part or exits a part onto a stop tab.[11] *Angular distortion* or *bowing* arises from through-the-thickness differences

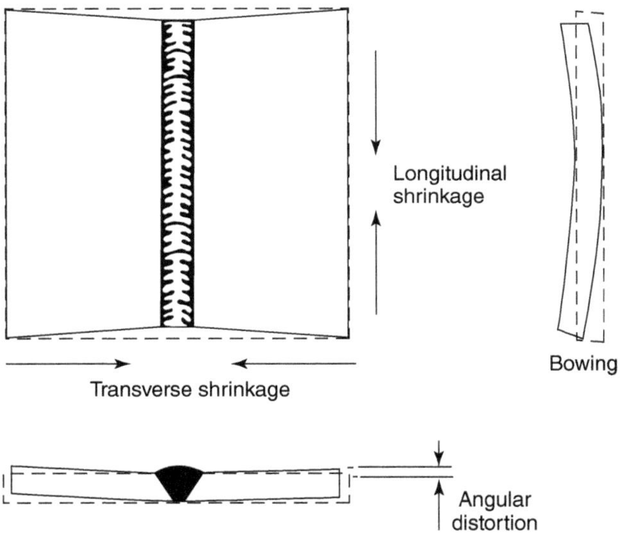

Figure 4.3 Schematic illustration showing the fundamental dimensional changes that can occur in weldments, including transverse shrinkage, longitudinal shrinkage, and angular distortion or bowing. Source: Allen (1981). Used with permission of TWI.

11 Once fusion welding has begun, and the heat source is moved at a constant speed along a part of uniform material and thermal mass (e.g. thickness), heat flow becomes quasi-steady state (i.e. the same everywhere along the weld path). Hence, thermally induced stresses only arise where there is some disruption to this quasi-steady state, i.e. where there is a transient condition of heat flow, similar to that at the free ends of the weldment. Tabs with the same composition as the base metal are sometimes added so that a weld can be started and stopped on these tables, rather than on the part. These are known as start and stop tabs. They prevent certain types of defects (see Chapter 6, "crater cracks").

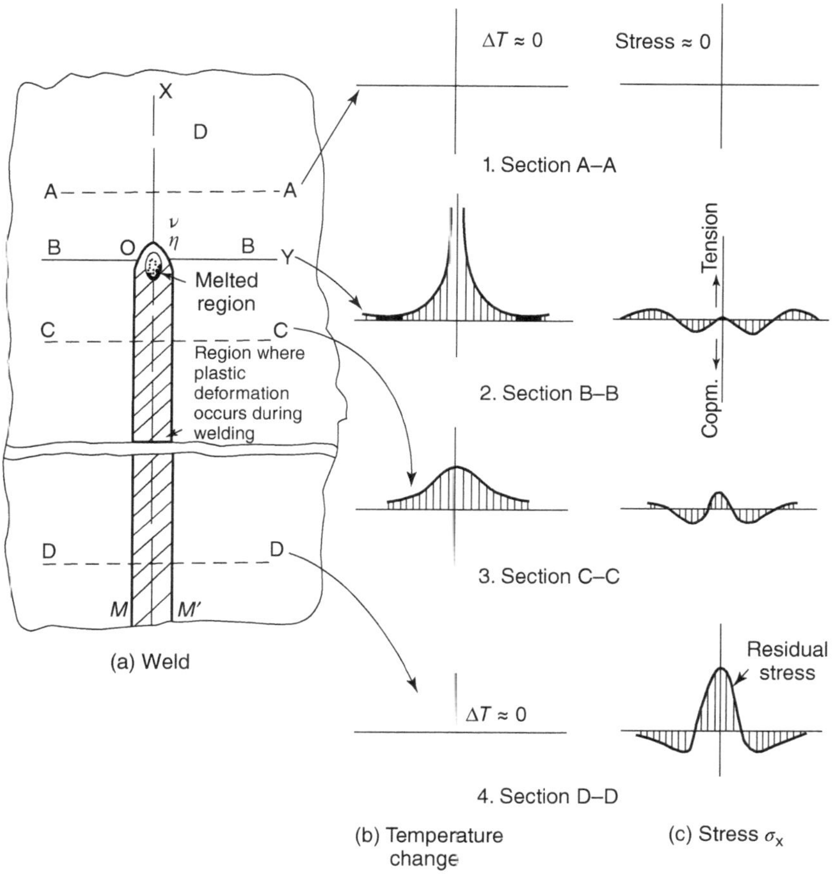

Figure 4.4 Schematic illustrations showing typical distribution of temperature and stress at several locations in a bead-on-plate weld. Source: Masabuchi (1981). Used with permission of The Minerals, Metals and Materials Society (TMS).

in shrinkage caused by welds that are wider at their face or crown than at their root (i.e. V-shaped), so that more shrinkage occurs at the crown than at the root because of the difference of the volume of the weld metal at each location. Of course, for a weldment to distort as illustrated, it must not be restrained from doing so, either by its sheer size, by surrounding components already welded on, or by fixtures.

Figure 4.4 schematically illustrates the typical distribution temperature and thermally induced stress at various locations in a bead-on-plate weld, in which a molten bead was made by a welding heat source moving along a plate without any joint therein.[12] Notice that induced tensile (i.e. pulling) and compressive (i.e. pushing) stresses (shown at the extreme right side of the figure) are – and must

12 *Bead-on-plate welds* are often made either to assess the inherent weldability of an unfamiliar base material or to zero-in on a set of welding parameters (i.e. voltage, current, and travel speed) to produce a weld of desired depth of penetration and crown and root width.

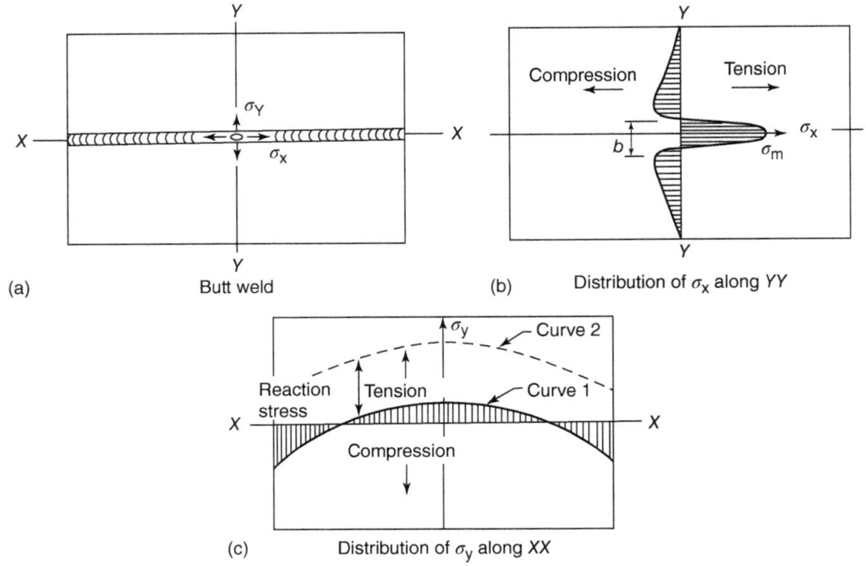

Figure 4.5 Schematic illustrations showing the typical distribution of residual stresses along (i.e. longitudinal stresses, c) and across (i.e. transverse stresses, b) a butt weld (a). Source: Connor (1987). Reproduced with permission from The American Welding Society.

always be – balanced (i.e. the shaded areas above and below the horizontal axis are equal), so the structure in in mechanical equilibrium. Thermally induced stresses that exceed the tensile strength of a material just below its melting (or freezing) point (or solidus temperature) can cause cracking of the newly solidified metal. Elsewhere in a weldment, residual stresses, unless removed by post-weld heat treatment (e.g. stress relief), can reach the yield (or flow) stress for the material.

Figure 4.5 schematically illustrates the typical distribution of residual stresses transverse to (in B) and longitudinally along (in C) a butt weld line. Once again, note that tensile and compressive stresses are balanced (i.e. shaded areas on each side of the line X–X and the line Y–Y are equal). Such stresses can easily lead to distortion of the weldment, if it is not restrained from moving, or, if it is so restrained, and lead to other problems from the locked-in stresses, as described in Section 4.6.

Finally, just to help with understanding the problems that can arise from thermally induced stresses, Figure 4.6 schematically illustrates the typical residual stress patterns for some simple welded shapes, including a T-joint made with opposing fillet welds (at the upper portion of the figure), an H-shape made using fillet welds (at the bottom left of the figure), and a box-beam shape made using corner welds (at the bottom right of the figure). One can only imagine the severity and complexity of stress patterns that can arise in geometrically more complex and larger weldments.

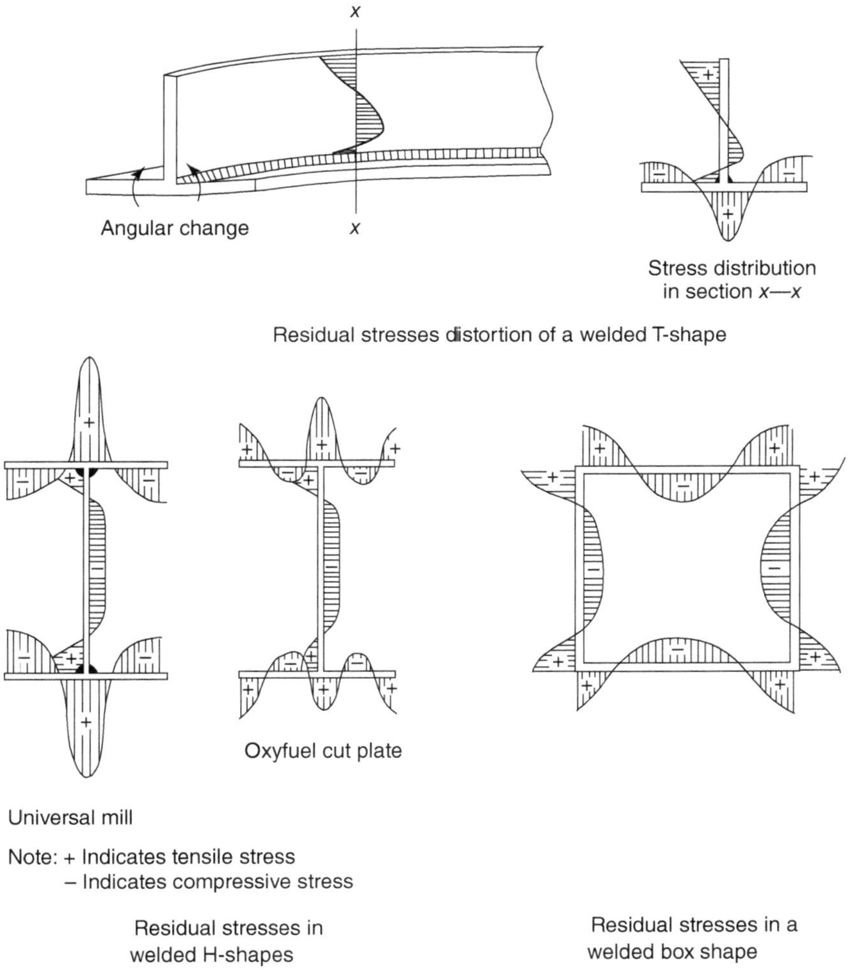

Figure 4.6 Schematic illustrations showing typical residual stress patterns for some simple welded shapes. Source: Connor (1987). Reproduced with permission from The American Welding Society.

4.5 Problems from Nonuniform Thermal Contraction and CTE Mismatches

Volumetric shrinkage and post-solidification thermal contraction can, as described, lead to dimensional errors (e.g. dimensions being too short), warpage or distortion to required shape, or cracking in the weld fusion zone, partially melted zone, or even very-high temperature HAZ.[13] However, other problems can arise from the CTE of a base material of the different CTEs of dissimilar base materials.

The CTE of a material is not a constant. Rather, it is a function – often a strong function – of temperature.[14] This can be seen in Figure 4.7, which shows the effect of temperature on the CTE of an austenitic stainless steel used for cryogenic applications. As a result of this factor and the fact that in the presence of a temperature gradient, as always found in the HAZ surrounding a fusion weld, the magnitude of contraction that occurs during post-weld cooling is greater for the hotter regions and less for the cooler regions. Worst of all, it is different everywhere across the HAZ, as the temperatures and temperature gradient are constantly changing as cooling progresses. The result is a further cause of thermally induced stresses around a weld because, in this case, of *nonuniform thermal contraction.*[15]

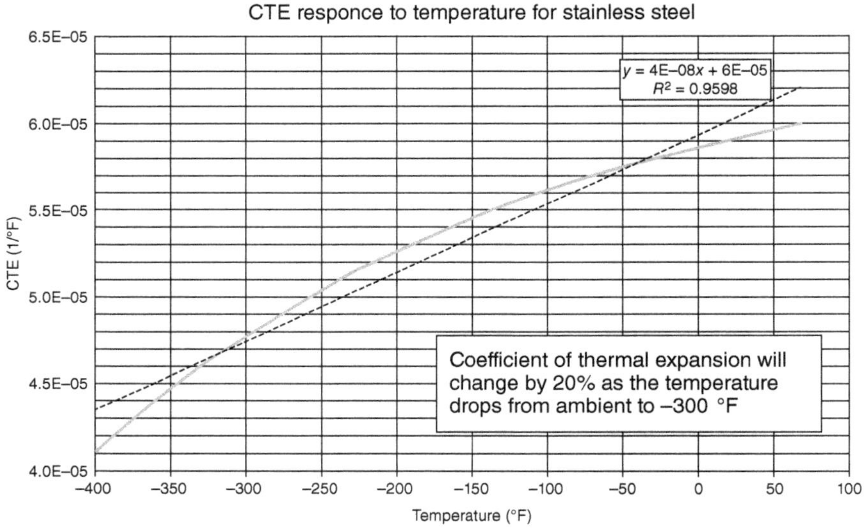

Figure 4.7 A plot showing the effect of temperature on the CTE (as in./in./°F) of an austenitic stainless steel (such as 304) commonly used in cryogenic applications. Note the sensitivity to temperature. Source: From the website of Wasatch Composite Analysis, LLC, Park City, UT, USA; used with permission of Wasatch Composite Analysis, LLC.

13 Cracking in the very-high temperature heat-affected zone arises from the highly nonequilibrium nature of fusion welding. The more cooling deviates from equilibrium (which must, by definition, take place infinitely slowly), the more phase transformation start temperatures are suppressed. The actual final temperature where the last liquid solidifies is always below the equilibrium solidus temperature and may be suppressed as much as 100 of more degree Celsius. Hence, subsolidus hot cracking is not uncommon.

14 Locating reliable data of the CTE of a material as a function of temperature can be quite difficult. Rather than plots, one is more likely to find tabulations that give the CTE for a material in different ranges, from near RT to higher and higher service temperatures.

15 Recognize that the metal that is progressively nearer the fusion zone has expanded progressively more on heating, so it contracts more progressively on cooling back to RT. Thermally induced tensile stresses develop when the contracting regions pull against less expanded regions.

A final source of problems that can arise from CTEs during welding is when different base metals are being welded and their CTEs are mismatched (i.e. are not the same). A general rule of thumb, not found in any specific reference but, rather, obtained from successful designers (as problems from CTE mismatch are common in structures comprising mixed materials not necessarily joined by welding), is that the CTE of abutting materials should not differ by more than about 20%. The more inherently brittle a material (i.e. the lower its ductility), the less mismatch is tolerable, with cracking being a possibility.

A modern example of where CTE mismatch becomes a particular problem involves the mounting of chips and various other electronic components onto circuit boards, heat sinks, etc., as seen for the IGBT module shown in Figure 4.8.

Table 4.1, described earlier, includes values of CTEs for commonly welded metals and alloys. The listed values are generally precise for the pure metal, whereas the values are given as a range for alloys where that metal is the base (solvent). The given range tries to represent the most common alloys for each base metal listed. Readers are encouraged to seek specific values for specific alloys.

The other properties listed in Table 4.1, including thermal conductivity k, density ρ, and elastic (Young's) modulus E, can be used in the calculations of heat flow, if desired, using the generalized equation of heat flow with simplifications in dimensionality, in particular.[16]

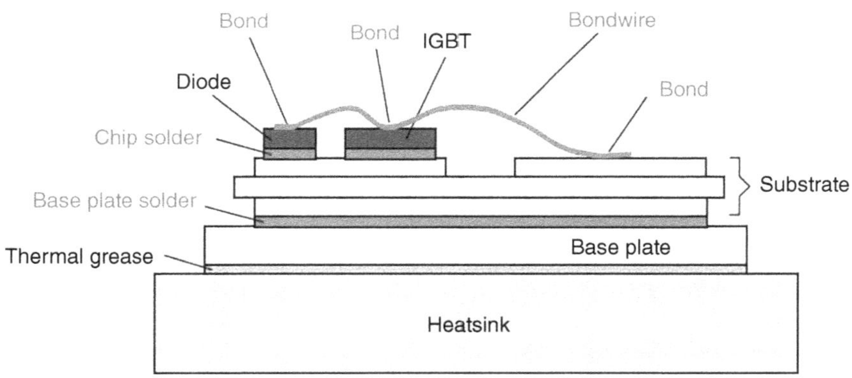

Figure 4.8 Illustration of a modern IGBT module showing the complex thermal–mechanical situation created by widely differing CTEs for the materials used in various portions of the module. Thermal grease is shown as the highly compliant interface between the base plate and the heat sink. A common FR-4 polymer-matrix composite circuit board has a CTE of 14.17 (in SI units), whereas pure aluminum used in the heat sinks has a CTE of 24 (in SI units). It is not unusual to find CTE mismatch between abutting materials of greater than 25–50%. Source: Mentor Graphics Corporation (2016). Used with permission of Mentor Graphics Corporation.

---

16 Two other thermophysical properties found in heat flow equations are specific heat C_p and thermal diffusivity α. Although the values for these properties are not listed in Table 4.1 (for brevity), an approximate value for the specific heat (if not found by a search) can be obtained from knowing the molar heat capacity of metals and alloys ranging between 22 and 24. Hence, an approximation of C_p can be obtained by dividing these numbers by the molar weight of a metal or alloy. A value for thermal diffusivity can be found from $\alpha = k/\rho\, C_p$.

4.6 Problems from Distortion and from Residual Stresses

The effects of distortion or warpage as a result of welding are much more obvious than the effects of residual stresses, but, for that reason, may be far less insidious. When linear or angular dimensions are not what they are supposed to be because of the shrinkage accompanying welding, such that functionality is compromised, detection, at least, is easily accomplished by inspection. Such linear and/or angular distortion becomes problematic when the resulting unintended alteration of geometry impedes fit or prevent function or compromises the level of the needed performance.[17]

The effect(s) of potential effect(s) of residual stresses is(are), regrettably, often unknown, underappreciated, or ignored. Locked-in or residual can lead to any of several problems, including the following:

- Movement of parts that require precise dimensions when machining removes the material and allows the part to move by the relaxation of such removal causes (i.e. geometric instability)
- Shortened fatigue life, as applied stresses add algebraically to unknown residual stresses
- Potentially compromised loss of impact toughness or fracture toughness for a given material
- Accelerated corrosion for a fixed environment
- Long-term delayed cracking due to the interactions between residual tension stresses and hydrogen (from any source) in untampered martensite in the fusion zone or, more commonly, the HAZ of steels
- Long-term shape instability, as long-term stress relaxation may occur in materials having lower melting points or materials/components operating at elevated temperatures (another manifestation of geometric instability).

To be safe, it is always advisable to stress relieve a weldment using post-weld heat treatment. The extra time and money it takes for the stress relief treatment is much less than the time and money lost when a weldment fails because of the failure to relieve weld-induced residual stresses.

4.7 Distortion Control and Residual Stress Reduction

Unless done thoughtfully and properly, attempting to control or reduce distortion can lead to residual stresses and, frustratingly, attempting to control or reduce

17 It is worth noting that thermally induced stresses can be used as an advantage, by introducing an intentional bowing into the longitudinal beams of a flat-bed trailer that causes it to have an upward hump when not loaded. The process for introducing the hump is known as heat shaping, while process for removing distortion is known as heat straightening. Both employ selective heating and rapid cooling (by water quenching) in localized regions of parts to be shaped or to have warpage removed.

residual stresses can lead to distortion. It is a Catch-22[18]; likened to the oft cited example "How is one supposed to gain the experience to get a license to drive a bus, if one must have a license in order to drive a bus to gain the experience?" Let us see why.

Controlling distortion in weldments begins with proper design and benefits from a proper setup and/or assembly procedures as well as from some techniques that can be employed during welding.

Proper design includes the following: (i) employing the minimum number of parts or components, thereby minimizing the amount of welding needed, (ii) selecting the proper joint configuration (especially for unrestrained butt joints) to minimize angular distortion (e.g. using double-V or double-U as opposed to single-V or single-U configurations to balance heat input and shrinkage), and (iii) using minimum-sized fillet welds. Proper setup and assembly procedures include (i) avoiding linear misalignment or mismatch at butt joints by checking with gauges and/or using tack welds to prevent movement, (ii) presetting joint members to compensate for expected angular distortion from shrinkage in single-sided welds with wider crowns than roots, and (iii) assembling the entire weldment so that it is nominally correct before welding and then using some form of restraint (e.g. fixtures and tack-welded strongbacks) to minimize distortion from welding. Techniques that can be used to help control distortion include (i) keeping net linear heat input to the minimum needed to produce a geometrically acceptable weld, (ii) sequencing welding to balance heat input and create offsetting shrinkage/distortion, and (iii), in some instances, preheating to help reduce distortion from nonuniform thermal contraction by minimizing temperature gradients in the weld zone.

Figure 4.9 schematically illustrates several techniques for preventing or minimizing distortion.

Care must be exercised with some of the preceding, as reducing distortion should not come at the expense of introducing locked-in or residual stresses, particularly if no post-weld stress relief treatment is planned. In restraining components of a weldment from distorting, residual stresses result. This especially includes the use of fixtures or temporary tack-welded strongbacks. The advantage of presetting components (e.g. in a butt joint as shown in Figure 4.9a) is that there are low or no residual stresses once the assembly has been welded and seeks its own final shape. Proper sequencing of welds (as shown in Figure 4.9c) certainly helps minimize distortion, but also results in residual stresses once the growing structure must be forced to fit together for the final weld(s).

Once distortion occurs and must be removed to allow proper fit and function of the weldment, there are two options: (i) mechanical straightening using presses or jacks and (ii) thermal straightening using localized heating and cooling to cause compensating expansion and contraction (in what is commonly called *heat straightening* or *flame straightening*, as oxy-fuel gas flame is usually used). Mechanical straightening is easier to understand and, thus, much more

18 A *Catch-22* is a paradoxical situation from which an individual cannot escape because of contradictory rules. The term was coined by author Joseph Heller (1923–1999), who used it in his 1961 novel Catch-22.

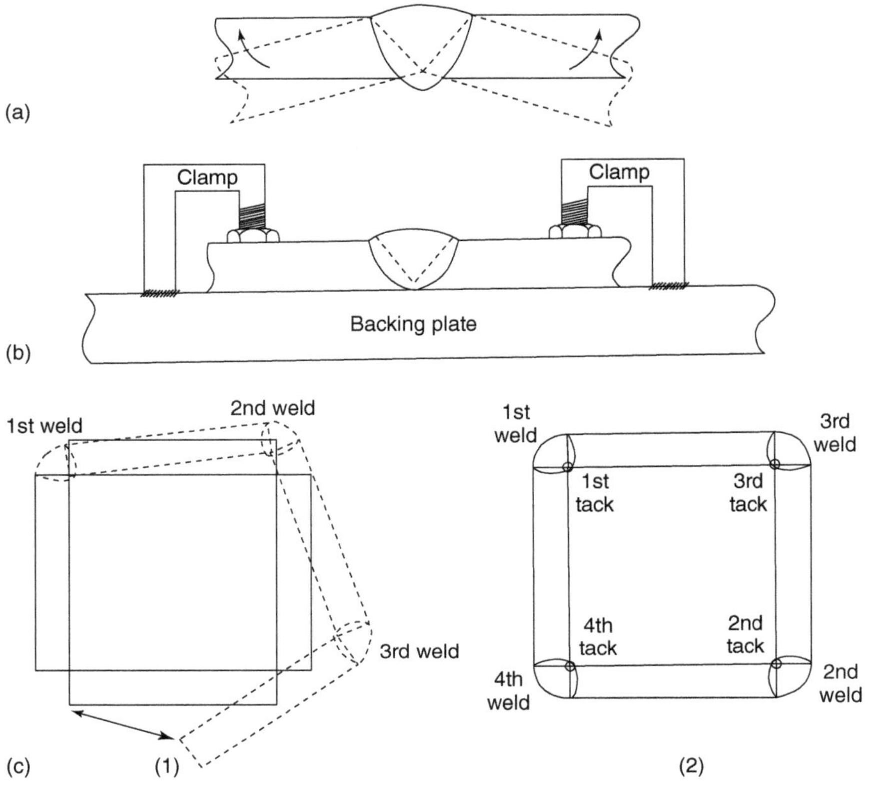

Figure 4.9 Schematic illustrations showing several techniques for preventing or minimizing distortion during welding, including (a) presetting elements of a butt joint, (b) rigid fixturing, and (c) sequencing welds (i.e. weld sequencing). Source: Messler (1999). Reproduced with permission from John Wiley & Sons, Inc.

commonly used in practice, unless and until structures get very large. The downside of mechanical straightening is that, when done cold and without being followed by some stress relief, residual stresses are introduced. Flame straightening is understandable in principle, but is more "arty" in practice. When done properly, flame straightening is applicable to even very large structures that may appear immoveable,[19] and because heating and cooling intentionally moves the structure, no (or low) residual stresses are introduced.

As for distortion, prevention (or avoidance) of residual stresses is better than attempting removal. In the case of residual stresses, prevention methods include (i) reducing the effects of heat during welding and (ii) allowing the structure to move in response to thermally induced stresses from shrinkage and/or

19 While working at Grumman Aircraft on LI, NY as a summer student, the author saw an experienced welding engineer straightening two 100-ft long, 4-ft high, 2-ft wide I-beams intended for building a moving overhead crane in an aircraft hangar. Using an oxy-acetylene torch for heating, a piece of chalk was used to mark the areas to be heated and cooled, and a sponge and bucket of water for rapidly cooling the just-heated areas, severe bows and twists were removed bit by bit until the beams were straight and true.

contraction. If it is an option, the most effective way of avoiding residual stresses is to avoid using heat altogether by employing a non-fusion welding process. Although a cold process is best, even processes involving heat (e.g. hot pressure welding, forge welding, roll welding, and friction welding) introduce far less residual stress than fusion welding processes. The reason is that no volumetric shrinkage occurs because no solidification occurs, and thermal contraction is usually less problematic because the peak temperatures reached tend to be well below those associated with fusion welding.

The approach for minimizing residual stresses during fusion welding is to minimize the volume of molten weld metal in every joint so as to minimize solidification shrinkage. This is best achieved by the judicious choice of joint preparation; U-groove versus V-groove, as less filler is required. Regardless of a joint groove's shape, its size should be kept as small as possible, while still providing adequate access for proper welding.

Another approach not often thought of, as most users of welding limit the choice of process to availability and/or prior experience, is to choose a fusion process with the highest melting efficiency possible. Higher energy density processes (such as electron-beam welding (EBW) or laser-beam welding (LBW) or plasma arc welding, PAW) minimize the width of a weld for a given penetration depth. Although this reduces the width of the fusion zone and the associated HAZ, thereby minimizing shrinkage and distortion, the steepness of the temperature gradient through HAZ can result in more severe residual stresses.

Reducing residual stresses following welding is virtually always prudent, even if not mandatory by a code (e.g. ASME Boiler and Pressure Vessel Code). The most common and only thoroughly proven approach is to employ a post-weld stress relief heat treatment (i.e. *thermal stress relief*). Heating lowers the yield strength of a base metal and, thereby, allows localized relaxation of residual stress, primarily by localized yielding and secondarily by slower acting creep. To prevent a weldment from moving and distorting as residual stresses are relaxed or removed, it is often advisable to perform the stress relief heat treatment with the weldment in a fixture (often the same one used to prevent distortion during welding). Figure 4.10 gives plots that show the effect of time at temperature and temperature of a thermal stress relief treatment on residual stresses.

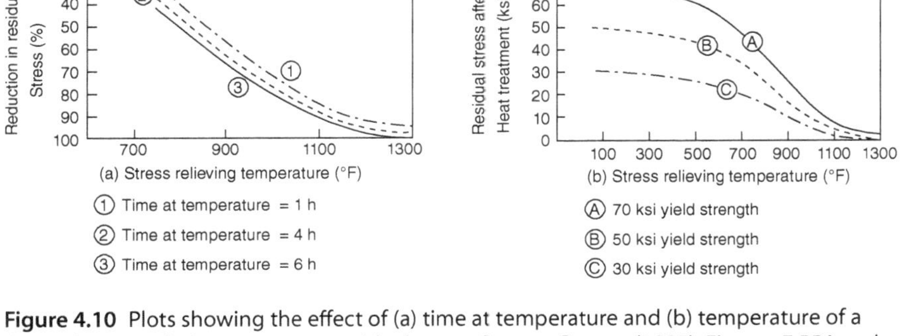

Figure 4.10 Plots showing the effect of (a) time at temperature and (b) temperature of a thermal stress relief treatment on residual stress. Source: Connor (1987), Figures 7.55A and 7.55B on p. 263. Reproduced with permission from The American Welding Society.

Table 4.2 Troubleshooting weld shrinkage, weldment distortion, and residual stress problems.

Problems with Weld Shrinkage:

o Shortening across weld	Excessive weld FZ size	Switch to autogenous with straight/square butt, redesign for smaller weld size, employ narrow-weld electron-beam (EB)/laser-beam (LB), employ tacked strongback, switch to non-fusion process
o Cracking in FZ	Excessive joint restraint	Reduce restraint, employ some preheat, use lower strength filler, switch to non-fusion process

Problems with Distortion:

	Angular joint misalignment with single groove	Preset joint elements, employ double groove, switch to EBW or LBW for narrow, parallel-sided welds
	Angular distortion with double-groove	Alternate welding to balance, switch to non-fusion process

General Guidelines to Avoid Distortion:

- ✓ Employ a minimum number of parts to reduce welding
- ✓ Select proper joint preparation/configuration
- ✓ Use minimum amount of weld metal (especially fillets)
- ✓ Keep net linear heat input to minimum needed for penetration
- ✓ Sequence welding to balance heat input and thermally induced stresses
- ✓ Possibly employ some preheat to reduce gradients/spread contraction

General Guidelines to Minimize Problems with Residual Stresses:

- ✓ Minimize restraint across welds; allow components to move, if possible
- ✓ Minimize severe temperature gradients across HAZ by using some preheat
- ✓ Minimize thermal contraction in HAZ using lowest acceptable heat input; consider using high energy density processes (PAW, LBW, and EBW)
- ✓ Avoid CTE mismatch between abutting joint elements exceeding about 10%
- ✓ Employ post-weld stress relief heat treatment whenever possible, always for fatigue-critical joints or joints exposed to possible impact or shock loading
- ✓ Consider hammer peening, shot peening, or planishing, if thermal stress relief is impractical and is fatigue- or fracture-critical

Despite claims to the contrary, it is virtually never impossible to perform some thermal stress relief on a finished weld. If the weldment is too large to bring to a heat-treating furnace, heat can be brought to the weldment using oxy-fuel gas torches or burners, quartz lamps, or resistance-heated blankets. Here is a case where anything is better than nothing!

Claims for *post-weld mechanical stress relief treatments* need to be accepted with caution. All methods rely on mechanical energy to relax and relieve or, at least, redistribute welding-induced residual stresses to reduce peak levels. Proven methods include hammer peening, shot peening, and planishing. With each method, the metal in regions with suspected high residual stresses is caused to plastically deform, thereby relieving the most severe (and potentially detrimental) locked-in stresses. Far less proven is the use of vibratory stress relief, in which it is alleged that elastic vibrations cause locked dislocations in regions of high residual stress to move, thereby relaxing or removing the worst stresses. Most data from reliable studies (done by other than the companies that manufacture and/or sell vibratory stress relief systems) have shown that it is necessary to vibrate the structure with a frequency that causes resonance in the structure. Doing this often does more structural damage than it does good.

As lawyers have said, even if not the Romans: Caveat emptor – let the buyer beware!

4.8 Troubleshooting Guide

Table 4.2 lists the various problems covered in this chapter: (i) problems with weld shrinkage, (ii) problems with weldment distortion, and (iii) problems with weld-induced residual stresses.

References

Allen, J.S. (1981). The effect of residual stresses on distortion. In: *Residual Stresses and their Effect* (ed. A.J.A. Parlane), 5. Cambridge: The Welding Institute.

Booyabazooka (2006). Water on Wikipedia.com, with image used under Creative Commons ShareAlike CC BY-SA 3.0. https://creativecommons.org/licenses/by/3.0.

Blyler, J. (2016). Increasing Power Density of Electric Motors Challenges IGBT Makers. Mentor Graphics a Siemen's Business.

Connor, L.P. (ed.) (1987). *Welding Handbook*, Volume 1: Welding Technology, 8e. Miami, FL: American Welding Society.

Masabuchi, K. (1981). Models of stresses and deformation due to welding – a review. In: *Modeling of Casting and Welding Processes* (ed. H.D. Brody and D. Apelian), 223. Pittsburgh, PA: Metallurgical Society of AIME, The Minerals, Metals and Materials Society (TMS).

Messler, R.W. Jr. (1999). *Principles of Welding: Processes, Physics, Chemistry, and Metallurgy*. New York, NY: Wiley.

Further Reading

Boley, B.A. and Weiner, J. (2011). *Theory of Thermal Stresses*. Mineola, NY: Dover Publications.

Campbell, J. (2008). *Castings*, 2e. Oxford: Elsevier Butterworth-Heinemann.

Messler, R.W. Jr. (2004/1999). *Principles of Welding: Processes, Physics, Chemistry, and Metallurgy*, Chapter 7. Weinheim, Germany/New York, NY: Wiley-VCH/Wiley.

Noda, N. (2002). *Thermal Stresses*. Boca Raton, FL: CRC Press.

Ueda, Y. (2012). *Welding Deformation and Residual Stress Prevention*. Oxford: Elsevier Butterworth-Heinemann.

5
Porosity

5.1 The Most Common Problem in Welds

There is probably no more common problem encountered in fusion welding than porosity in welds. The bad news is that more than 20 potential causes have been identified, which challenges troubleshooting. The good news is that most causes are easy to remedy, even if not easy to pinpoint, and there are definitely some metals and alloys that are prone to porosity.

In short, *porosity* results from the contamination of the molten metal or alloy (i.e. the fusion zone, FZ) by trapped evolved gas. The four main reasons for the gases that cause porosity can be identified as follows:

1. Dirty base metal
2. Water (in any form) of joint surfaces or in or on consumables
3. Insufficient or improper shielding
4. Incorrect welding conditions or techniques or faulty equipment.

Three of these (i.e. 1, 2, and 4) demand better housekeeping. The rules for welding without problems from porosity are reducible to the following:

- Make it clean and keep it clean
- Check equipment
- Check shielding (especially gas flow)
- Check workspace conditions.

As "Location, Location, Location" is the mantra for success in real estate,[1] the guiding principle (even if not the mantra) for success in welding (at least as it affects porosity formation) is Cleanliness, Cleanliness, Cleanliness.

Evidence of porosity almost always comes in the form of rounded holes or pores. Some pores may be open to the surface (i.e. *surface-breaking porosity*),

1 Language expert William Lewis Safire (1929–2009), American author, syndicated columnist, journalist, and presidential speechwriter, responded to a reader's inquiry on the origin of the phrase in an essay in the 29 June 2009 issue of *Time Magazine*, shortly before his passing in December of that year. He wrote that although many attribute it to Lord Harold Samuel (1912–1987), a real estate tycoon in Britain, when he wrote "There are three things that matter in property: location, location, location" when he was about 44 years old (i.e. around 1955–1956). However, further research found the "Location, Location, Location" appeared in a classified ad for real estate in the *Chicago Tribune* in 1926 by an unknown author, when Lord Samuel would have been just 14 years old!

A Practical Guide to Welding Solutions: Overcoming Technical and Material-Specific Issues,
First Edition. Robert W. Messler, Jr.

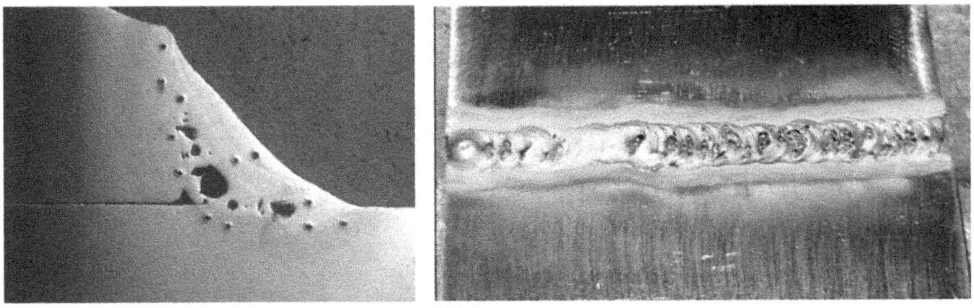

(a) (b)

Figure 5.1 Photomacrographs showing (a) very severe internal porosity in a weld made in aluminum and (b) surface-breaking porosity in a weld made in steel. The presence of surface-breaking porosity is often an indicator of severe internal porosity. Source: From the website of welding experts Ed and Emily Craig, i.e. www.weldreality.com. Used with the kind permission of Emily Craig.

usually at the weld's face or crown, and, thus, may be detectable during the visual inspection that should always follow welding. On using a dye or a fluorescent penetrant, for ferromagnetic metals and alloys,[2] magnetic particle inspection (MPI) may enhance detection. The majority of porosity, however, is normally internal to the weld (i.e. *internal porosity*), in the fusion zone. Thus, most porosity is only detectable using X-radiography or ultrasound.

Figure 5.1 is a photograph showing both forms of porosity, i.e. open to the surface (at the right) and internal to the weld (at the left). It is generally agreed that surface-breaking porosity usually indicates that large amount of distributed porosity is present within the weld.

Porosity is often classified by its shape and distribution within the weld. The shape is usually *spherical porosity* (also, known as *globular porosity*), whereas the distributions can be

- random, uniformly scattered (or distributed porosity)
- cluster porosity
- linear porosity.

An elongated form can occur in either of two types, often referred to by the same terms, although the cause(s) is(are) usually different, i.e.:

- Wormhole porosity or cavity
- Piping porosity.

A third form of elongated pore not necessarily caused by the gas trapped in a molten weld metal is known as a "crater pipe" (Figure 5.2). A *crater pipe* forms in the final solidified weld pool. The defect results from the volumetric shrinkage that accompanies the solidification of virtually all metals and alloys (see Section 4.3) but leaves a cavity at the termination of a weld because the weld pool at that point solidifies from all directions, not just at the trailing

2 Ferromagnetic metals include, in descending order of magnetic strength, Fe, Ni, and Co and some of their alloys. Notably, austenitic stainless steel (AISI-SAE 300-series) are generally not ferromagnetic, although ferrite in some compositions imparts some ferromagnetism.

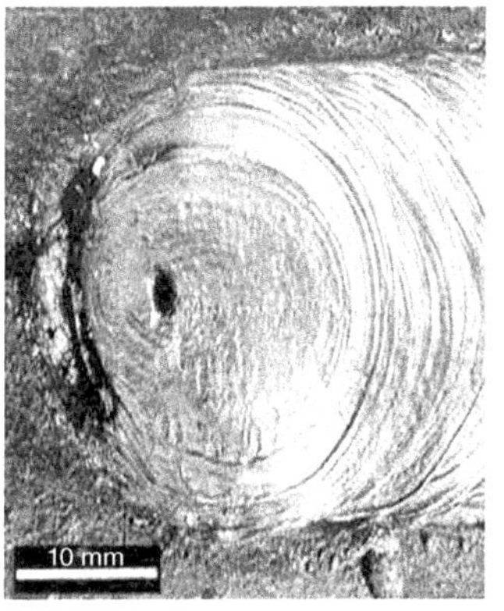

Figure 5.2 Close-up photograph showing a crater pipe at the termination of a fusion weld. Crater pipes need not be the result of gas porosity and often are simply the result of volumetric shrinkage in a stop crater. Source: Yusof (2003). Used with permission of TWI.

end of the weld puddle. This results in the weld pool getting shut off from any supply of molten metal, thus resulting in a shrinkage cavity. Any condition that exaggerates the liquid-to-solid volume change will promote crater pipe formation, the most common being abruptly switching off the welding current and forcing rapid solidification of a large weld pool. Prevention of crater pipes include the following methods:

- Use a runoff tab (stop tab) with butt joints so that the weld is terminated from the weldment properly
- Progressively reduce the welding current to reduce the weld pool size:
  - ✓ by having the welder slowly withdraw the arc, while using a circular motion in the crater
  - ✓ by employing a programmable ramp-down (i.e. downslope) of current at the end of a weld available on modern power supplies
- Add a filler to compensate for the shrinkage of the final weld pool.

The other forms or types of porosity are discussed in the next section.

Some (arguably too many!) claim, with little technical support, that not all porosity is unacceptable, unless explicitly forbidden by a standard or code.[3] Many claim (somewhat correctly) that porosity does not degrade the strength of a weld unless it becomes very severe. This may be true for static strength, which is a material's response to a constant, nonfluctuating load. The reason is that the most commonly encountered porosity has a minute effect on reducing the load-bearing area of a welded joint. Only extremely severe porosity would result in any significant reduction in load-bearing area. On the other hand, there are three situations of loading for which porosity in welds (or castings) could be a problem: (i) in cyclic- or fatigue-loaded structures (whether involving high- or low-frequency dynamic loading), (ii) in structures designed to fracture

3 Many codes, notably the American Society of Mechanical Engineers' (ASME's) Boiler and Pressure Vessel Code (including its extension to nuclear containment vessels and piping) and the American Petroleum Institute's (API's) code and standards applicable to petrochemical vessels and piping, rigidly control the level of porosity permitted to exist in welds. The most critical welds cannot have any detectable porosity!

toughness criteria (involving sustained loading under a sustained stress), and (iii) in structures designed to tolerate shock or impact loading.

Under fatigue loading, any manufacturing-induced flaw (e.g. crack, pore, low-density nonmetallic or high-density metallic inclusion, machined undercut, and severe surface scratch) or even a geometric notch in a design (e.g. sharp radius groove or section transition and fastener holes) concentrates and amplifies stress such that the flaw or notch that acts like a flaw can grow over time, reach a critical size, become unstable, and result in complete (often catastrophic) failure of the part or structure by overload fracture. Under fracture toughness design criteria, a manufacturing-induced flaw or a designed-in geometric notch cannot be permitted to remain in a given material under a maximum allowable sustained stress if that flaw or notch that acts like a flaw exceeds a certain (critical) size.[4] Under impact or shock loading, any flaw or flaw-like geometric notch can lead to catastrophic failure by rapid overload. For these loading situations, porosity can be problematic.

The above notwithstanding, the problem with accepting porosity is that even if a particular pore is acceptable (because of its size or fortuitous location), the presence of pores indicates the lack of process control. Hence, good luck results in an acceptable pore – by chance. Bad luck results in unacceptable pores – by chance. The question is whether one can succeed in engineering by chance?

5.2 Types of Weld Porosity

Figure 5.3 schematically illustrates various types of porosity, whereas Figures 5.4–5.6 show the radiographs or photographs of some of the types.[5] The various types are described below:

It is not impossible for a *single pore* to occur in a weld, although this is probably the exception. When they are found, single pores tend to be found more often in solder joints made in microelectronic assemblies, as such joints are often very small in volume. Because low-melting-point solder joints are not made under shielding of any kind, the source of such pores is usually from air entrapped at the joint. Nothing more will be said about the single pore here, as if there was really only one pore in a fusion weld, (i) it would likely never be detected, and (ii) if a single pore was detected, the cause would be of little concern or, in most cases, consequence.

Uniformly scattered porosity (also known as *distributed porosity*) almost always consists of large numbers of randomly distributed small spherical pores

4 In fracture mechanics, under fracture toughness design, the criteria for fracture tolerance (when conditions of plane strain prevail in a sufficiently thick part) is given by the equation $K_{IC} = Y\sigma_0(\pi a)^{1/2}$, where K_{IC} is the material's fracture toughness, Y is a geometric constant (about 1, but available in handbooks for various geometric conditions), σ_0 is the nominal applied stress, and a_c is the maximum allowable (i.e. critical) length of a surface flaw or half-length of an internal flaw. No flaw can be permitted that exceeds a_c at open to a surface or $2a_c$ internally.

5 Endless examples of images of porosity in welds can be found Internet, but one must be careful not to become discouraged or depressed by the most horrific examples, as these are showing extremes!

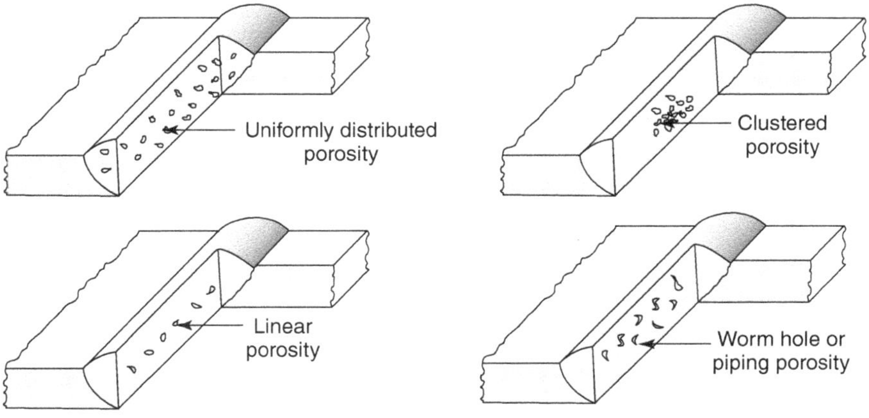

Figure 5.3 Schematic illustration showing various types of porosity found in fusion welds, including uniformly scattered porosity, cluster porosity, linear porosity, and wormhole or piping porosity.

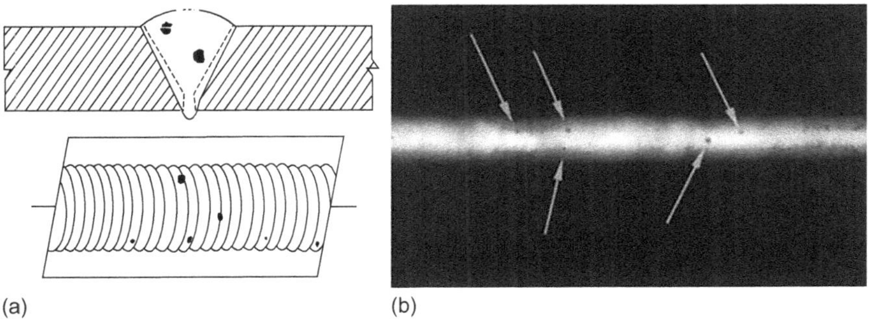

Figure 5.4 (a) Schematic and (b) radiograph showing scattered porosity in a weld.

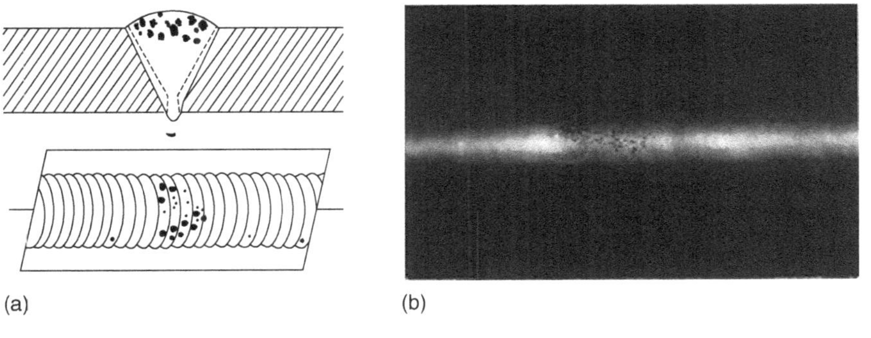

Figure 5.5 (a) Schematic and (b) radiograph showing cluster porosity in a weld.

Figure 5.6 Photograph showing severe wormhole porosity because of excessive shielding gas flow rate (at 60 ft^3/min versus 30, as recommended) during gas-shielded FCAW. Source: From the website of welding experts Ed and Emily Craig, i.e. www.weldreality.com. Used with the kind permission of Emily Craig.

(Figure 5.4). The most common source of scattered porosity is improperly or inadequately inert-gas-shielded arc welding, although it can also result from improperly cleaned base metals (e.g. rusted steel, improperly cleaned Al or Mg, or Al or Mg alloys). There is some evidence that running a weld with an arc gap that is too short can cause scattered porosity, as operating an arc welding process at very low amperage.

Cluster porosity, as the name implies, consists of pockets of many, generally small pores at intermittent discrete locations along a weld (Figure 5.5). It is common to find such clusters at equal spacing, suggesting that wherever the gas comes from, it builds up to some level within the molten weld metal and, when it exceeds some solubility limit, comes out as bubbles that become trapped. The most common cause of cluster porosity is moisture (water) on the surface of joints, adsorbed on or absorbed in the coatings of shielded metal arc welding (SMAW) electrodes, or associated with rust on steel filler wires used with gas metal-arc welding (GMAW) or flux-cored arc welding (FCAW). Another common cause is *arc blow*, in which a direct current or DC arc is erratically deflected off the weld line by induced electromagnetic fields from improperly positioned ground connections. Loss of shielding leads to porosity. Cluster porosity tends to occur in the bulk of the fusion zone, as opposed to near the edges of the FZ.

Linear porosity, as the name implies, occurs as discrete spherical pores strung out in lines, either along the centerline of the weld (where solidifying metal from opposite sides of the weld impinge) or near the edges of the fusion zone, near the fusion boundary. Linear porosity along the weld centerline suggests that the causative gas stays dissolved until the final solidification occurs (i.e. solubility of the gas is high in the weld metal). Linear porosity along the edges of the weld suggests that the solubility of the causative gas is low and decreases as the temperature of the molten weld metal decreases (being lowest near the fusion boundary or edge of the weld pool). Dissolved hydrogen, from dissociated water or hydrocarbon contaminants, is often the culprit.

Wormhole porosity suggests gross contamination, as evolving gas forms large pores that get dragged along as the weld puddle moves with the welding heat source, creating an elongated cavity behind (Figure 5.6). The most common cause of wormhole porosity is the air that was trapped in crevices under the vertical leg of T joints or where joint elements meet in corner joints. This being the case, wormhole porosity is more commonly encountered with fillet welds than with butt welds.

Piping porosity refers to cavities that are longer than their width and typically (but not only) oriented along the weld line. The most common cause is insufficient protection of the molten weld during its production. Another common source, however, is from dragging air trapped at the root of a weld made in improperly fit-up tee joints or corner welds.

5.3 Gases in Molten Weld Metal

All gases, except the inert gases (i.e. He, Ne, Ar, Kr, Xe, and Rn), dissolve in liquids, including molten metals. They do so by having gas molecules, nascent atoms, and/or ions occupying the relatively-large spaces among the atoms of the metal in the liquid form. During fusion welding, nitrogen, oxygen, and hydrogen (the latter produced by the dissociation of water molecules in an arc) are often drawn into the weld pool from the surrounding air, surface oxides on joint elements, hydrocarbon contaminants, or water in any form on any surface.[6] The amount of any diatomic gas (e.g. N_2, O_2, and H_2) that can dissolve in a molten metal is a function of the partial pressure of the gas species above the liquid, expressed by Sievert's law as

$$k = [\text{gas}]/P^{1/2}{}_{\text{gas2}}$$

where k is the equilibrium constant, [gas] is the concentration of a particular gas in the molten metal as weight percent (wt%), and P_{gas2} is the partial pressure of the particular gas in the diatomic form.[7] That the law works is shown by the plot for N_2 in Figure 5.7.

The amount of nitrogen, oxygen, and hydrogen that can be dissolved in a molten metal almost always increases with increasing temperature of the liquid.[8] Plots showing the equilibrium solubility of nitrogen and oxygen in pure iron and pure nickel (at 1 atm pressure) as a function of temperature are shown in Figure 5.8, whereas plots of the equilibrium solubility of hydrogen in pure aluminum, pure copper, pure iron, and pure nickel (at 1 atm pressure) as a function of temperature are shown in Figure 5.9. These plots also show how the

6 When H_2S or various oxides of nitrogen, carbon, or sulfur are present (as they are in the environments of many chemical- and petroleum-processing plants), these can also dissolve in molten metals.
7 Atoms and ions of these gases also dissolve in molten metals, but the amounts dissolved are not predicted by Sievert's law. As a result, the amount of a particular gas dissolved in a molten metal is almost always greater than predicted by Sievert's law, even though the time for which the metal is molten in the weld pool is very short for such a highly nonequilibrium process as fusion welding.
8 This is not necessarily to be expected, however, as the amount of oxygen (O_2) and carbon dioxide (CO_2) that can dissolve in water decreases with increasing temperature, as evidenced by the bubbles that form inside glasses as cold water or soda sits and warms to room temperature.

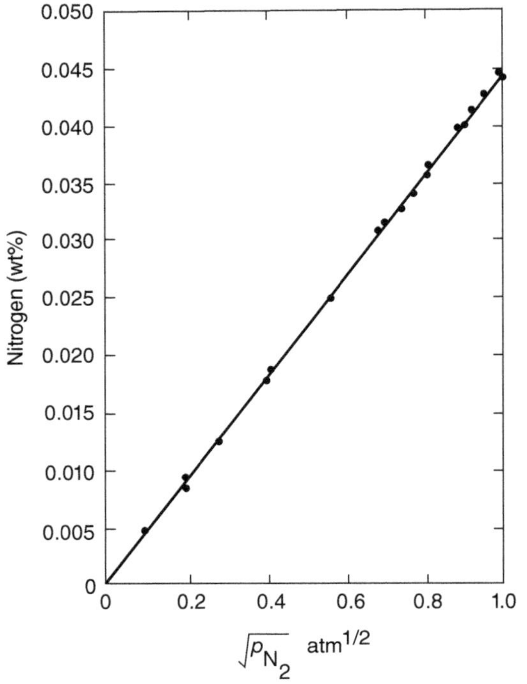

Figure 5.7 Plot showing the solubility of nitrogen in molten iron (at 1600 °C) obeys Sievert's law, increasing with the square root of the partial pressure of nitrogen above the melt. Source: Pehlke and Elliott (1960). Reproduced with permission from The Minerals, Metals and Materials Society (TMS).

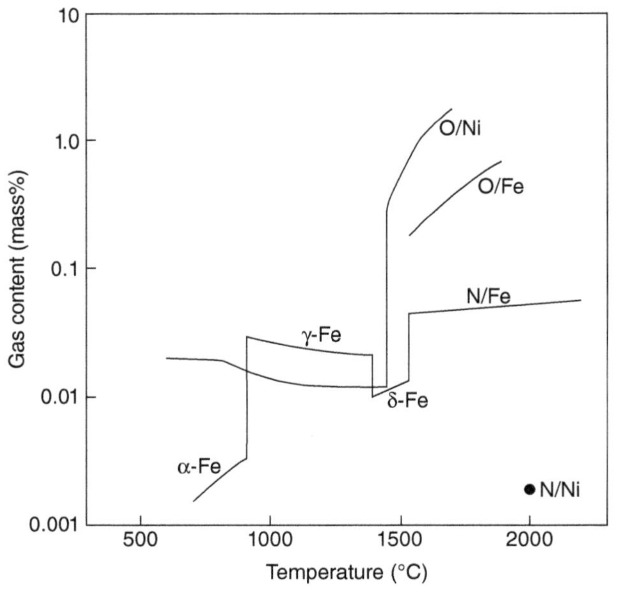

Figure 5.8 Plots showing the equilibrium solubility of nitrogen and oxygen in pure iron and in pure nickel at 1 atm pressure. Source: Lancaster (1993). Reproduced with permission from SCSC.

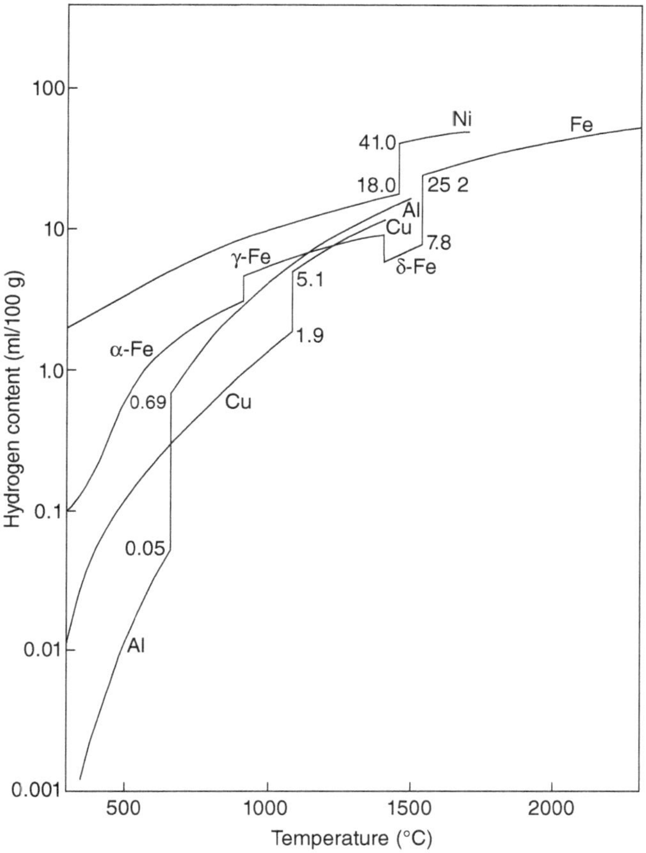

Figure 5.9 Plots showing the equilibrium solubility of hydrogen in pure aluminum, pure copper, pure iron, and pure nickel at 1 atm pressure. Source: Lancaster (1993). Reproduced with permission from SCSC.

solubility of these gases drops dramatically in the solid as soon as solidification occurs. This is because there is far less space among the atoms of a metal in its crystalline form than in its liquid form. Surprisingly, the solubility of these gases is greater in the close-packed face-centered cubic (FCC) γ-iron than in the non-close-packed body-centered cubic (BCC) α-iron and δ-iron forms.[9]

The solubility of a gas in a molten metal is affected by the presence of different alloying elements (i.e. solutes), so the susceptibility of alloys to porosity can be different from that of the pure metal. The reasons are complex and not simply related to the available space among atoms of the solvent (host) metal, sometimes relating to the affinity of the gas to come out of the solution to form a compound with the solute element (e.g. oxide, nitride, or hydride). Figure 5.10 gives plots for the equilibrium solubility of nitrogen in binary iron alloys at 1600 °C under 1 atm

9 Calculation reveals larger and a greater number of sixfold octahedral interstitial sites in FCC than in BCC iron.

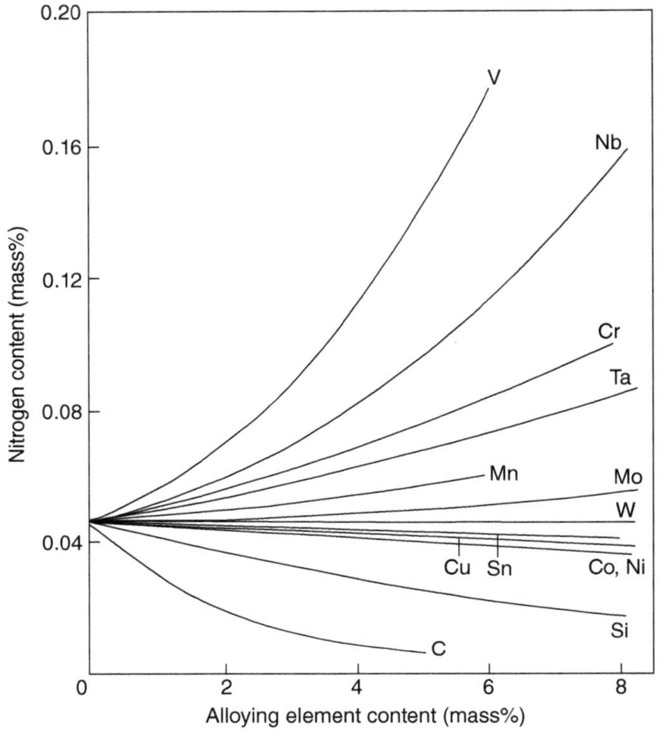

Figure 5.10 Plots showing the effect of various solutes on the solubility of nitrogen in binary iron alloys under equilibrium conditions at 1600 °C and 1 atm pressure. Source: Lancaster (1993). Reproduced with permission from SCSC.

nitrogen pressure, whereas Figure 5.11 presents plots showing the influence of alloying elements on the oxygen content of weld metal, after solidification.

Once a gas (such as nitrogen, oxygen, or hydrogen) has been taken into a molten metal, it can lead to one or more of several things: (i-a) they can remain in the solution to cause interstitial hardening in the solid state; (i-b) they can remain in the solution and stabilize a particular phase of a metal exhibiting different crystalline forms (i.e. allotropic forms), as dissolved nitrogen stabilizes the FCC γ form of Fe; (ii) they can be rejected from the melt upon solidification when solubility in the solid drops precipitously, as it often does, leading to the formation of porosity; or (iii) they can react with either the host metal or some solute to form brittle compounds (e.g. hydrides in Ti, oxides, and nitrides in Al). Table 5.1 gives the general solubility of hydrogen, nitrogen, and oxygen in various liquid metals at their melting points.

As it pertains to discussion in this chapter, the formation of porosity from dissolved gases is the concern, as porosity formation follows the sequence: Gas (or gases) dissolve > solidification takes place > solubility decreases > gas bubbles form > bubbles may not escape (despite violent pool stirring or convection) pores

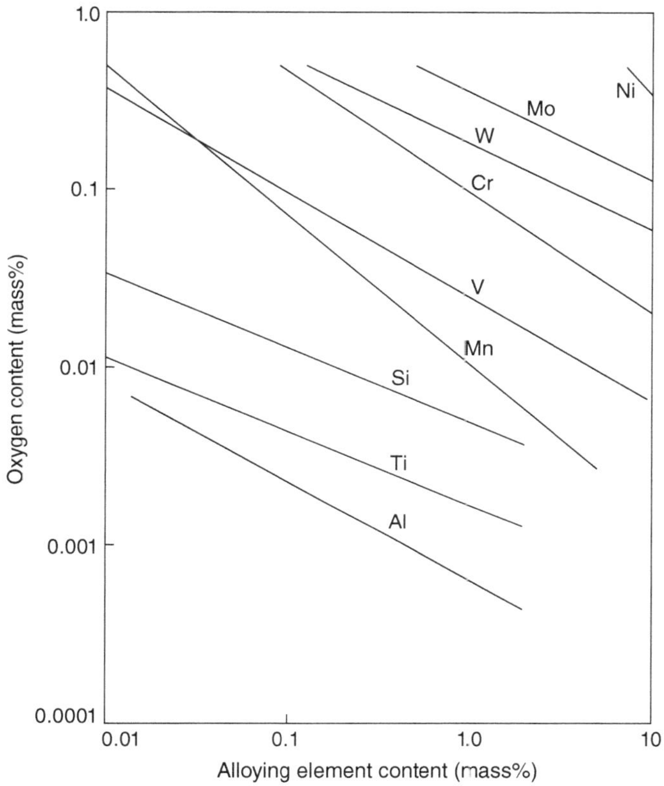

Figure 5.11 Plots showing the influence of alloying elements on the oxygen content of iron weld metal following solidification. Source: Lancaster (1993). Reproduced with permission from SCSC.

result.[10] It will be seen in Section 5.4 that there are many possible sources of porosity in fusion welds, but the most common, by far, is lack of cleanliness of the joint elements. Table 5.2 lists the principal gas(gases) responsible for porosity in the most commonly welded engineering metals and alloys. Also listed is the suggested method for cleaning each to minimize the likelihood of porosity.

Although they will be easier to understand (or rationalize) after referring Section 5.4, Figures 5.12 and 5.13 present interesting bar graphs showing the levels of oxygen and nitrogen (Figure 5.12) and of hydrogen (Figure 5.13) found in welds made in steel as a function of the process employed, as well as, in Figure 5.13, some process variables.

Therefore, with this knowledge about gases in molten metals and alloys, let us look at the many possible causes of porosity in fusion welds.

---

10 Section 5.5 will discuss how the various forces that can give rise to convection in a molten weld pool exacerbate or, potentially, reduce the likelihood of gas bubbles becoming trapped in a weld as porosity.

Table 5.1 Solubility of hydrogen, nitrogen, and oxygen in various liquid metals at their melting point.

		H	N	O
No compound formed (endothermic solution)	Gas soluble	Ag, Al, Be, Cd, Co, Cr, Cu, Fe, Mg, Mn, Mo, Ni, Pb, Pd, Pt, Rh, Ru, Sn, W, Zn	—	Ag
	Gas insoluble	Au, Hg	Rb, Cs, Cu, Ag, Au, Zn, Pb, platinum metals	Au, platinum metals
Compound formed (exothermic solution)	Compound very soluble	Sc, Y, rare earths, Ti, Zr, Hf, Th, V, Nb, Ta, U	Ti, Zr, Hf, Th, V, Nb, Ta, U	Ti, Zr, Hf, Th, V, Nb, Ta
	Compound moderately soluble	Alkali and alkaline-earth metals: Li, Na, K, Rb, Cs, Ca, Sr, Ba	Co, Cr, Fe, Mn, Mo, W	Alkali and alkaline-earth metals: Li, Na, K, Rb, Cs, Ca, Sr, Ba, Cu, Co, Cr, Fe, Mn, Mo, Ni, Pb, Sn, W
	Compound insoluble	—	Li, Na, K, Be, Mg, Ca, Zn, Al	Al, Mg, Be, Ca, Sr, Ba, Zn, Cd

Source: Lancaster (1987). Used with permission of Allen & Unwin Book Publishers.

Table 5.2 Principal gases causing porosity in the most commonly welded engineering metals and alloys.

Material	Gas	Recommended cleaning method
Carbon steels	Hydrogen	Remove all hydrocarbon contaminants with solvents; eliminate water and/or moisture from all surfaces and shielding fluxes and inert gases; remove rust by wire brushing; use low-hydrogen SMAW electrodes for C_{eq} (for C steels) over 0.50 to avoid hydrogen-assisted cracking
	Nitrogen	Grind to remove the mill or heat treating scale coatings
	Oxygen	Grind to remove the mill or heat treating scale coatings; beware semi-killed steels
Low-alloy steels	Hydrogen	Eliminate all sources of hydrogen (as for carbon steels, above); use low-hydrogen SMAW electrodes for C_{eq} (for low-alloy steels) over 0.40
Stainless steels	Hydrogen	Degrease + wire brush + degrease (especially austenitic and martensitic types)

(Continued)

Table 5.2 (Continued)

Material	Gas	Recommended cleaning method
Aluminum and alloys	Hydrogen	Chemically clean + wire brush + degrease + scrape just before welding
	Oxygen	Wire brush + degrease + scrape just before welding
Copper and alloys	Hydrogen, nitrogen	Degrease + wire brush to remove tarnish + degrease
Nickel and alloys	Hydrogen, nitrogen	Degrease + wire brush + degrease
Titanium and alloys	Hydrogen	Degrease + pickle in aqua regia + scrape just before welding; vacuum degas if incoming hydrogen exceeds 150 ppm
	Nitrogen, oxygen	Pickle in aqua regia to remove the mill surface and α-case + degrease

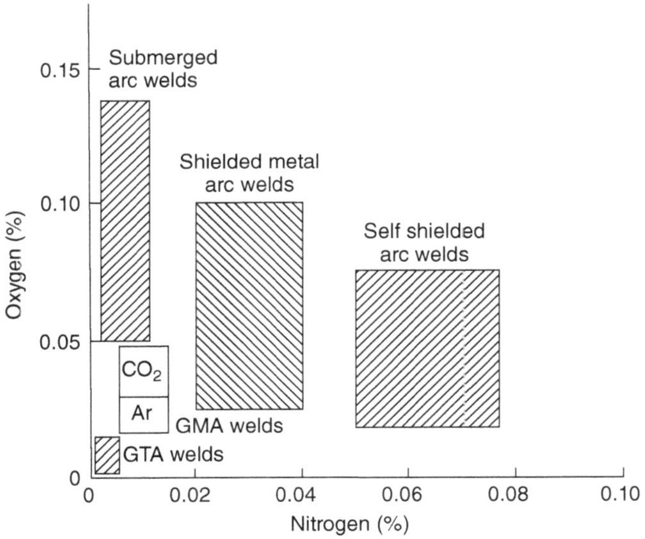

Figure 5.12 Bar graphs of oxygen and nitrogen levels expected in steel after fusion welding using any of the several different arc processes shown. Source: Rein (1974). Published by and used with permission of The American Welding Society.

5.4 The Many Possible Causes of Porosity in Welds

Knowing that porosity is formed as gas (typically, nitrogen, oxygen, or hydrogen) forms bubbles that become entrapped in the new solidified weld before they had any chance of escaping raises the question: From where does this gas originate? Herein is the challenge in attempting to resolve a problem with porosity

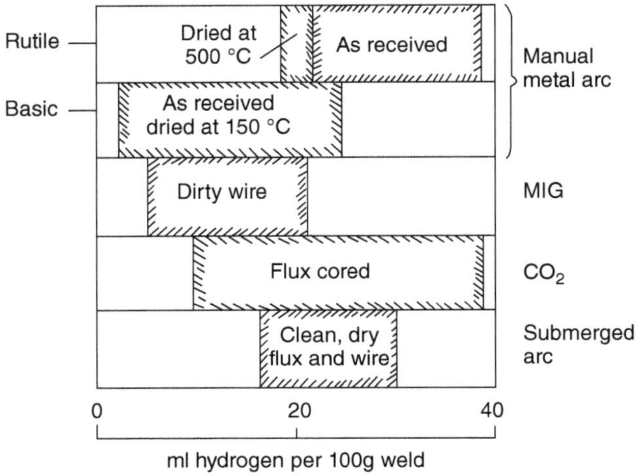

Figure 5.13 Bar graphs of the amount of hydrogen found in welds in steel as a function of the process and certain process conditions. MIG, metal inert gas. Source: Easterling (1992). Used with the kind permission of Easterling family.

in fusion welds. To be successful, either the root cause[11] of the gas pickup must be identified *or* some technique during welding must be found to sweep forming bubbles out of the molten weld pool before solidification entraps them forever. The former is almost always more successful, although the latter will be discussed in Section 5.5.

So, let us consider the many possible sources of the occurrence of porosity in fusion welds, essentially in the order of the frequency each is encountered, numbered for convenient organization and referral, and indicating the most prevalent materials affected (when applicable). Correction of the problem is either self-evident or will be given.

1. Contamination of the surface of joint elements:
   a. Mill scale (on carbon- and low- and medium-alloy steels), rust (on C and low-alloy steels, as well as on cast irons), surface oxide (on Al/Al alloys, Mg/Mg alloys, austenitic stainless steels, Ni/Ni alloys, and Zn alloys), α-case[12] (Ti/Ti alloys), and tarnish layers (Cu/Cu alloys) can result in porosity from released oxygen or from hydrogen dissociated from water commonly adsorbed onto the oxidized surface. Chemical removal by acid or alkaline cleaning (as recommended for the material) is acceptable for

11 A *root cause* is an initiating cause of either a condition or a causal chain that leads to an outcome or an effect of interest or concern. The term denotes the earliest, most basic, deepest cause for a given outcome. The idea is that an error is only seen by its manifest signs, which can be multitudinous, widespread, and convoluted. The root cause leading to these is often a lot simpler. Without finding the root cause, a problem tends to persist.

12 Dissolution of oxygen into reactive Ti and Ti alloys results in the stabilization of the lower temperature HCP α phase as a hard, brittle, fatigue-prone surface layer typically only a few thousandths of an inch (under 0.1 mm) thick. Such a layer may be present from the mill or, more likely, is created by subsequent heat treatment.

thin layers. Some form of mechanical cleaning[13] may be needed before chemical removal can be effective for thicker layers.

b. Hydrocarbon contaminants such as machining lubricants (as well as coolants), lubricating or corrosion-preventing oils, greases, paint, wax crayons and other markers (not intended for use with welding), adhesives/glues, tapes (e.g. masking tape, rubberized tapes, electrical tape, and Scotch Tape™), maskants, etc., can all lead to porosity from hydrogen.[14] Dirty work gloves or dirty cleanup rags must be avoided! Complete removal of any hydrocarbon contaminant, usually by liquid solvent or solvent vapor degreasing, is recommended. Handling (except with clean gloves) after degreasing is discouraged. Care must be taken to prevent welding electrodes and wires from becoming contaminated with oil, grease, paint, glue, or tape.
c. Zinc plating (such as galvanize coatings on steels) should be stripped from the area to be fusion welded (usually by thorough wire brushing), including the immediate surrounding area of any heat-affected zone (HAZ), as zinc has a low melting point (419.53 °C/787.15 °F) and boiling point (907 °C/1665 °F) and a high vapor pressure, all of which leads to it forming Zn vapor in electric arcs. Although porosity can result, the more serious problem is toxicity to the welder!

2. Contamination of the welding environment by water, in any form:
   a. Joint element surfaces must be dried before welding and kept dry during welding in order to prevent the formation of porosity from hydrogen, in particular (especially with steels prone to forming martensite in their HAZ). Dew, condensation (below 10 °C/50 °F), frost, snow, etc., must be removed, perhaps by heating the area to be welded using an oxy-fuel gas torch. Rain, sleet, or snow must be kept off the weld, preferably using some form of shelter. Humidity (for high-humidity locations) should be considered, perhaps employing an inert-shielded glove box or dry box for inherently reactive metals and their alloys (e.g. Al, Mg, Ti, and Zr). Another option is to employ preheat above the reverse polarity (RP) of water to exclude water pickup.
   b. Welding consumables must be free of water, including no water of hydration associated with rust (on steel FCAW, GMAW, or submerged-arc welding or SAW wires and bare gas tungsten-arc welding or GTAW wires or rods), no absorbed water in the coatings of SMAW electrodes[15] or in the flux employed with SAW, no water (or hydrogen) produced by SMAW electrodes with cellulosic coatings (for steels), and no water vapor in bottled

13 Mechanical cleaning methods to remove oxidation/tarnish layers include abrasive grinding, blasting with abrasive grits, draw filing, scraping with a sharp blade, or brushing with a *clean* metal wire brush, best used only on one type of material, to prevent cross-contamination. Care must be taken with softer metals and alloys to preclude the embedding of particles of any abrasive, as these can result in either porosity or nonmetallic inclusions in welds.
14 It is not out-of-the-question for inert shielding gases to be contaminated with oils when pressurized cylinders are filled at facilities where proper procedures are not followed. Filters in the gas lines help!
15 Although moisture can be picked up by the core of FCAW wires, it is rare; provided wires are properly stored and not allowed to get wet.

shielding gases (by purchasing gases from a reputable supplier), preferably employing drying filters in the gas lines where they exit the cylinder(s).

3. Contamination of the weld by air from the atmosphere:
   a. Moving air must not be permitted to disturb the delivery of and coverage by inert shielding gas during welding, or porosity can result from nitrogen or oxygen, or even hydrogen (from humidity). Overhead fans and floor fans should be kept more than 20-ft (6 m) away from the point of welding. Drafts (from open doors or windows or vents) greater than 4–5 mph (6.5–8 kpm) should be avoided. Welding must be clear of air discharges from machines.
   b. Cylinders of compressed welding gases cannot be allowed to run out of gas during welding, as protective shielding from the atmosphere would be lost. The amount of gas remaining in a cylinder should be checked just before beginning to weld.
   c. Shielding gas nozzles (for GTAW, GMAW, or FCAW with auxiliary shielding gas) must not be permitted to get too far from the weld pool, or diminished coverage by shielding gas could result in porosity from nitrogen or oxygen, or even hydrogen from humid air.
   d. For GMAW, the gun angle must be held so that shielding gas is not spread so much that air is drawn back into the weld opposite the nozzle's travel direction. An angle of 5–15° off 90° is best.
4. Contaminated or defective welding equipment:
   a. GMAW and FCAW wire-feed cable liners, gun contact tubes, and wire feeder toothed or knurled wheels must be kept free of dust, grease, oil, rust pickup from rusted wires, etc., any of which could contaminate the weld pool and cause porosity, usually from hydrogen.
   b. Shielding gas hose fittings must be kept tight and, periodically, have O-rings checked.
   c. Gas feed lines or hoses must be regularly checked for contamination from oil pickup from hose fittings or contaminated gas cylinders. Hose must also be checked to ensure that they are free of bugs, nests, or webs, as these can lead to porosity in a contaminated weld pool. Also, gas solenoid valves should be regularly checked for proper operation.
   d. GMAW nozzles must be kept from plugging, most often by weld spatter associated with certain metals or alloys or molten metal transfer modes (e.g. short-circuit welding).
   e. Shielding gas hoses must be kept free of kinks (especially when run distances exceed about 20 ft (6 m) and must be checked for pinching or crushing by other equipment or material, as well as for cuts or burned areas (always a risk in a welding environment).
5. Contamination from improper welding procedure or technique:
   a. Care must be taken to ensure that air is not being sucked into the root of a weld by the joint being too open (especially with GTAW and GMAW).
   b. Care must be taken to ensure that a GMAW gun nozzle is not at an awkward angle on corner welds or edge welds, as resulting turbulence causes air pickup and possible porosity from nitrogen or oxygen, or even hydrogen for humid air.

   c. Gas flow rates (for GTAW, GMAW, or gas-shielded FCAW) must not be run too high. Flow rates should be about 20 cfh (9.5 l/min) for GTAW and 30–45 cfh (14–21 l/min) for GMAW/FCAW.
   d. Improper use of anti-spatter compounds, sprays, or gels should be avoided, as too much of use can lead to contamination of the weld pool and porosity.
6. Contamination from within the base metal:
   a. On rare occasions, a gaseous element is intentionally added to a metal or an alloy either to improve the strength via interstitial solid solution alloying or to stabilize one allotropic form (i.e. crystalline solid phase) over another or to improve corrosion resistance. The only known examples involve the intentional addition of nitrogen to (i) austenitic stainless steel to (a) offset the loss of strength when the carbon content is reduced to prevent sensitization (e.g. 304LN grade) or to (b) stabilize the austenitic phase over the ferritic phase in alloys with both phases present (according to constitution diagrams[16]) or (ii) to 316 to improve corrosion resistance in nuclear applications (as 316NG) and (iii) to certain Ni–base superalloy compositions to increase the strength via "nitrogen enhancement." With any of these alloys, nitrogen can be released to form porosity. The problem can often be overcome by adding some dry nitrogen gas to an inert shielding gas (e.g. Ar) or a gas mixture (e.g. Ar + He), as the increased partial pressure of N_2 above the molten pool tends to keep nitrogen in the solution via Sievert's law.
   b. Oxygen, nitrogen, and hydrogen can be present in some metals and alloys as an impurity associated with the metal's production by melt metallurgy. Some examples include, oxygen in pure copper (except oxygen-free high-conductivity (OFHC) copper), oxygen in steels that have not been fully deoxidized (i.e. "killed"), usually using Si or occasionally with Al, and hydrogen in Ti and Ti alloys. Fortunately, welding of OFHC is rare and virtually all modern steel is "killed." For Ti and Ti alloys, problems with hydrogen porosity are more common with electron-beam welding (EBW) than with any other process, as the hard vacuum employed with EBW (to keep the beam from dispersing) causes dissolved gas to be released under the low partial pressure of the vacuum. Preventing such porosity demands hydrogen content to be maintained below about 150 ppm, which might be attainable by the mill or may require vacuum degassing before EB welding.
   c. Oxides (as inclusions in Al and Mg and their alloys), nitrides (especially in nitride grades of steel), and hydrides (especially in inherently reactive Ti and Ti alloys) can all generate porosity as they thermally decompose in high-energy arcs, plasmas (in plasma arc welding, PAW), or beams (in EBW or laser-beam welding (LBW)). Prevention of porosity from such decomposition is difficult to impossible, although small additions of either oxygen (at 1–2 vol.%), nitrogen (up to 10 vol.%), or hydrogen (at 2–4 vol.%) to inert shielding gas (using a glove box) can be tried.

16 Such diagrams (including Schaeffler diagram, DeLong diagram, and WRC diagram) plot the microstructural constituent(s) found in a stainless steel as a function of the ferrite-stabilizing chromium equivalent (Cr_{eq}) and austenite-stabilizing nickel equivalent (Ni_{eq}).

Although there are many potential sources of porosity formation in fusion welds, it is hoped that the organization used above helps. The use of the term "contamination" in each category of potential sources of porosity should evoke the mantra for all quality welding: Cleanliness, Cleanliness, Cleanliness.

5.5 Attempting to Minimize Porosity Formation in Fusion Welds

Because the only fully effective way of eliminating porosity once it occurs in a weld is by machining, grinding, or arc-gouging out the defective weld metal and rewelding with a new filler, prevention is the key. Whenever porosity occurs, the following should always be done, in the order given below:

1. Review to ensure that precleaning of joint elements (to remove oxides or tarnish layers, coatings, and hydrocarbon contaminants) was done properly (i.e. as recommended by best practices of the American Welding Society, by material producers, or using reliable online references).
2. Review to ensure that all sources of water, in any form, were prevented (e.g. by shelters, dry boxes, or low preheating to about 250 °F/125 °C and by baking coated SMAW electrodes and SAW fluxes).
3. Check that all consumables (e.g. electrodes and wires) were correct and free of rust or contaminants, all shielding gases were correct and dry, and all gas cylinders had gas present throughout welding.
4. Check that all support equipment was correct and operational: all gas hoses, wire guide tubes, torch or gun nozzles, and contact tubes were clean; all hose fittings were tight; and no gas hoses were kinked, crushed, or burnt.
5. For SMAW, GTAW, and GMAW, verify that no excessive drafts were present from open doors or windows, fans, improperly operated fume eliminators, or nearby machine tool cooling air.
6. Review welding procedures (and welders and welding operators) to ensure that nozzles for inert-shielded processes were properly positioned, gas flow rates were correct, weld root gaps were not excessive, and backing bars or strips were properly placed (if employed).
7. No gases were introduced by impurities within or additives to the base materials.

Special attention should be given to the development of welding procedures for porosity-prone metals and alloys, such as Al and Al alloys, Mg and Mg alloys, Ti and Ti alloys, Zn-galvanized steels, and cast irons that had been in service (as such inherently porous cast alloys become impregnated with oil that must be removed by torch burning before welding).

Another potential opportunity for preventing problems from porosity in welds involves taking advantage of the ever-present convection forces that combine to complex flow patterns in weld pools. Four forces act to cause convection: (i) the slow movement-inducing (1 cm/s) buoyancy force of gravity acting to pull the more dense cooler molten metal near the edges of the pool downward and

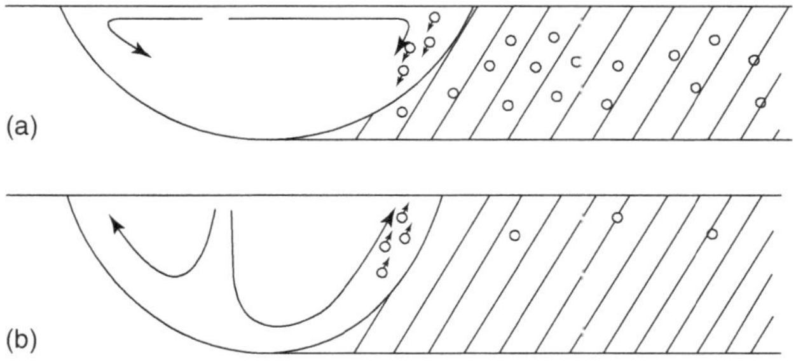

Figure 5.14 Schematic illustration of the two possible effects of different convection patterns on forming gas bubbles in a molten weld pool; one pattern (in a) sweeping forming bubbles downward to increase the likelihood/ they will become entrapped as pores and the other pattern (in b) sweeping forming bubbles upward to increase the likelihood they will escape from the surface of the weld puddle. Source: Kou (1987). Used with permission of John Wiley & Sons.

allowing the less dense hotter molten metal at the center of the pool to rise; (ii) a fast movement-inducing (10–100 cm/s) surface tension gradient (Marangoni) force that pulls the hotter molten metal from the center of the pool to its edges; (iii) a fast movement-inducing (10 cm/s) electromotive (Lorentz) force associated with arc, plasma, and electron-beam processes that forces the molten metal down at the center of the pool, to turn upward upon encountering the boundaries of the pool; and (iv) a generally slow (1 cm/s), but variable (to 10 cm/s), movement-inducing impingement force (or arc-, plasma-, or electron-beam friction force) that pushes the molten metal down at the center of the pool, to turn upward upon encountering the boundaries of the pool.

As shown in Figure 5.14, the ever-present (on Earth) buoyancy force and ever-present Marangoni forces act to sweep gas bubbles forming in the cooler portions of the weld pool down into the pool, so as to increase the likelihood they will be trapped as pores in the solidified weld. On the other hand, as also shown in Figure 5.14, the ever-present variable impingement force and the strong (and variable) Lorentz force associated with arc, plasma, and EB processes act to sweep gas bubbles forming in the cooler portions near the edges of the weld pool up and out of the pool, as they burst at the pool's surface. Thus, by increasing the current in arc, plasma, and EB welding processes, the Lorentz force is increased, and the sweeping action is enhanced as a result. Likewise, the like-acting impingement force can be enhanced by increasing the welding current for arc or plasma processes or, if it is an option, by switching to higher energy density EBW, for which the impingement force is the greatest for all fusion-welding processes. Although not sure, manipulating these two forces may help reduce the occurrence of porosity.

5.6 Troubleshooting Porosity Problems in Welds

Table 5.3 lists the various ways through which problems from porosity can arise during fusion welding and in fusion welds.

Table 5.3 Troubleshooting problems with porosity in welds.

Contaminated Joint Element Surfaces:	
• Mill scale, rust, surface oxide/tarnish, Ti α-case (1a)	Remove as recommended
• Hydrocarbon contaminants (1b)	Remove by degreasing or solvent wiping
• Zn-galvanize (1c)	Remove by stripping area to be welded
Contamination of Welding Environment by Water:	
• Wet joint elements (2a)	Dry and keep dry; consider low preheat
• Moisture in flux coatings/cores (2b)	Bake SMAW electrodes + store properly; properly store FCAW wires
• Moisture in SAW flux (2c)	Store properly; bake if necessary
Contamination by Air from Atmosphere:	
• Moving air disrupts shielding gas (3a)	Remove or reduce draft or source
• Gas cylinders empty during welding (3b)	Check gas level before welding
• GTAW, GMAW, FCAW nozzles too far away (3c)	Move nozzles closer for better coverage
• GMAW nozzle angle wrong (3d)	Correct to 5°–15° off 90
Contaminated or Defective Welding Equipment:	
• Dirty feed-wire liners, feeder wheels, contact tubes (4a)	Check; clean as required and periodically
• Loose or faulty shielding gas fittings (4b)	Check; tighten or change
• Dirty or old and porous gas hose (4c)	Check; clean or change, as needed
• Dirty GMAW nozzles (4d)	Check of spatter; change
• Blocked shielding gas flow (4e)	Check for kinks, crushing, burnt areas; change
Contamination from Improper Technique:	
• Air sucked in at root with GTAW, GMAW, gas-FCAW (5a)	Reduce root gap; consider back-shielding
• Awkward GMAW or gas-FCAW nozzle angle (5b)	Correct for T and edge joints
• Severe wormhole porosity; GMAW, gas-FCAW (5c)	Reduce shielding gas flow rate
• Problems with anti-spatter materials (5d)	Reduce amount of compound, spray, or gel
Contamination from Base Material Chemistry:	
• With 304LN, 316NG, or N-enhanced Ni–base alloys (6a)	Consult material producer; consider adding up to about 10 vol.% dry N_2 to shielding gas
• Impurity gases such O_2 in steel not "killed" (6b)	Consult material producer
• Decomposition of oxides, nitrides, or hydrides (6c)	Consult material producer; try adding small amount of offending gas to shielding gas

Always consider the positive effect of favorable weld pool convection using higher current for more Lorentz force and higher current or higher energy density process for more impingement force.
The #/letter in parentheses ( ) after each potential cause refers to the #/letter in the text.

References

Easterling, K. (1992). *Introduction to the Physical Metallurgy of Welding*, 2e. Oxford, UK: Butterworth-Heinemann. As referenced in Robert W. Messler Jr. (1999). *Principles of Welding: Processes, Physics, Chemistry, and Metallurgy*, Figure 11.9, p. 325, Wiley, New York, NY.

Kou, S. (1987). *Welding Metallurgy*, 1e. New York, NY: Wiley.

Lancaster, J.F. (1987). *Metallurgy of Welding*, 4e. London: Allen & Unwin.

Lancaster, J.F. (1993). *Metallurgy of Welding*. Springer.

Pehlke, R. and Elliott, J.F. (1960). Solubility of nitrogen in liquid iron alloys, 1: thermodynamics. *Transactions of the Metallurgical Society AIME* 218: 1088–1101.

Rein, R.H. (1974). *Proceedings of a Workshop on Welding Research Opportunities* (ed. B.A. McDonald), 92. Washington, DC: Office of Naval Research Eager, T.W. (1978). "Sources of weld metal oxygen contamination during submerged arc welding," *Welding Journal*, 57(3), 1976s–1980s.

Yusof, F. (2003). *Welding Inspection – Defects/Repairs*" (ref. WIS 5), as Slide #15/55. Cambridge, UK: TWI.

Further Reading

Devletian, J.H. and Wood, W.E. (1983). *Factors Affecting Porosity in Aluminum Welds – A Review*. New York, NY: Welding Research Council.

Evans, R.M. (1964). *Porosity in Titanium Welds*, vol. 194. Columbus, OH: Battelle Memorial Institute, Defense Materials Information Center.

Fast, J.D. (1976). *Gases in Metals*. London, UK: Macmillan Publishers.

Fast, J.D. (1965). *Interaction of Metals and Gases*. Cambridge, MA: Academic Press.

Messler, R.W. Jr. (2004/1999). *Principles of Welding: Processes, Physics, Chemistry, and Metallurgy*, Chapter 11. Weinheim, Germany/New York, NY: Wiley-VCH Verlag/Wiley.

Westwood, L.W., Carney, D.J., and Sims-Smith, C.E. (1953). *Gases in Metals: A Series of Four Lectures*. Metals Park, OH: American Society for Metals.

6

Cracks

6.1 The Most Dreaded Defect in Welds

To the detriment of the process of welding, there are a host of physical defects that can adversely affect the quality of a weld and, thereby, a weldment. Geometric joint mismatch, excessive dimensional or geometric distortion, insufficient fusion, undercut, porosity, inclusions, and cracks all compromise weld quality and weldment performance. Of these, cracks are universally considered the worst and, rightfully, are the most dreaded of all weld discontinuities (see Table 3.2). All welding codes and most welding standards[1] show zero tolerance for cracks, whereas other defects are generally tolerated within certain limits.[2]

A *crack* may be defined as a local break in the physical continuity of a material (i.e. a physical discontinuity). There are three requirements for cracks to form and grow in a material: (i) a stress-raising defect, (ii) a tensile stress field, and (iii) a low tolerance to fracture.[3] Welds provide all of the requirements. Stresses in and around a weld are inherent with the fusion welding operation, as the volumetric shrinkage accompanying the solidification of all commonly welded engineering metals and alloys and the thermal contraction in the heat-affected zone (HAZ) surrounding the fusion zone (FZ) both result in tensile stresses, balanced by compressive stresses outside this region (see Figures 4.4–4.6). Discontinuities in welds, as well as in base metal, and the geometric notches typically associated with a weldment's design all act as stress raisers (or stress risers) by concentrating thermally induced and/or externally applied mechanical

1 Welding codes (such as those of the ASME Boiler and Pressure Vessel Code) were developed to protect the public from death or injury from catastrophic failures attributable to faulty welds. Codes impose design rules and limit allowable design stresses and, as it pertains to welding, stipulate welding practice and procedures, including those that may be necessary to make repairs. Welding standards are generally similar, but, unlike codes, are not developed by committees of experts working as volunteers. Rather, welding standards tend to be developed within a company for its own use and only rarely by representatives of companies for an industrial sector.
2 In fracture-critical design based on a material's inherent resistance to the growth of a flaw under a sustained stress intensity, defects below a certain critical size are allowable, and need not be removed or repaired.
3 "Low tolerance to fracture" in a material results from a low value of *fracture toughness*, which is not necessarily an indication of low tensile strength, but, rather, a complex property involving a combination of plastic flow strength, ductility, and the energy associated with the formation of surfaces by a crack.

A Practical Guide to Welding Solutions: Overcoming Technical and Material-Specific Issues,
First Edition. Robert W. Messler, Jr.

stresses two to five times the nominal stress ($\sigma_{nominal}$) in the structural member. The crack sensitivity of a material is associated with its chemistry (i.e. composition) and/or its susceptibility to the formation of some inherently low fracture toughness constituent (e.g. inherently brittle phase, such as as-quenched untampered martensite in steels, inherently brittle ceramic or intermetallic compounds, such as hydrides in Ti alloys or σ phase in high-Cr stainless steels, or brittle low-melting eutectics, such as those formed between Fe and either S or P in steels). When present together, these three situations can spell TROUBLE.

Ingel's equation for calculating the maximum stress that develops around a defect reveals that it is the width much more than the length of a defect that controls the amplification by:

$$\sigma_{max} = \sigma_{nominal}\left(1 + \frac{2\sqrt{a}}{\rho}\right)$$

where a is the length of a flaw open to a surface or the half-length of a flaw at the interior of a material, and ρ is the radius at the tip of an assumed elliptical flaw (analogous to the flaw's width). For a crack, ρ can approach atomic dimensions (i.e. 0.25–0.4 nm) in an inherently brittle material in which no plastic yielding occurs in the region surrounding the crack-tip to blunt the crack. Thus, nominal stress ($\sigma_{nominal}$) can be raised to extremely high values, more than sufficient to cause growth. To make matters even worse – as if that were necessary – the state of stress in the region of a sharp-tipped flaw is triaxial, precluding any relief by the Poisson effect of lateral contraction with axial extension that occurs with uniaxial or biaxial stress states. This leaves a material with no option but to fracture![4]

Therefore, cracks are always bad but are usually worse in fusion welds when highly nonequilibrium solidification often results in microstructures with inherently low fracture toughness, only to be aggravated by the concentration of already high thermally induced tensile stresses by geometric notches.

6.2 Classification of Cracking and Cracks in Welds and Welding

Whether weld cracking is the result of the poor fit-up of parts comprising a weldment, excessive restraint, rapid cooling of hardenable steels, any of a variety of possible contaminants from the atmosphere, inadequate cleaning of parts before welding, or the composition and structure of the base metal and/or weld filler metal, it carries with it significant consequences for any welding operation. Not only do weld defects adversely affect the integrity of the completed weldment but they also require significant time and money to rectify. In the best cases, unacceptable defects, most notably cracks, must be removed by arc gouging, abrasive grinding, or machining, and the affected area rewelded. In the worst cases, the defective weldment must be rejected and scrapped. However, the cost does not end here.

4 Solid materials have two options when loaded, i.e.: (i) to yield and deform without fracturing or (ii) to fracture, either partially (i.e. cracking) or completely. This, in fact, is a key differentiator in the analysis of failures (i.e. in failure analysis).

In critical applications, the cost for inspecting welds that may ultimately be rejected is incurred. In addition, once reworked, rejected and repaired weldments require a second round of inspection that inevitably increases the fabricator's costs. In those situations that necessitate post-weld heat treatment (PWHT),[5] hours or even days can pass between the completion of rework and evaluation of the reworked weld(s) or weldment. This adds to costs through increased downtime and lost productivity.

The good news amidst all of this bad news is weld cracking does not have to be an enigma, nor does it have to be a severe drain on a fabricator's productivity and profits. Understanding the basic types of weld cracks and their causes greatly helps in preventing them in the first place. As usual, prevention is better than remediation.

Cracks associated with welding are most commonly classified into one of two types: *hot cracks* and *cold cracks*. As described in Section 6.3, *hot cracks* develop at elevated temperatures (i.e. between the liquidus temperature and the solidus temperature of an alloy), commonly form during solidification of the weld metal (i.e. fusion zone or partially melted zone [PMZ], for some alloys) as the result of volumetric shrinkage, and are recognizable as they propagate between grains of the material.[6] As described in more detail in Section 6.4, *cold cracks* develop after solidification of the weld (i.e. subsolidus) as the result of thermal contraction stresses, and are recognizable as they propagate both between and through grains. These two types of weld cracks (i.e. hot cracks and cold cracks) can alternatively be classified as cracks in the weld metal and cracks in the base metal, respectively.

Although the preponderance of cracking and cracks associated with welding and welds is hot cracks and cold cracks, cracks can also form in and near a weld during service, beyond delayed cold cracking (see Section 6.4). These other cracks are often not considered weld cracks, as such, even though they are most certainly the result of changes brought about by welding and would not have occurred had there been no welding. The two types are due to (i) fatigue or (ii) corrosion and will be addressed in Section 6.5.

Figure 6.1 schematically illustrates the various forms of hot cracks and cold cracks found in fusion welds.

6.3 Hot Cracking and Cracks

The most commonly encountered type of cracking found in fusion welds is, by far, *hot cracking*, with the result being generically known as *hot cracks*.[7]

---

5 A case will be made many times in the rest of this book that PWHT to relieve almost inevitable residual stresses introduced during fusion welding is *always* prudent, even if not mandated

6 "... between the grains ..." means *interdendritic* in a weld's fusion zone; and could mean along any boundary between growing dendrites (e.g. along the weld's centerline or between equiaxed dendrites, columnar dendrites, or cellular dendrites).

7 On an absolute basis, cold cracking and cold cracks are the most prevalent in sheer frequency and number; the reason being they occur in hardenable carbon and alloys steels. These alloys are, by far, the most widely welded by tonnage, including: steel structural frames for buildings; bridges;

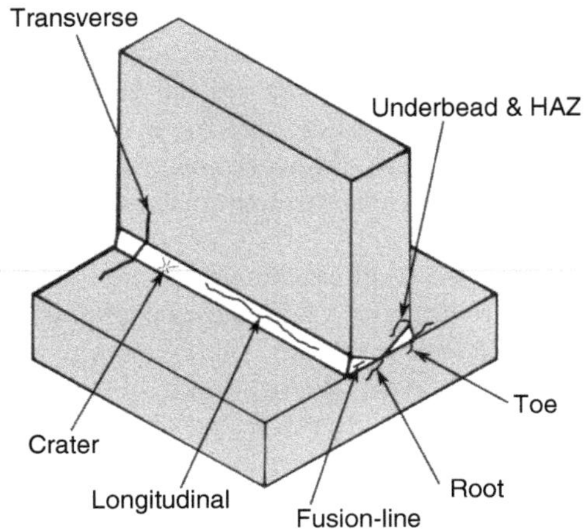

Figure 6.1 Schematic illustrations showing various forms of hot cracks and cold cracks in fusion welds (here, fillet T-welds), including hot centerline (or longitudinal) cracks, crater cracks, and fusion-line cracks (usually at the plane of impingement of opposing solidification structure growth) and cold toe cracks, root cracks, underbead cracks caused by hydrogen, and transverse cracks caused by severe restraint. Source: Wikipedia "Welding defects". Freely used under Creative Commons ShareAlike CC BY-SA 4.0.

This type of cracking is found in many commonly welded engineering metals and, especially, alloys, including, in alphabetical order, Al and Al alloys, cast irons welded using Ni–Fe and pure Ni filler, Cu and Cu alloys, Mg and Mg alloys, Ni and Ni alloys; steels, including carbon steels, low-alloy steels, austenitic stainless steels, and tool steels; and Ti alloys, particularly β- and β-rich α–β alloys. Hot cracks occur most often in the weld metal (i.e. fusion zone), with three forms prevailing in descending order of frequency: (i) *longitudinal cracks* (often called *centerline cracks*); (ii) *crater cracks*; and (iii) *transverse cracks*. Another particularly difficult-to-prevent form occurs in the PMZ of certain age-hardenable alloys, most notably, heat-treatable Ni alloys, known as *reheat cracking*. In some alloys, especially in multiple-pass welds, a phenomenon called *liquation cracking* is observed, most notably in certain Al alloys and certain Ni–base superalloys. Both types are addressed in detail in Chapter 15 in Part III but are briefly described later in this section.

Longitudinal weld (hot) cracks run in the same direction as the weld and are almost always confined to the center of the joint for single-pass welds, whether a butt or a fillet type. For this reason, they are commonly referred to as *centerline cracks*. In the case of multi-pass welds, where several beads per layer may be deposited, the crack may not be in the geometric center of the joint, although it will always be at the center of a bead. These cracks form as the last molten metal remaining in the weld solidifies and contracts to create tensile stresses that pull still weak interdendritic boundaries apart.

This form of hot crack may be an extension of a crack that originated at the end (i.e. termination) of a weld but can also initiate elsewhere during welding. Longitudinal cracks (especially centerline cracks) are usually visible at the crown bead of a butt weld or the face of a fillet weld, but can be entirely subsurface.

Figure 6.2 shows a typical example of a longitudinal/centerline crack open to the crown bead of a fusion weld (here an autogenous bead-on-plate weld in an Al alloy). Figure 6.3 shows centerline cracks open to the crown and contained within a butt weld in steel.

---

pressure vessels; natural gas, oil, water, and steam piping and pipelines; ships; agricultural, earthmoving, mining, and construction equipment; etc.

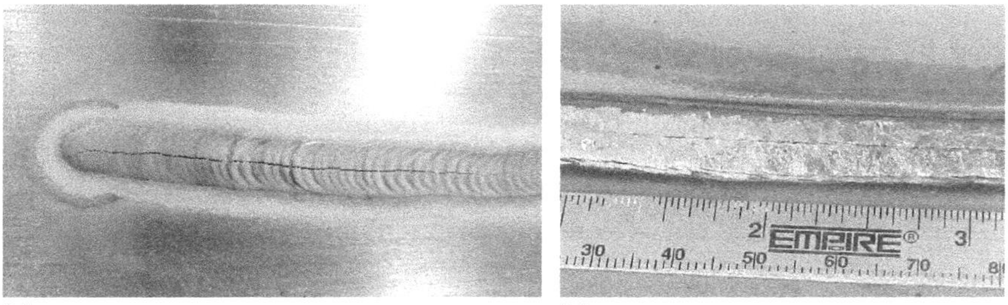

(a) (b)

Figure 6.2 Photographs showing (a) a longitudinal crack commonly referred to as a "centerline crack" in a bead-on-plate weld made on plate of 6061-T6 using the gas metal-arc (GMA) process at high I and low TS (at top) and (b) in a fillet weld made in steel using the GMA process. Source: (a) Pfaller (2015). Used with permission of Welding Productivity Magazine. (b) Hensley (2011). Used with permission of Fabricating and Metalworking Magazine.

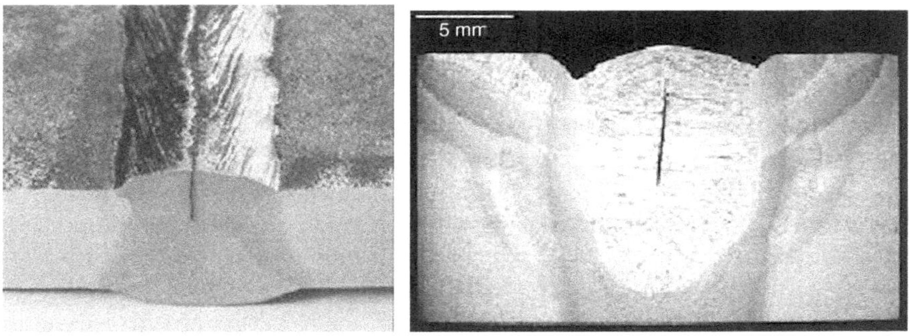

(a) (b)

Figure 6.3 Photomacrographs showing centerline cracking (a) open to the surface and (b) buried at the interior of butt welds in steel. Source: TWI (1999). Used with permission of TWI.

Centerline cracking can result from any of three phenomena: (i) segregation-induced cracking, (ii) bead shape-induced cracking, or (iii) surface profile-induced cracking. Any of these three causes produce the same form of crack, sometimes making it difficult to identify the root cause for the cracking. To make matters worse, two or even three of the causative phenomena often interact to lead to cracking. To help troubleshoot a root cause, let us consider each mechanism.

Segregation-induced cracking occurs when a low melting point constituent (e.g. compound or eutectic) forms in the melt and is forced to the center of the joint where epitaxial growth from base metal along each edge of the weld leads to impingement of dendritic grains or subgrains. Normal rejection of *all* solutes from the solvent (i.e. alloy base) at the advancing solid–liquid interface leads to the enrichment of the last remaining liquid and lowers the solidus to a minimum for the system. Rising tensile stresses from progressive solidification shrinkage act to pull the newly formed solid network apart, forming a crack.

To prevent this cause for centerline cracking, all sources of contamination by sulfur, phosphorus, selenium (in some austenitic stainless steels), lead, zinc (usually from galvanize coatings), and trace amounts of copper must be eliminated by scrupulous cleaning of all joint elements before welding, proper shielding during welding, and checking base metal certifications for excessive residual or tramp elements. Once contaminants have been eliminated as much as possible, welding current should be lowered as much as possible for the desired weld, as this refines solidification structure and, by increasing interdendritic boundary area, spreads any offending constituent around, perhaps to a level below the threshold at which hot cracking results. It is also possible to disrupt an obvious line of impingement along the centerline of a weld by employing transverse weaving, arc oscillation, or current pulsing.

The second variation of centerline cracking is *bead-shape-induced cracking*. This variant tends to occur with deep penetrating processes, including submerged-arc welding (SAW), CO_2-shielded flux-cored arc welding (FCAW), and, especially, electron-beam welding (EBW) and, to a lesser extent, laser-beam welding (LBW). The cause is the pronounced line of solidification structure impingement that is exacerbated by weld profiles where the depth is greater than the width of the fusion zone. For such growth situations, grains growing in opposite directions from the base metal may not achieve sufficient cohesion across the joint. This cause is greatly exacerbated when low-melting segregates are present in the melt, regardless of their source.

The obvious solution is to change the width-to-depth ratio to approach 1 : 1 to 1.5 : 1 for arc welding processes. This may require a change in the design of certain joints. For arc welding processes, lower welding current (i.e. amperage) and/or larger diameter electrodes may help. For high energy density beam processes (such as LBW and, especially, EBW), narrow, deep penetration single-pass welds are usually a prime factor in choosing such a process in the first place. In this case, special care must be taken to ensure that (i) there are no sources of contamination by tramp elements (e.g. S, P, Pb, Se, As, and trace amounts of Cu with P) and (ii) restraint that adds to thermally induced stresses acting across the joint should be minimized. The latter may not be problematic as the narrow fusion zone and HAZ associated with high energy density beam processes are one of the advantages of such processes.

The final variant of centerline cracking is *surface profile-induced cracking*. Concave weld surfaces (whether on butt welds or fillet welds) result in internal shrinkage stresses that place the weld metal at the crown or face surface in tension, thereby tending to pull the newly forming solid apart at the centerline. As such contours are frequently the result of high arc voltages, a slight decrease in arc voltage will tend to cause the bead to return to a slightly convex profile, greatly reducing the cracking tendency by the compressive shrinkage stresses that occur at the surface of such welds. Reducing welding or travel speed increases the amount of fill and, hence, tends to create a convex profile. Finally, vertical-down welding (i.e. moving downward on a vertical plane) tends to generate crack-sensitive concave profiles, whereas vertical-up welding tends to remedy the problem by producing a more convex bead profile.

Crater cracks can form during the abrupt termination of a fusion weld if a crater (representing the final position of the molten weld pool) is left with insufficient weld metal such that the volumetric shrinkage of the isolated puddle creates an often star-shaped crack. Although such cracks initially only extend to the edge of the crater, they can (and often do) propagate into longitudinal centerline cracks. Crater cracks are almost always visible at the crown or face of a weld.

Figure 6.4 shows a typical crater crack in a gas tungsten arc weld made in an Al alloy.

The cure for crater cracking is straightforward: do not allow an isolated weld pool with insufficient weld metal to form at the termination of a weld. This can be achieved in two primary ways: (i) by using end tabs or stop tabs at the end of a planned joint,[8] such that the weld is not terminated on the actual part or weldment but, rather, on a disposable extension, or (ii) (a) by having a manual welder slowly withdraw the arc from the work while using a slight circular motion and watching the pool gradually shrink or (b) by employing a programmable downslope of the welding current with a mechanized or automated arc welding process using a modern (e.g. solid-state) power supply.

The least frequently encountered form of hot cracking is *transverse cracking*, also called *cross cracking*, with the resulting cracks being called *check cracks*. As the name implies, transverse cracking is characterized by cracks within the weld metal that run perpendicular to the direction of welding (i.e. transverse to the welding direction). Because cracks occur under an opening-mode tensile stress, it should come as no surprise that transverse cracking is the result of thermally induced shrinkage stresses that act along the weld (i.e. longitudinally).

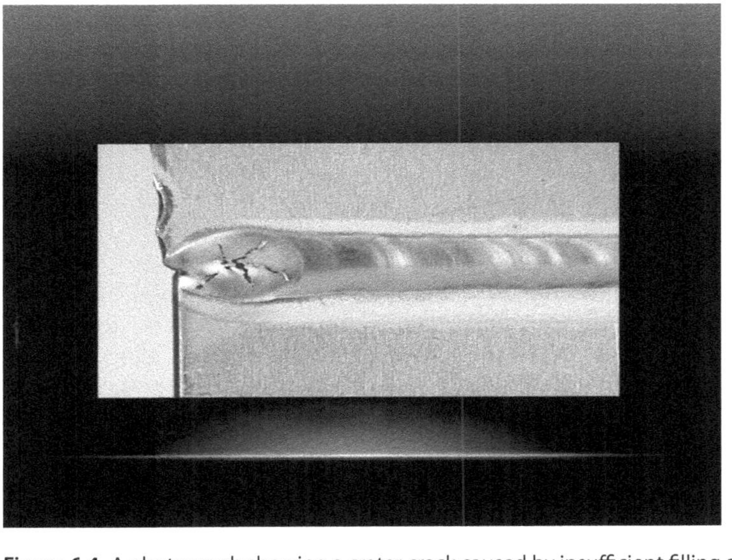

Figure 6.4 A photograph showing a crater crack caused by insufficient filling at the end of a weld. Source: TWI (2015). Used with permission of TWI.

8 Such tabs are often used at both the start and the stop end of a weld, so as not to allow either an arc strike on the work at the start of welding or an abrupt removal or extinction of the arc at the termination of welding. Tabs can be integral to the workpiece and later machined off or like-material pieces can be tack welded and removed, usually by an impact with a hammer and finish grinding, but, perhaps, by machining.

These longitudinal stresses develop because the metal in the weld fusion zone undergoes volumetric shrinkage in all three orthogonal directions during solidification but is constrained in the longitudinal direction by the surrounding, ever-solid base metal. This constraint gives rise to high tensile stresses that lead to cracking.

Although transverse cracking usually occurs during solidification, ergo is a form of hot cracking; it is also known to occur well after a weld has completely solidified and cooled to room temperature. In this case, it is a form of cold cracking which, like all forms of cold cracking, is caused by hydrogen pickup by the weld, HAZ, or unaffected base metal (UBM). More is said about this in Section 6.4.

Almost without exception, transverse cracking is associated with weld filler metal that is higher in strength – and almost always far less ductile and less tough and, as a consequence, less strain tolerant – than the base metal, that is, significantly overmatching the base material.[9] It is the higher strength, greater hardness, and lower ductility and/or toughness of such overmatched welds that commonly lead to check cracks in hard-facing beads. Such check cracks act to relieve some of the high tensile residual stresses that would otherwise be present in the weld.[10]

Overcoming a problem with transverse cracking first demands a determination of whether cracking was observed to occur upon solidification of the weld (often by the welder) or well after the weld has solidified and cooled. In the former case, the simplest and most effective solution is to reduce the strength of the filler metal. Of course, this depends on the intended purpose of having employed an overmatched filler in the first place. Because restraint exacerbates hot cracking of all forms, reducing restraint from fixtures or surrounding structure may help. Another effective approach is to employ preheating. Doing so will (i) reduce some of the troublesome longitudinal tensile stresses causing transverse hot cracking and, should the cracking actually be cold cracking, will serve to prevent the formation of susceptible martensite in the FZ and HAZ and will assist in diffusing problematic hydrogen, particularly in hardenable steels.

Figure 6.5 shows an example of transverse cracking.

From the foregoing, it should be clear that (i) microsegregation during the nonequilibrium solidification of fusion welds is inevitable and (ii) the tensile stresses that arise from the volumetric shrinkage of molten metals/alloys to solids pull on the last liquid regions to solidify. Hot cracking tends to result when the range over which solidification takes place in an alloy (i.e. the liquidus-to-solidus freezing range) is large, as here, shrinkage stresses are greater. In short, trapped solute-rich, low-melting boundary regions tend to pull open under thermally induced volumetric shrinkage stresses.

---

9 Overmatched filler metal means richer in key alloying additions (or solutes), so that (i) the weld metal will be stronger or (ii) harder or (iii) more corrosion resistant than the base metal. In the case of strength and corrosion resistance, this is usually done to offset the loss of strength and corrosion resistance always present with cast versus wrought alloys of the same nominal composition. In the case of hardness, this is done for improving resistance to wear on a surface overlaid by a weld metal in surface welding or hard facing.

10 Check cracks act to self-relieve residual stresses in a weld, thereby preventing distortion in the weldment and compromised performance in service later.

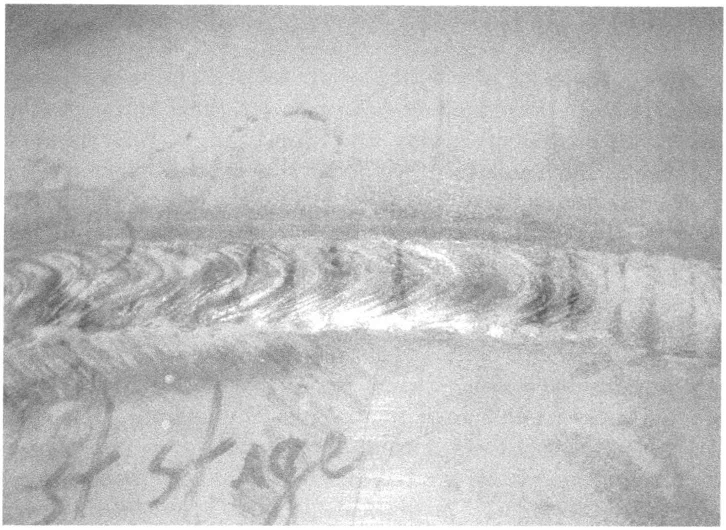

Figure 6.5 Photomacrograph showing transverse cracks in a circumferential weld made in steel. Source: website www.ndt.net, appearing in a response to a question from Juan Amado by Deston Henson, NDT Inspector of Canada, on 21 January 2011. Used with kind permission of Deston Henson.

A final form of hot cracking is *liquation cracking*. As the name implies, localized liquation, or melting, occurs either in segregate-containing grain or dendrite boundaries or around the particles of sulfides, phosphides, carbides, nitrides, or silicides or silicates, depending on whether the phenomenon occurs in the PMZ (above the equilibrium solidus temperature of the alloy) or in the highest temperature portion of the HAZ below the equilibrium solidus but above the depressed solidus temperature caused by prevailing nonequilibrium conditions. In the later situation, partial dissolution of particles leads to the formation of a low-melting composition that leads to cracking. This form of liquation cracking is most accurately called *constitutional liquation cracking*, as it is the result of a localized change in the constitution of the region.

Shrinkage stresses increase as the volume of shrinking metal increases, i.e. as welds are larger because of higher heat input (typically, excessive welding amperage). Large size welds and deep penetrating arc welding processes and procedures increase the shrinkage strains and, thus, the tendency for welds to hot crack. Somewhat ironically, the stresses that induce these strains are increased when higher strength base metal and, especially, filler metals are involved. With higher yield strength-material weld materials, higher residual stresses will also be present.

Under conditions of high restraint, extra precautions must be taken to overcome cracking tendencies. Attention needs to be given to joint design, filler metal selection, base metal and filler metal alloy freezing range and known susceptibility to cracking, net linear heat input, welding technique, welding sequence, preheat and interpass temperature, and PWHT for stress relief.

So much for hot cracking. What is *cold cracking*?

6.4 Cold Cracking and Cracks

Cracking that occurs in the HAZ of a weld is characterized by the separation that occurs immediately adjacent to the weld bead or fusion zone. Although such cracking is related to the process of welding, cracks occur in the base material, not in any filler or melted and resolidified weld material. A further characteristic associated with such cracking is that it is transgranular rather than intergranular in nature. That is, cracks tend *not* to track dendrite, grain, or subgrain boundaries, but, rather, cross such boundaries. This process is called is *cold cracking*, with the result being *cold cracks.*

While cold cracking can occur in metals and alloys other than steels and cast irons, such occurrence is extremely rare and usually associated with exceptional brittleness. For this reason, discussion here is restricted to cold cracking in steels.

Cold cracking/cold cracks are known by several other names, based on their location relative to a weld bead, including *toe cracking/toe cracks, root cracking/root cracks,* and *underbead cracking/underbead cracks.* As shown by the schematic illustration in Figure 6.1, these are located, respectively, where the crown of a butt weld or face of a fillet weld meets the unmelted HAZ, where the root of a weld meets the unmelted HAZ, and along a line in the high-temperature portion of the HAZ separated from but running parallel to the weld's fusion zone boundary.

Figure 6.6 shows examples of toe cracking and root cracking in fusion welds.

In order for HAZ cold cracking to occur, three conditions must be present simultaneously: (i) there must be a sufficient level of hydrogen present in the material, (ii) there must be a sufficiently susceptible microstructure of the base material involved, and (iii) there must be a sufficiently high level of residual or applied tensile stress. The good news is that adequate reduction or elimination on any one of these three factors will usually prevent HAZ cold cracking. The best practice in welding is to limit two of the three variables, namely (i) the level of hydrogen present from any source(s) and (ii) reduce the susceptibility of the HAZ microstructure. More will be said about these later in this section.

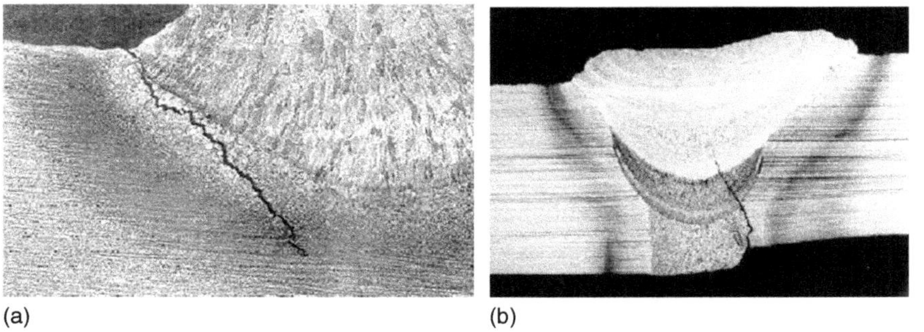

(a) (b)

Figure 6.6 Photomacrographs showing (a) toe cracking form of hydrogen-induced cold cracking in the coarse-grained region of the HAZ of a weld in a hardenable steel and (b) root cracking form initiating in the root HAZ and propagating into weld metal in steel. Note the linear mismatch of the joint elements in (b). Source: (a) TWI (2000) and (b) Hart (1999). Used with permission of TWI.

Given that all HAZ cold cracking is the result of hydrogen, cold cracking is commonly referred to as hydrogen embrittlement, although this is technically incorrect. A much more appropriate and technically accurate term is *hydrogen-assisted cracking* or *hydrogen-induced cracking*.

Hydrogen embrittlement properly refers to the creation of some inherently brittle condition or situation in a metal or alloy. Three major possibilities exist: (i) formation of an inherently brittle hydride (i.e. a nonmetallic compound of hydrogen and a metal, such as TiH_4 in Ti and Ti alloys), (ii) formation of microscopic voids filled by diatomic hydrogen under extraordinarily high hydrostatic pressure, which can lead to cracking by void extension and appear as shiny opened pores, known as "fisheyes" (Figure 6.7), in fracture surfaces (as in hardenable steels), or (iii) the lowering of the surface energy of microcracks by adsorbed atomic hydrogen, such that unstable crack growth is promoted in susceptible microstructures (most notably, as-quenched, untampered ferrous martensite). The last two phenomena are attributed to lead to hydrogen-assisted or hydrogen-induced cold cracking. The problem is most prevalent – and most serious – in steels having a strong tendency to form metastable martensite during rapid cooling, such as that routinely occurs in the high-temperature region of the HAZ surrounding fusion welds made in room temperature or colder material, especially in thicker section weldments.

Although there is no free hydrogen (i.e. H_2) in the Earth's atmosphere, there is water (H_2O), as humidity, fog, rain, dew, sleet, snow, or frost. Water molecules dissociate into atomic hydrogen (i.e. $H_2O \rightarrow 2H + O$) in a welding arc or plasma (from plasma-arc welding [PAW], EBW, or LBW). The tiny atoms diffuse very

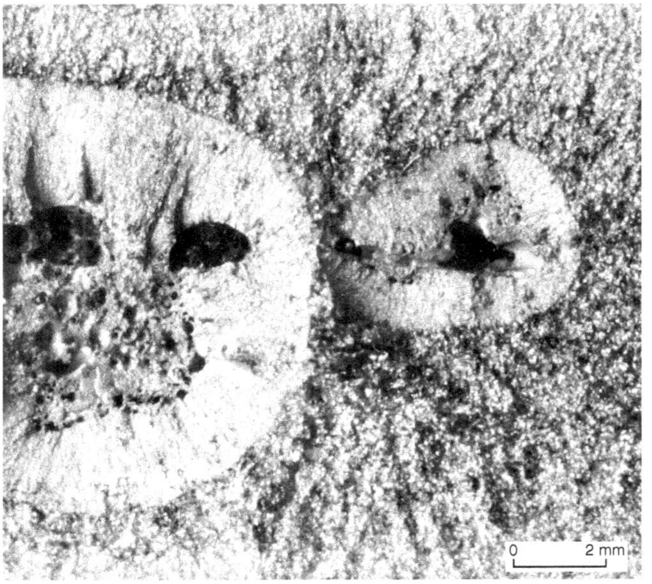

Figure 6.7 Photomacrograph showing characteristic "fisheyes" in the surface of the fracture, indicative of hydrogen-induced cracking. Source: Moeser (1984). Used with permission of Mr. Moeser.

rapidly through the crystal lattice of all metals, including steels (i.e. essentially, Fe). Other potential sources of water in the welding environment include water absorbed into the flux coatings of shielded metal arc welding (SMAW) electrodes, the granular fluxes used with SAW, and even within the cores of FCAW wires; waters of hydration associated with rust on steels, oxides (e.g. on Al and Mg and their alloys), and on tarnish layers (e.g. patina on Cu and Cu alloys); moisture contained in shielding gases (without drying filters in gas lines); moisture adsorbed on the inside of gas lines or aspirated into the shielding gas stream through loose fittings or porous hoses; hydrocarbon contaminants such as oils, grease, paint, and marking crayons and pens; and hydrogen in base metals (especially, Ti and Ti alloys). To prevent hydrogen cold cracking, all of these potential sources should be checked regularly! Remember: Cleanliness, Cleanliness, Cleanliness.

The second necessary condition for cold cracking is the presence of a susceptible microstructure. The most prevalent, by far, is martensite that has formed by rapid quenching and not been tempered to relieve some of the tremendous lattice strain that gives this metastable phase its high strength and hardness but, most significantly, its extreme brittleness. The conditions that promote the formation of brittle martensite in steels are high cooling rates (exceeding the alloy's critical cooling rate) and high hardenability. Suitably high cooling rates are routinely present in the high-temperature region of the HAZ of steels, especially when the ambient temperature is below around 10 °C/50 °F and/or the joint and/or surrounding structure is thick (typically over 50 mm/2 in.). Cooling rates also tend to be higher for lower net linear heat inputs (notably from high welding speeds). Higher hardenability levels in steels result from, in order, greater carbon contents (e.g. over about 0.40–0.45 wt% in carbon steels and over 0.30–0.35 wt% in low-alloy steels) and additions of Mn, Cr, Mo, Ni, and Si.[11]

For steels for which hardenability (as indicated by C_{eq} values) indicates susceptibility to martensite formation in the HAZ, the most effective way to reduce cooling rate and avoid potential problems is to employ preheat. By raising the temperature of the material in and around the joint to be welded, the cooling rate is reduced, limiting the formation of martensite. Suggested temperatures for preheat are available in a variety of sources, most notably the American Welding Society (AWS, www.aws.org). Preheating can be done using furnace soaks just before welding, localized heating by oxy-fuel gas torches or burners, banks of high-intensity quartz lamps, resistance-heated blankets, or induction heating coils. Regardless of the method used, the temperature can be checked using TempilStik™ specially formulated wax temperature indicator sticks.

The residual stresses that almost inevitably arise during fusion welding can be reduced through post-weld thermal stress relief treatments, with recommended practices available from the AWS or from applicable codes, as appropriate. Although arguments are commonly made as to why post-weld stress relief heat treatments are (i) impractical or (ii) time consuming and (iii) unnecessarily

11 The hardenability of a carbon or low-alloy steel is best assessed using empirical equations for carbon equivalent, C_{eq}. A commonly used equation for carbon steels is: C_{eq} = wt% C + wt% Mn/4 + wt% Si/4 < 0.45. A commonly used equation for low-alloy steels or carbon steels by Deardon and O'Neill is: C_{eq} = wt% C + wt% Mn/6 + (wt% Cr + wt% Mo + wt% V)/5 + (wt% Cu + wt% Ni)/15 < 0.35 excellent, 0.36–0.40 very good, 0.41–0.45 good.

expensive, worldwide experience has demonstrated that, over and over again, PWHT is always possible, if there is sufficient motivation, and the time and cost for PWHT becomes incidental if a failure occurs in weldment because of cold cracking. The most insidious aspect of hydrogen-induced cold cracking is that it need not occur as the weldment cools to room temperature but can be delayed by hours, days, weeks, months, or even longer. Hence, the other common name *delayed cracking*.

Considering what has been presented above concerning cold cracking in the HAZ of steels, in particular (but not uniquely!), the best practices for avoiding problems include the following:

1. Checking the hardenability of a carbon or low- or medium-alloy steel to be welded for the first time using (i) appropriate empirical equations for calculating C_{eq} using the minimum and maximum possible values of alloy additions permitted by specifications and (ii) referring to Jominy end-quench hardenability curves
2. Using low-hydrogen-type SMAW stick electrodes or FCAW wire electrodes
3. Storing coated SMAW electrodes, FCAW wires, and SAW fluxes in proper dry boxes, using baking as recommended by the manufacturer
4. Eliminating or reducing all conceivable potential sources of hydrogen in the welding environment
5. Employing preheat and/or post-weld thermal stress relief treatment as recommended by best practices of applicable codes or the AWS.

6.5 Other Weld-Related Cracking and Cracks

Some alloys are susceptible to cracking that is absolutely related to the material having been welded, but cannot properly be considered weld cracking or weld cracks. The two examples are weld-related (i) *corrosion cracking* and (ii) *fatigue cracking*.

Stainless steels are made "stainless" – free of rusting – because they have a naturally forming protective (i.e. passivating) layer of chromium oxide (Cr_2O_3) on their surface that keeps oxygen (from the atmosphere) out and iron in (away from oxygen in the atmosphere). To be effective, however, at least 13 wt% Cr must be in solution in the alloy. As a result of fusion welding, temperatures in the HAZ in the range of 400–800 °C/750–1500 °F can cause some of the Cr in the solid solution phase to react with C in the same phase to form chromium carbide particles (mixtures of Cr_3C_2, Cr_7C_3, and/or $Cr_{23}C_6$) to form in grain boundaries. As a result, Cr is depleted in the matrix phase in the so-called "denuded zones" around particles. This depletion leads to sensitization of the alloy to intergranular corrosion attack and *sensitization cracking*. As usual, actual cracking is more likely to occur when assisted by tensile stresses at or near the surface of the weldment, whether residual stresses or applied stresses.

Options for avoiding sensitization include (i) using stainless steel grades with less than around 0.045 wt% C (which tends to be the level found in most modern grades by specification), (ii) employing a so-called stabilized grade of stainless

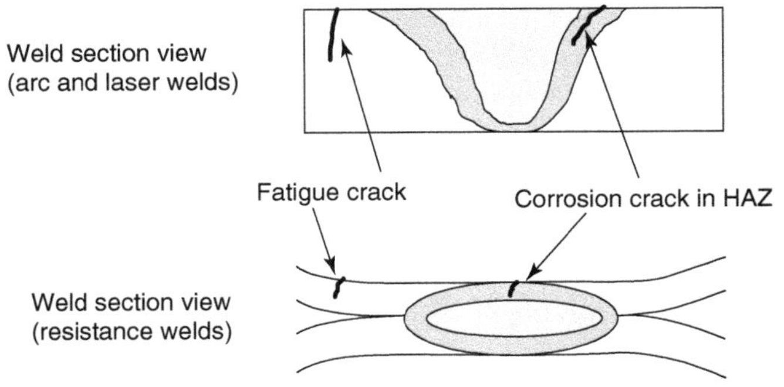

Figure 6.8 Schematic illustrations of fatigue and corrosion cracking related to welding. Source: Kelkar (2014). Used with permission of WJM Technologies.

steel, such as 321 (with added Ti to preferentially form TiC to Cr carbides) or 347 (with added Nb to preferentially form NbC to Cr-carbides), or (iii) post-weld solution, heat treat the entire weldment at 1040 °C/1900 °F or above for two to four hours, which is usually not practical.

Residual stresses left in a weldment following welding can lead to stress corrosion cracking in some alloys, notably Al alloys in the presence of chlorides, Cu alloys in the presence of ammonia, and ferritic and austenitic stainless steels in the presence of chlorides.

Fatigue cracks can form in regions near a weld because of the left-in residual stresses that assist in the formation and growth of such cracks in the presence of otherwise acceptable cyclic (fatigue) loading. Such cracks have been observed in locations beyond the HAZ where residual surface tensile stresses were produced by welding and not removed by post-weld stress relief. The best option is post-weld stress relief heat treatment, but, short of this, localized hammer or shot peening or planishing (all of which introduce some compressive stress to offset tensile stresses) is better than doing nothing.

Figure 6.8 schematically illustrates the typical locations of weld-related corrosion and fatigue cracks in and near welds.

6.6 Crack-Prone Metals and Alloys

The AWS has defined *weldability* as "… the capacity of a material to be welded under the imposed fabrication conditions into a specific suitably designed structure and to perform satisfactorily in the intended service." By this definition, some metals and alloys may exhibit good weldability under some conditions, for some intended service, but poor weldability under other conditions or for other intended service.

Although weldability depends on many factors, including the process being used, the operating parameters (most particularly, the net linear heat input), the procedure followed and techniques employed, the degree of restraint on a joint, and conditions in the welding environment (especially, the presence of hydrogen in any form from any source), the most important factor of all is the chemical composition and nature (e.g. microstructural condition) of the base metal or alloy. Composition determines inherent weldability in some metals and alloys, with some being inherently weldable, some being inherently more difficult to weld, and some being essentially unweldable by fusion processes, at least. With cracking being the most dreaded and universally unacceptable of all weld discontinuities, it only makes sense to consider which among the most commonly welded engineering metals and alloys are prone to cracking by one mechanism or another.

Let us consider crack-prone systems in alphabetical order.

Cracking in aluminum alloys: Although notorious for problems with porosity during fusion welding, aluminum alloys are correctly considered weldable by arc, plasma, and electron-beam processes without cracking-related problems *provided* a proper filler (typically, 4043 or 4047) is used with appropriately developed and tested welding procedure. As a rule, the non-heat-treatable grades (3xxx and 4xxx) are most weldable, followed by the only heat-treatable Al alloy to have been specifically developed for weldability, i.e. 6061, being a close second (or third). This all said, hot cracking (i.e. cracking in the presence of a liquid phase) does occur in restrained joints in heat-treatable Al–Cu (2xxx series) alloys with 2.0–3.2 wt% Cu, in non-heat-treatable Al–Mg (5xxx series) alloys with 0.5–2 wt% Mg (notably, 5052), and in heat-treatable Al–Mg + Si (6xxx series) alloys with 1.0–1.5 wt% Mg + Si. Alloys in the heat-treatable Al–Zn–Mg (7xxx series) tend to be free of hot cracking, unless small amounts of Cu have been added for additional strengthening, as in 7075 (i.e. an Al–Zn–Mg–Cu alloy).

In all of the alloys above, the operative mechanism is constitutional liquation cracking in the PMZ or very-high-temperature HAZ, where nonmetallic particles (usually not the operative precipitate phase) begin to dissolve and create a surrounding low-melting region that melts under the highly nonequilibrium conditions of heating and cooling. Most (although not all) of these particles tend to be in grain boundaries, so cracking tends to be intergranular.

Liquation hot cracking in Al alloys prone to the problem can generally be welded by applying one or more of the following principles:

- Avoid the most extremely crack-sensitive alloys, if possible
- Use a suitable (manufacturer-recommended) filler with a solidification point close to or below that of the base material
- Use the highest welding speed that will provide needed penetration
- Use assembly techniques and welding sequences and techniques that minimize restraint and minimize thermally induced stresses
- Apply a compressive force across the joint during welding to counteract the pulling force that leads to cracking.

Problems with hydrogen embrittlement observed in some high-strength Al alloys tend to be from stress corrosion, which is weld related through welding-induced residual stresses that were not relieved following welding.

Cracking in copper alloys: The greatest challenge in fusion welding copper and copper alloys is the high thermal conductivity of these base materials. Welding current and the type of shielding gas employed must maximize heat input to the joint without producing burn through. Hot cracking can occur in alloys with wide liquidus-to-solidus temperature (freezing) ranges, such as Cu–Sn (bronze) and Cu–Ni (Monel) alloys. Hot cracking tendency can be minimized by reducing restraint during welding, preheating to slow the otherwise inherently high cooling rate and reduce the magnitude of thermally induced stresses, and reducing the width of the root opening and increasing the size of the root pass.

Hydrogen cold cracking is not observed in copper alloys.

Cracking in nickel and nickel alloys: Nickel and its alloys can generally be reliably welded by most welding processes, except forge welding and oxy-fuel gas welding. Wrought alloys tend to exhibit similar welding characteristics to austenitic stainless steels. Cast grade of Ni alloys, especially those with high Si contents, present difficulties. Non-heat-treatable grades are the most weldable, whereas the heat-treatable, precipitation-hardenable grades are best welded in the solution-treated condition.

It is imperative to remember that Ni and Ni alloys are extremely susceptible to hot cracking from additions or residual amounts of S, P, Pb, and other low-melting-point elements (e.g. Cd, Zn, Sn, Ag, B, and Bi).[12] These elements are often present in oils, greases, paints, marking crayons and inks, machining and forming lubricants, shop dirt, and many processing chemicals. Hence, cleaning (using alkaline cleaners containing sodium carbonate) is best following normal solvent cleaning. Cracking in the weld metal or FZ usually results from contamination by the aforementioned elements. Cracking in the HAZ is usually the result of intergranular penetration of contaminants from the base metal surface, especially S for cutting oils.

Weld metal hot cracking can be reduced by strict process control centered around cleaning. Although weld metal cracking may be the result of excessive welding current, centerline cracking is usually the result of a tear-drop shaped weld puddle, and backward-pointing V-shaped ripples on the crown bead. In the latter case, welding speed should be reduced.

Weld cracking is also promoted in Ni and Ni alloys by undue restraint. Techniques for reducing cracking from this source include reducing width-to-depth ratio if cracks are open to the surface, reduce depth-to-width ratio if cracks are internal to the weld, and be sure that all welds have convex crowns (for butt welds) or faces (for fillet welds).

Liquation cracking, specifically constitutional liquation cracking, is known to occur in some age-hardenable alloys intended for elevated temperature service (i.e. Ni–base superalloys, such as Inconel 718/Alloy 718). These cracks will be found in the PMZ or very-high-temperature HAZ and should be dealt with as described earlier.

Ni–base alloys are immune to hydrogen-induced cracking as well as to stress corrosion cracking.

Steels: All carbon, low- and medium-alloy steels are susceptible to hot cracking because of contamination from residual (i.e. tramp) S and/or P, as well as other low-melting elements found nearby in the Periodic Table of Elements (notably, Cd found in electroplate, and Pb or Se added to improve machinability).

---

12 Many of these latter low-melting-point metals are present in brazing and soldering filler metals.

Obviously, efforts must be expended during setup before actual welding to remove all potential sources of S by appropriate cleaning.

The most prevalent form of cracking found in steels, by far, is hydrogen-induced cold cracking in the HAZ. The problem is worse as the value of C_{eq} increases above about 0.45–0.50 for C steels and 0.4 for low- and medium-alloy steels (see Section 6.4).

High-alloy (greater than 10 wt%) tool steels designed for special purposes (e.g. impact shock resistance, hot or red hardness, and high-temperature deformation resistance) are often prone to hydrogen cold cracking and generally require preheat in order to be made weldable.

Stainless steels grades fall into four major types: (i) ferritic stainless steels, which are resistant to hot cracking and immune to cold cracking, (ii) austenitic stainless steels, which are immune to cold cracking but susceptible to hot cracking unless 5–10% ferrite is present in the alloy (using constituent diagrams, such as the Schaeffler diagram or DeLong diagram), (iii) martensitic stainless steels, which tend to be resistant to hot cracking but very prone to hydrogen cold cracking until and unless they are post-weld tempered, and (iv) precipitation-hardening (PH) stainless steels, which can exhibit liquation cracking. Duplex grades with ferrite + austenite tend to weld well, whereas grades with martensite tend to be susceptible to hydrogen cold cracking, which is often delayed.

Titanium and titanium alloys: Based on inherently reactive titanium, care must be taken to avoid contamination by C, O, N, and H. Oxygen leads to the formation of an extremely brittle, fatigue-crack prone α case (oxygen stabilizing the α form of Ti). The other elements all tend to lead to the formation of brittle compounds (i.e. carbides, nitrides, and hydrides), which can promote fatigue cracking. The all-β and β-rich $\alpha + \beta$ Ti alloys are susceptible to hydrogen embrittlement, which may involve a mix of brittle hydride formation and hydrogen-assisted cracking.

6.7 Troubleshooting Cracking Problems in Welds

Table 6.1 that lists the various ways problems from cracking can arise during fusion welding and in fusion welds.

Table 6.1 Troubleshooting weld cracking problems.

Problems with Hot Cracking and Cracks:	
o Longitudinal cracking/centerline cracking (FZ)	
✓ Segregation induced	Reduce/eliminate low-melting solutes and impurities, disrupt growth with pronounced centerline impingement
✓ Bead shape induced	Reduce W/D ratio to 1 : 1 to 1 : 1.4
✓ Surface profile induced	Reduce welding voltage, reduce TS to produce convex bead contours

(*continued*)

Table 6.1 (Continued)

o Crater cracking (FZ)	Reduce abruptness of arc or beam weld termination by withdrawing slowly or ramping down current; add stop tabs
o Transverse cracking/cross-check cracking (FZ)	Reduce restraint, reduce strength, and especially hardness of the filler metal
o Liquation cracking/constitutional liquation cracking (PMZ/high-temperature HAZ)	Reduce strength of filler metal, lower heat input via lower I
Problems with Cold Cracking and Cracks:	
o Hydrogen-induced cracking/ hydrogen-assisted cracking	Reduce/eliminate all potential sources of hydrogen, use preheat for C steels with $C_{eq} = 0.50$ and low-alloy>0.40, post-weld temper or stress relieve
Problems with Other Weld-Related Cracking:	
o Corrosion cracks (HAZ)	
✓ Sensitization cracking	Use low-C grade (<0.045 wt%), use stabilized grade (321 with Ti or 347 with Nb), post-weld solution treat
✓ Stress corrosion cracking	Post-weld heat treat for stress relief
o Fatigue cracks (low T HAZ or UBM)[a]	Post-weld stress relieve thermally, if possible, using hammer or shot peening or planishing, if not

a) Fatigue cracks can and often do initiate at geometric notches created by weld crown or root beads.

References

Hart, P.H.M. (1999). Hydrogen cracking – its causes, costs and future occurrence, figure 3. www.TWI-global.com.

Hensley, T. (2011). Basic tips for understanding (and preventing) common types of weld cracking. Posted on 9 November 2011, www.fabricatingandmetalworking.com.

Kelkar, G.P. (2014). Weld cracking – an engineer's worst nightmare, figure 3.

Moeser, M. (1984). Fractography and mechanism of hydrogen cracking – the fisheye concept. www.martin-moeser.de.

Pfaller, T. (2015). Unlocking aluminum welding. www.weldingproductivity.com.

TWI (1999). Defects – solidification cracking, figure 1. Job Knowledge 44, www.TWI-global.com.

TWI. What is hot cracking (solidification cracking)? Figure 3. Frequently Asked Question, www.TWI-global.com.

TWI (2015). Welding inspection and quality control, Task #2, Flaw/Discontinuity/Defect, Cracks, slide #6. www.slideplayer.com.

TWI (2000). Defects – hydrogen cracks in steels – identification, figure 2. Job Knowledge 45, www.TWI-global.com.

Further Reading

Bailey, N., Coe, F.R., Gooch, T.G. et al. (1993). *Welding Steels Without Hydrogen Cracking*, Woodhead Publishing Series in Welding and Other Joining Technologies, 2e. Cambridge: Woodhead Publishing Ltd. (now part of Elsevier Ltd.), Kindle edition.

Böllinghaus, T. and Herold, H. (2008). *Hot Cracking Phenomena in Welds II*. New York, NY: Springer Publishing Company.

Nagumo, M. (2016). *Fundamentals of Hydrogen Embrittlement*, 1e. New York, NY: Springer Publishing Company.

7

Nonmetallic and Metallic Inclusions

7.1 Solid Versus Gas Inclusions

Chapter 5 described the occurrence of *gas inclusions* in fusion welds, the two major forms being spherical or globular porosity (or simply, porosity) and elongated porosity (commonly referred to as either wormholes/wormhole porosity or pipes/pipe porosity). The underlying cause of all gas inclusions is the entrapment of gas within the solidified weld. Gas formation may arise from any of several root causes, such as inadequate shielding from the atmosphere to allow absorption of nitrogen, oxygen, or hydrogen from dissociated water; excessive moisture from any source; contamination of joint element surfaces by hydrocarbons, oxides, or tarnish layers; improper welding procedures or techniques; or gaseous impurities in base materials. However, there is another type of inclusion that can occur in fusion welds, i.e. *solid inclusions*.[1]

Solid inclusions that occur in the fusion zones of welds are the result of solid foreign substances that become trapped in the weld metal during welding. Four predominant types of inclusions can occur:

- Flux inclusions
- Slag inclusions
- Oxide inclusions
- Metallic inclusions.

The first three types are all *nonmetallic inclusions*, involving nonmetallic materials (i.e. ceramics or glasses) having densities lower than the base and/or the filler metal. The fourth types are a *metallic inclusions*, which can, and almost always do, have a higher density than the base and/or filler metal. The significance of an inclusion's density relative to the weld metal relates to its ease of detection by X-radiography and its appearance in radiographs.

A fifth type of inclusion can actually occur, although it is rare compared to the other types. This fifth type involves a foreign material that falls into the molten weld pool and becomes trapped. The most common examples are particles of dirt or other debris from the welding environment (e.g. air or work surfaces).

1 Solid inclusions can, in fact, also occur in welds made by non-fusion welding processes, where no melting and solidification is involved. The source is virtually always oxides or other tarnish-layer non-metallic compounds, with the resulting defect consisting of particles of that/those compounds trapped at the original interface between joint elements, often lines of inclusions.

A Practical Guide to Welding Solutions: Overcoming Technical and Material-Specific Issues,
First Edition. Robert W. Messler, Jr.

The first three nonmetallic inclusion types listed above can be further classified by self-evident terms as

- o Isolated
- o Linear
- o Clustered.

Metallic inclusions are virtually always isolated in their occurrence.

The type of inclusions found (i.e. flux, slag, oxide, metallic, or foreign debris), together with how they appear (i.e. as isolated, linear, or clustered forms), gives clues to their root cause. According to the American Society of Mechanical Engineers (ASME), causes of welding discontinuities (as listed in Table 3.2) can be broken down as follows: 41% because of poor process conditions (e.g. contaminated joint surfaces, consumables, or base materials), 32% to operator error (e.g. incorrect welding parameters, failure to remove the slag between welding passes, and sticking an electrode into the weld pool), 12% to the wrong technique (e.g. allowing molten metal to flow over the molten slag), 10% to incorrect consumables (e.g. wrong choice for out-of-position welding), and 5% to improper weld groove choice or preparation (e.g. insufficient accessibility). Solid inclusions can be the result of most of these, so, as usual when troubleshooting a problem from inclusions, the root cause must be found.

7.2 Nonmetallic Inclusions in Welds

The three principle types of nonmetallic inclusions (i.e. flux inclusions, slag inclusions, and oxide inclusions) have two different sources. *Flux inclusions* and *slag inclusions* both come from slag-protected fusion welding processes, which include shielded metal arc welding (SMAW), flux-cored arc welding (FCAW), submerged arc welding (SAW), and electro-slag welding (ESW).[2] Oxide inclusions come from the surfaces of joint elements that have not been properly cleaned of oxide layers (e.g. mill scale on steels, naturally forming oxides on Al and Mg alloys, and stainless steels). Let us find out how.

In the primary extraction and purification of metals from ores and the subsequent production of alloys from these refined metals, molten compounds of acidic, basic, and neutral oxides and halides (or oxide–halide mixtures) are used to remove unwanted nonmetallic impurities by oxidation–reduction reactions. The starting ingredients, often obtained from natural sources, such as minerals, are properly called a *flux* (short for fluxing agent). Once such a flux has reacted with impurities from within a metal or alloy (predominantly dissolved gases, sulfur, and phosphorus) and with gases from the atmosphere (most particularly, oxygen), the resulting viscous, glassy melt and subsequent glasslike solid are properly called a *slag*. It is not uncommon that the term slag is used regardless of

2 It is possible for slag inclusions in welds to come from re-melting of slag inclusions in certain base materials containing significant amounts of such inclusions. The most commonly encountered example is wrought iron, which routinely contains up to 2% by weight of fibrous slag inclusions, which make the material stronger.

the stage of the extraction–refining–alloying process in production metallurgy being referred to.

With fusion welding actually being a close cousin to primary metal and alloy production, as well as to casting, not surprisingly, fluxes and slags are also used in fusion welding. In fact, in fusion welding, these materials perform many functions. Firstly, dry flux ingredients that make up the largest proportion of the coatings in stick electrodes for SMAW and in the cores of tubular wire electrodes for FCAW, and almost all of the granular flux used with SAW and ESW,[3] melt in the heat of the arc (or by I^2R heating in ESW) to become extremely chemically reactive and protect or shield the molten weld pool from oxidizing gases in the surroundings. They do this by "gettering" (scavenging) gases. Secondarily, the reactive molten fluxes remove nonmetallic impurities (e.g. S, P, and Si) from the molten weld metal by scavenging or gettering these from within the weld pool, aided by convection.

After potentially contaminating gases from the welding environment and nonmetallic impurities in the weld pool have been scavenged, the resulting molten material becomes a viscous glass (of mixed oxides and halides), which then hardens into a friable, glasslike shell over the weld. Although molten, the slag, being less dense than the molten weld metal, floats to the top of the weld pool and provides cover that protects (shields) the reactive molten metal from oxidation by the surrounding air. As the weld metal cools, the molten slag cools, and, eventually, both become solid; the weld being a crystalline metal or alloy, the slag being an amorphous glass. The glasslike shell of hardened slag protects the cooling but still reactive weld metal from oxidation and, depending on its formulation, can help contain the weld metal as a liquid and/or slushy mixture of solid and liquid and help shape the solidifying metal to form the crown bead.

Adding to the complexity of formulating fluxes for coating SMAW electrodes, filling FCAW tubular wires, and, only to a slightly lesser degree, protecting the arc and molten weld metal in SAW and ESW, is the fact that it is the balance between the surface tension and the viscosity of the molten flux and, then slag, that influences how easily either can become trapped within a weld pool to form a nonmetallic inclusion. Surface tension influences how well the molten flux wets the underlying metal, whereas viscosity influences how easily the molten flux can flow. These two properties need to be balanced to help protect and shape the weld, without becoming g trapped beneath the still molten metal. Different chemical makeups of fluxes produce very different characteristics.

Just as an example, an oxidizing flux containing iron oxide produces a low surface tension weld pool with a more concave weld crown bead profile that promotes wetting of the parent metal to preclude overlap defects. A high silicate flux produces a glasslike slag that is easy to remove and is often self-detaching. Fluxes with a lime content are free of hydrogen but produce an adherent slag

---

3 The coatings of SMAW consumable electrodes and the cores of FCAW consumable wires also contain ingredients that thermally decompose to produce non-oxidizing shielding gases, release ions to stabilize the arc, and, in some cases, add alloying additions to a core wire to adjust the composition of the weld metal or increase the deposition rate of weld metal. SAW fluxes are sometimes enhanced with metallic elements (e.g. Mn and/or Cr) to offset evaporative losses of these elements in an electric arc (i.e. compensating fluxes).

that is difficult to remove and prone to producing slag inclusions, especially in multi-pass welds.

With the preceding as background, *flux inclusions* result when unfused flux becomes entrapped in welds during welding. This can occur when SMAW electrodes have damaged coatings or are run too hot (at an amperage above that recommended by the manufacturer) and coating pops off to fall into the weld pool. It can also occur from faulty FCAW tubular wires and with spoiled SAW flux. The imperfection is irregular in shape and tends to be isolated. Problems can be avoided using well-maintained consumables and by employing proper welding techniques to produce deposits with good, smooth crown bead contours.

Slag inclusions result when molten slag becomes entrapped during welding or in multi-pass welding when incompletely removed slag from an earlier pass becomes trapped by a subsequent pass. The imperfection is irregular in shape and can appear as isolated defects or in clusters or as elongated stringers and is much more common that flux inclusions. When molten slag becomes trapped between molten metal and the side wall(s) of a prepared joint groove, the result is often incomplete fusion at the wall(s). Prevention is facilitated by employing the proper electrode angle to preclude slag overflow, which impairs the arc characteristics. Well-maintained consumables must be used and correct welding current, voltage, travel speed should be employed to produce weld beads with smooth contours. Weld joint groove preparations must allow adequate access to the root of the weld, in particular, as most slag inclusions occur near the root. Most importantly, welders must be trained to be observant, an alert to soon-to-be-trapped flux particles or molten slag.

Figure 7.1 schematically illustrates slag inclusions in a weld (a) along with the appearance of these low-density defects in X-radiographs (b). Figure 7.2 shows

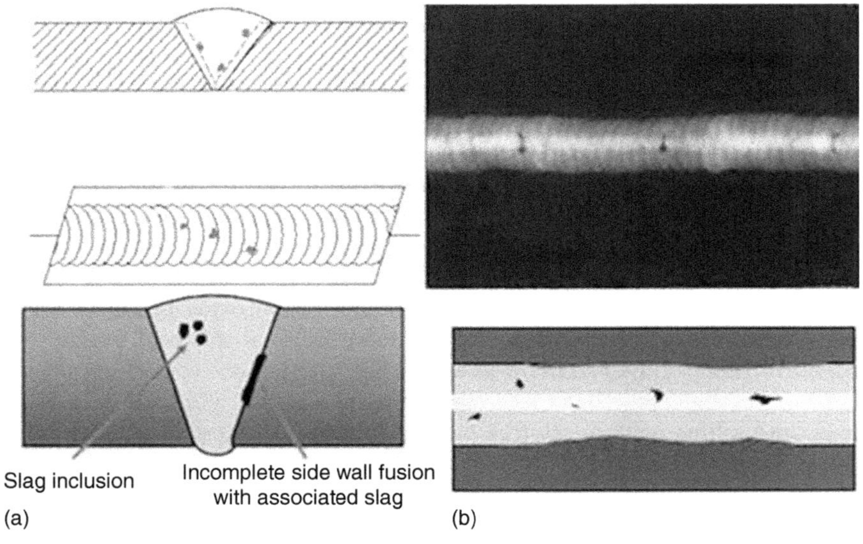

Figure 7.1 Schematic illustrations (a) showing various locations and forms of slag inclusions in fusion welds, along with X-radiographs (b) showing the appearance of such defects during nondestructive evaluation of welds. Source: NDT Resource Center (2009). Used with permission of the Center for Nondestructive Evaluation, Iowa State University.

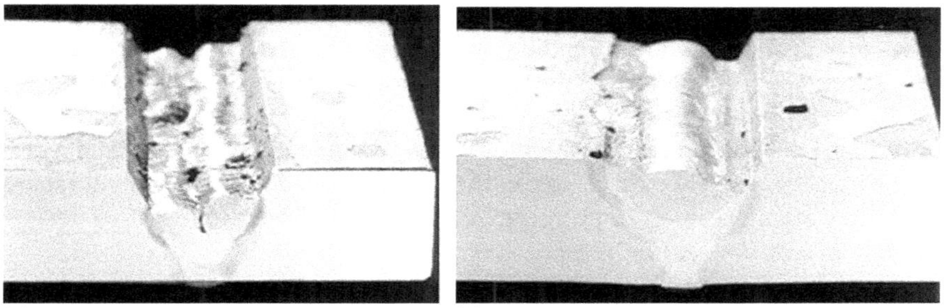
(a) (b)

Figure 7.2 Photographs showing the weld crown surfaces of (a) a poorly made weld and (b) a properly made weld, showing the susceptibility for slag entrapment (perhaps from difficult removal between passes) with rough, irregular contours. Source: Lukas (1999). Used with permission of TWI.

photographs of welds made with poor technique (in a) and proper technique (in b). Irregular surface contour (as in a) tends to make complete removal of hardened slag difficult, increasing the likelihood of entrapment as inclusions after subsequent passes are made. Smooth surface contours (as in b) make hardened slag removal easier.

The third principal type of nonmetallic inclusion, *oxide inclusions*, is the result of surface oxides that were not properly removed from joint element surfaces before welding becoming entrapped within the weld metal. Many oxides (e.g. Al_2O_3 and MgO on Al alloys and Mg alloys, respectively, and Cr_2O_3 on stainless steels) are highly refractory and do not melt in a weld pool. Most appear on the surface of the welds, especially with Al alloys (Figure 7.3), although some can become entrapped as irregular solids.[4] Oxides on the surface of other metals and alloys, notably Cu and Cu alloys and steel, readily melt in weld pools,

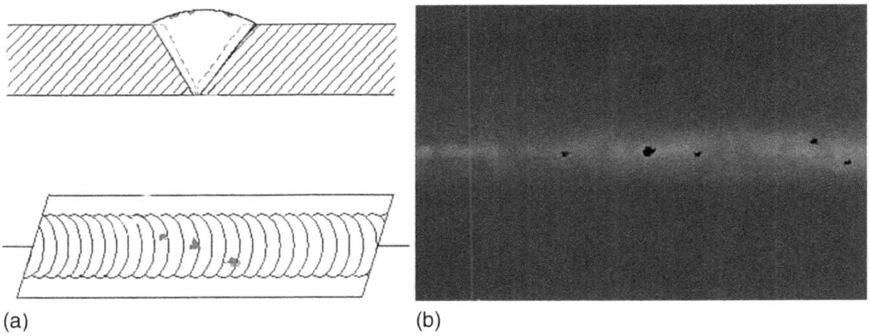
(a) (b)

Figure 7.3 Schematic illustrations (a) showing the possible locations of oxide inclusions on or in fusion welds, along with an X-radiograph (b) showing the typical appearance of such defects during NDE by X-ray. Source: NDT Resource Center (2009). Used with permission of the Center for Nondestructive Evaluation, Iowa State University.

4 The melting points of refractory Al_2O_3, MgO and Cr_2O_3 are 2072 °C/3762 °F, 2825 °C/5117 °F, and 2435 °C/4415 °F, respectively; tending not to melt in welding arcs. The melting points of Fe-oxides and Cu-oxides are 1539–1565 °C/2802–2849 °F and 1232 °C/2250 °F, respectively; easily melting in welding arcs.

thereby tend to result in globular oxide inclusions. In seeking the root cause of a particular problem with nonmetallic inclusions, simple element identification using EDS (energy-dispersive X-ray spectroscopy) or, possibly, chemical tests on unetched samples can help.

Although various sulfide inclusions (i.e. FeS and MnS) and silicate (mixed metal SO_4) stringers routinely exist in steels, these seldom lead to inclusions in fusion welds. They are more likely to promote hot cracking and base metal (BM) delamination (e.g. lamellar tearing), respectively.

7.3 Metallic Inclusions in Welds

Anytime foreign metals (fundamentally different than the base material)[5] are taken into the molten weld, metallic inclusions or metal inclusions may occur. The problem most frequently occurs in gas tungsten arc (GTA) welding where small pieces or particles of tungsten from the permanent, nonconsumable electrode enters the weld pool. The high-density inclusions are strong absorbers of X-radiation, therefore appear as distinct intense white indications on radiographs. The cause of these inclusions may be overheating of the tungsten electrode because of excessive current, most particularly when the process is being operated in direct current (DC) reverse polarity, with the W electrode positive. As the electrode becomes too hot, some melting can occur, with droplets falling into the weld pool. Another common cause is welder incompetency, as they allow the tip of the tungsten electrode to dip into the weld pool or stick to the workpiece as an arc is initiated. A possible source can be certain shielding gas mixtures, particularly those with small additions other than inert gases (i.e. active gases). Finally, although rare, the tungsten electrodes themselves can be inferior, especially if not produced by reputable, established manufacturers.

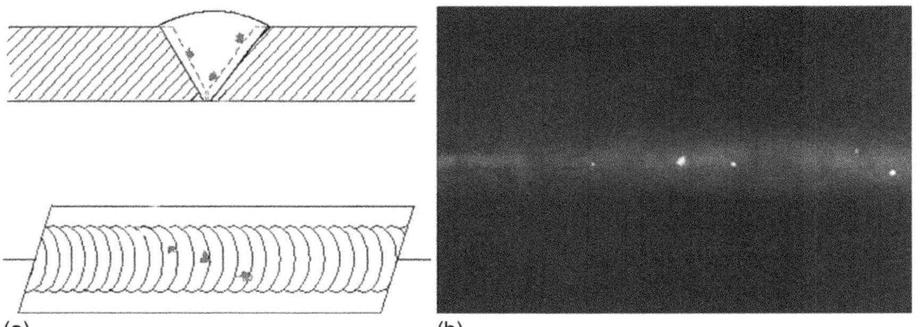

(a) (b)

Figure 7.4 Schematic illustrations (a) showing the possible locations of tungsten inclusions in fusion welds made by the GTAW process, along with an X-radiograph (b) showing the typical appearance of such defects during NDE by X-ray. Note the intensely white and sharp indications from these high-density inclusions. Source: NDT Resource Center (2009). Used with permission of the Center for Nondestructive Evaluation, Iowa State University.

5 It is possible for a piece of or incompletely-melted particle from a filler metal to enter the weld pool and become entrapped in a fusion weld, thereby leaving an incompletely melted and assimilated metallic inclusion.

Some examples have radically higher hardness than most weld metals in which they can become entrapped, the risk of such inclusions is as initiators of fatigue in service.

Although tungsten inclusions are the most common, other metals can also become entrapped to form metallic inclusions. The examples are copper from overheating (and melting) of the contact tube in FCAW, gas metal arc welding (GMAW), or SAW and metal/alloy from a beam strike of a welding fixture during electron-beam welding (EBW) or laser-beam welding (LBW).[6]

Figure 7.4 schematically illustrates tungsten inclusions in a weld, along with such inclusions as they appear as intense white indications in an X-radiograph.

7.4 Troubleshooting Problems with Inclusions in Welds

Table 7.1 lists the various ways problems from inclusions can arise during fusion welding and in fusion welds.

Table 7.1 Troubleshooting problems with inclusions in welds.

Nonmetallic Inclusions:		
o Flux inclusions	Contamination by flux off damaged or poorly maintained SMAW or FCAW electrodes or SAW flux	Use well-maintained consumables Employ proper parameters and techniques
o Slag inclusions	Molten slag in slag-protected processes becomes entrapped or slag from earlier passes is not fully removed	Hold electrodes at proper angle Use well-maintained consumables Employ proper grooves for access Train welders to be alert to slag Thoroughly remove slag between passes during multi-pass welding
o Oxide inclusions	From oxide or tarnish from joint element surfaces, base materials with excessive oxide inclusions (e.g. Al alloys)	Properly and thoroughly remove mill scale, oxide layers, and tarnish layers from all joint surfaces Beware of high oxide inclusions in BM
Metallic Inclusions:		
o W inclusions	From W electrodes in GTAW	Do not overheat W electrodes by too high I, especially with DCRP Do not allow electrode tip to dip into weld pool Beware of active gases added to inert shielding gas
o Cu inclusions	From contact tubes and nozzles	Maintain; do not overheat

6 It is good practice to construct fixtures from the same alloy as that being welded by EBW, if at all possible.

References

Lukas, B. (1999). Defects/imperfections in welds – slag inclusions, figure 2a,b. Job Knowledge 43 (with help from Gene Mathers and Colin Eileen). Modified from an original in *Contact*, September/October 1999, www.twi-global.com.

NDT Resource Center (2009). Radiographic interpretation – welds. NDT Resource Center at Iowa State University, Ames, IA, www.nde-ed.org.

Further Reading

Mandal, N.R. (2016). *Welding Defects*. SpringerLink (on line).

Messler, R.W. Jr., (2004/1999). *Principles of Welding: Processes, Physics, Chemistry, and Metallurgy* (Chapter 11). Weinheim/New York, NY: Wiley-VCH Verlag/Wiley.

Preto, M. (2013). *Welding Defects*. Rome: Aracne Publishing.

8

Weld Appearance

8.1 Can You Judge a Book by Its Cover? Is Beauty Only Skin Deep?

Two sayings come to mind as one considers *weld appearance*; one an idiom, the other a proverb. "Don't judge a book by its cover" and "Beauty is only skin deep."[1] Although both may be true for books or beauties, neither is usually true for fusion welds. Quite the contrary. The visual appearance of a weld in all likelihood is very indicative of its inner quality and integrity.

Welding, in general, is a process and fusion welding, in particular, is often a manual process. Both depend for their quality on the following, in a sequential order: quality of base materials, quality of joint preparation, quality of pre-weld cleaning, quality of consumables (e.g. coated electrodes, cored or solid wire electrodes, fluxes in any form, and shielding gases), quality of welding equipment (e.g. power supplies, electrode holders in shielded metal-arc welding (SMAW), welding guns or torches, cables, and gas hoses), quality of weld procedure development, quality of welder training and skill, and quality of process control. Good materials, good equipment, good parameters and procedures, good operators, and good techniques are, together, the key to good-looking *and* good-performing welds.

Experience has shown, over and over and over again, a weld's appearance speaks volumes to the care and pride that went into its production. Although there have surely been good-looking welds that lacked internal quality, these are rare. On the other hand, the overwhelming majority of bad-looking welds have not had good internal quality.

While it is true that some unattractive features of welds are just that, i.e. unattractive, others can indicate potentially more serious consequences in service. An attractive weld indicates not only skillful welding but also the pride of workmanship by the welder and the pride of ownership by design engineers, process engineers, quality assurance (QA) personnel, and management, to the highest level. Contrarily, if a welder produces welds and weldments that are poor

1 An *idiom* is a group of words established by usage as having a meaning not deducible from the use of the individual words, whereas a *proverb* is a short pithy saying in general use, stating a general truth or piece of advice. "Don't judge a book by its cover" came from *The Mill on the Floss* by George Elliot's (1860), whereas "Beauty is only skin deep" was written as "All the carnal beauty of my wife, is but skin deep" by Sir Thomas Overbury in 1613.

A Practical Guide to Welding Solutions: Overcoming Technical and Material-Specific Issues,
First Edition. Robert W. Messler, Jr.

in appearance and these are acceptable to QA; plant and production managers; engineering, sales, and marketing; and top-level executives, quality has taken a backseat to other motivations. One only has to look at the welds on big, shiny, yellow earthmoving equipment to know that everyone at the company, from top to bottom, takes pride in their product!

So, let us look at weld appearance – no pun intended.

8.2 Weld Crown Bead Faults

The first thing one sees when looking at a weldment is the crown bead of a butt weld and/or the face of a fillet weld, as these are the top surfaces of welds. Several features of these crowns or faces reveal a great deal about the conditions that prevailed during their production. These features include (i) the width of the weld, (ii) the contour (or profile) of the bead's top surface, (iii) the uniformity of the weld, (iv) the shape and uniformity of ripples on the weld, (v) undercut adjacent to beads, and (vi) spatter alongside welds.

The width of a fusion weld is primarily controlled by welding arc or electron-beam current (or laser-beam intensity); the higher the current, the wider the weld. Welds (both butt welds and fillet welds) should be no larger (i.e. wider) than necessary to fully fuse joint elements and keep the volume of the weld metal to a minimum for heat control and cost efficiency. The contour of a weld crown bead should be slightly convex and blend reasonably smoothly to the base material at each edge (without a re-entrant angle and no rollover), whereas a fillet weld's face should have very slight concavity (just below the hypotenuse of an imaginary triangle drawn from the end of each leg) and transition smoothly to the base materials. Crown beads tend to become too high and roll under as linear heat input gets too low, more often because welding speed is too fast but, possibly, because welding amperage is too low. The ripple marks (or ripples) found on most fusion welds should have circular or elliptical shape, and not be V-shaped, as these arise from the shape of the back end of the moving weld pool and indicate the welding speed (see Section 8.4). Welds should never exhibit undercut, which tends to occur with either excessive welding voltage or so-called "digging arcs" (see Section 3.4). Excessive spatter around welds indicates improper welding procedure, with an arc length that is too long, amperage that is too high, or operation of gas–metal arc welding (GMAW) in a spatter-prone molten metal transfer mode (e.g. short-circuiting or bridging mode). All welds (butts or fillets) should have uniform width, contour, and ripples, as these all indicate good process control.

Besides being able to assess whether welds are being made at faster-than-ideal welding speeds based on V-shaped ripples (especially if other defects, such as centerline cracking, are being encountered), very narrow and irregular crown beads can indicate inconsistent penetration and, in the worst cases, intermittent loss of penetration. Most telling of all is when truly poor appearing welds (e.g. with undercut, excessive spatter, or burned-on slag from overheating) are permitted to remain, as this suggests a lack of concern for workmanship and, potentially, quality.

Figure 8.1 schematically illustrates top and profile views of the crown beads of butt welds made under different conditions, as indicated. Figure 8.2 shows

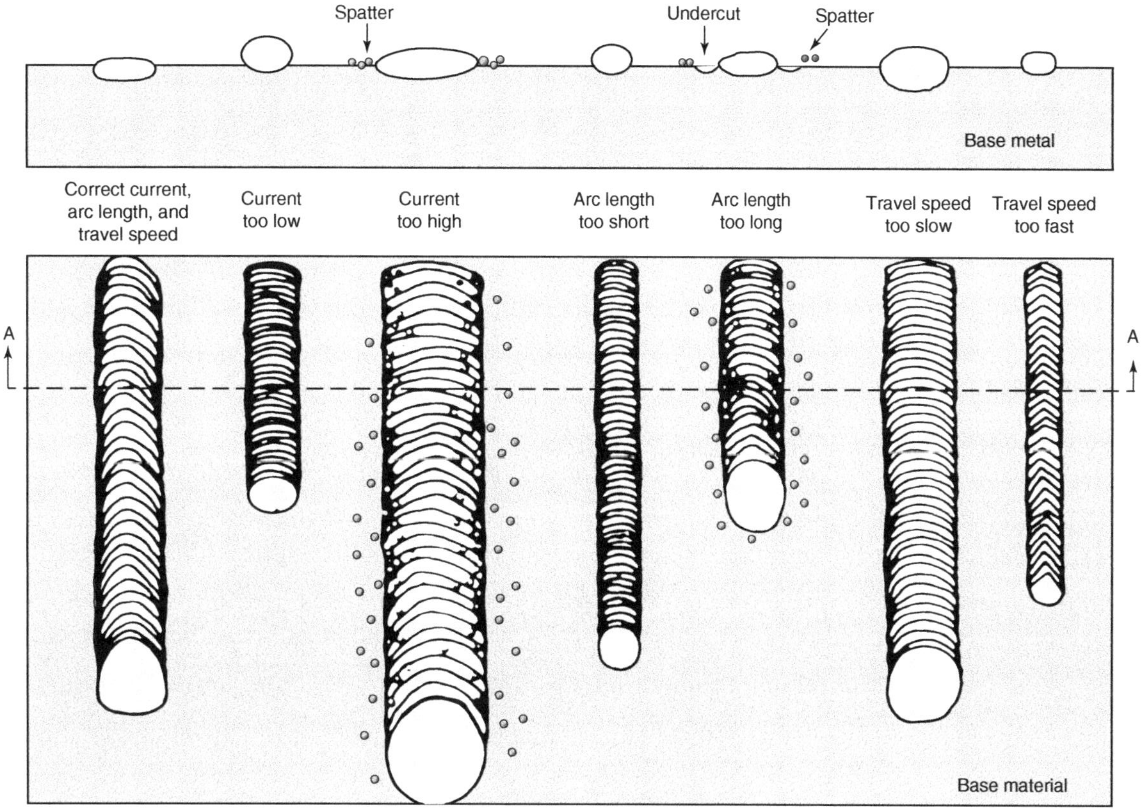

Figure 8.1 Schematic illustrations showing the appearance of the crown bead surface and profile for welds made under various conditions of current, arc length, and speed using SMAW, some of which are acceptable and some of which are not.

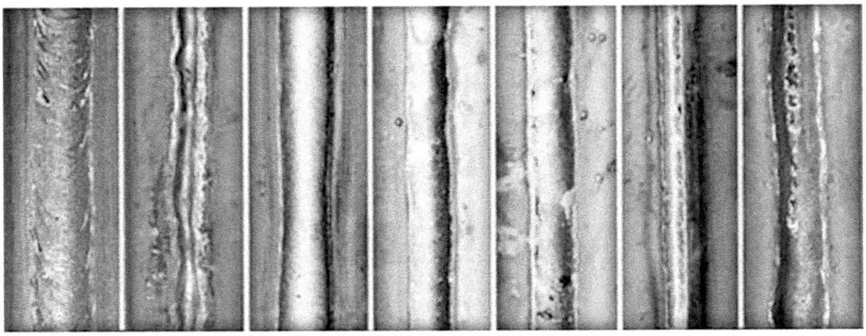

Figure 8.2 Photographs showing arc welds made under various conditions are as follows: A, good weld; B, travel too fast; C, travel too slow; D, voltage too low; E, voltage too high; F, amperage too low; G, amperage too high. Based on a manufacturer's standards, some of these welds would likely be unacceptable. Source: Miller Electric Manufacturing Company. Used with permission of Miller Electric Manufacturing Company.

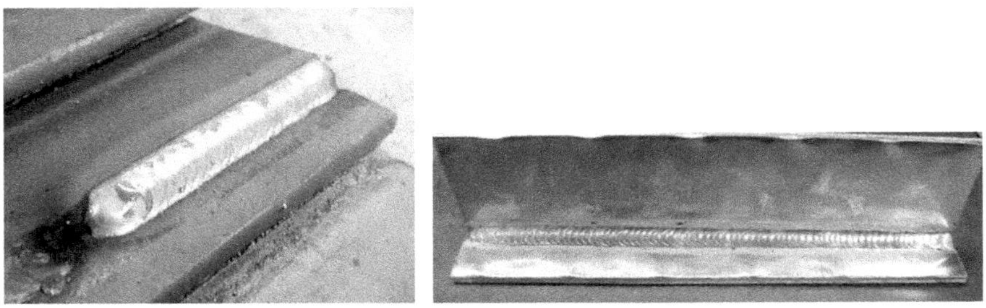

(a) (b)

Figure 8.3 Photographs showing good-looking manual welds made by GMAW/metal inert gas (MIG) on steel (a) and by GTAW/tungsten inert gas (TIG) on aluminum (b). Source: (a) Collier (2010a) and (b) Collier (2010b). Used with permission of Jody Collier.

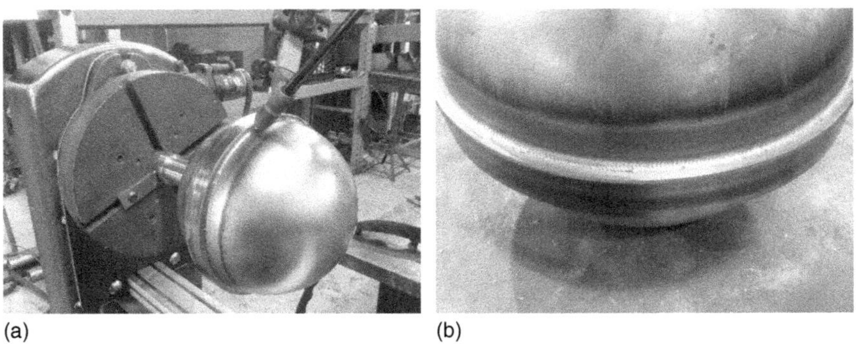

(a) (b)

Figure 8.4 Photographs showing a GTAW/TIG weld being made (a) and in a close-up (b) using a semiautomated weld positioner and a pulsed-current setup. Source: Miller Electric Manufacturing Company (2011). Used with permission of Miller Electric Manufacturing Company.

photographs of welds made on steel under different conditions using the SMAW process. Figure 8.3 shows photographs of truly attractive manual welds in steel and an Al alloy, whereas Figure 8.4 shows a photograph of a weld being made with an automated system, rather than by a manual welder.

8.3 Weld Root Bead Faults

Although often not as readily visible on many weldments, unless there is access to the underside of a joint and one makes the effort to look, the appearance of the root bead in full-penetration butt welds made without backing bars or strips is also an indicator of good or bad welding parameters and/or technique. The root bead of well-made welds should have slight convexity or reinforcement and uniformity along the weld length.

The most obvious and detrimental defect with root beads for full-penetration welds is intermittent loss of penetration, which can also result in lack of fusion defects (Figure 8.5). This problem is the result of insufficient (usually, marginal) heat input, which can be either due to current that is too low or, more often, welding speed that is too fast. The tip-off is the shape of ripples in the crown bead; V- and opposed to oval-shaped ripples suggest that welding speed is too high. The possible consequence of lack of penetration and/or lack of fusion is the stress raiser effect that can compromise performance in fatigue, especially if loading acts to open the unfused portions of the joint.

A problem that too often tends to be ignored is highly irregular convexity of root beads. These are referred to by welders as "icicles" or "grapes" or, in the vernacular, "dingle-berries" (Figure 8.6). They are indicative of one of the following: (i) excessive welding current (with possible burn through) or (ii) a root gap that is too wide or opens during welding because of thermally induced stresses.

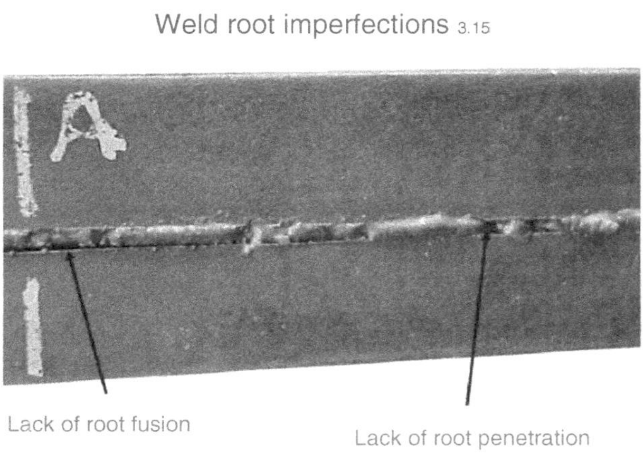

Figure 8.5 Photograph showing lack of root fusion defect (near the left end of the weld) and lack of root penetration (near the right end of the weld) in fusion butt welds made in steel. Source: Yusof (2003). Used with permission of TWI Ltd., Cambridge, UK, holder of the copyright.

Figure 8.6 Photograph showing excessive root penetration in a fusion weld to create what are variously called "icicles," "grapes," or, in the vernacular, "dingle-berries." Source: Khichade (2015). Used with permission of KEC International Ltd., Mumbai, India.

The most common method for preventing a problem involves the use of backing bar or strip or ring (as in pipeline welding). Although such irregular root beads can promote fatigue, the more immediate concern is the use of improper welding parameters.

Another problem encountered in the welding of stainless steel pipes, in particular, is known as "sugaring" (Figure 8.7a). *Sugaring* is actually the result of severe oxidation around a fusion weld when it is exposed to oxygen in the air. The black crusty oxide that forms is unattractive in itself, but it is more problematic because it indicates that the backside of the weld was not properly shielded during welding. The solution is proper purging of the inside of the pipe with inert gas (usually argon) before welding. Figure 8.7b shows the effect of reducing oxygen content

(a) 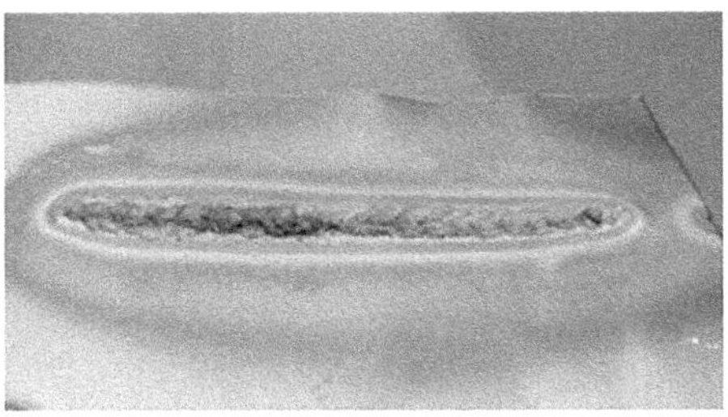

Figure 8.7 (a) A photograph showing "sugaring" on the root side of a full-penetration fusion weld made in stainless steel with insufficient inert shielding gas coverage; (b) a photograph showing the importance of proper argon purging of pipes or tubes before full-penetration fusion welding. Note: A color version of this latter photograph can be found online using Google image search for "heat tint on stainless steel from improper argon purge." Source: (a) Miller Electric Manufacturing Company. Used with permission of Miller Electric Manufacturing Company, Appleton, WI, USA. (b) Dunbar (2000). Used with permission of Dave Dunbar of Intercon Enterprises, Inc., Blaine, WA, USA.

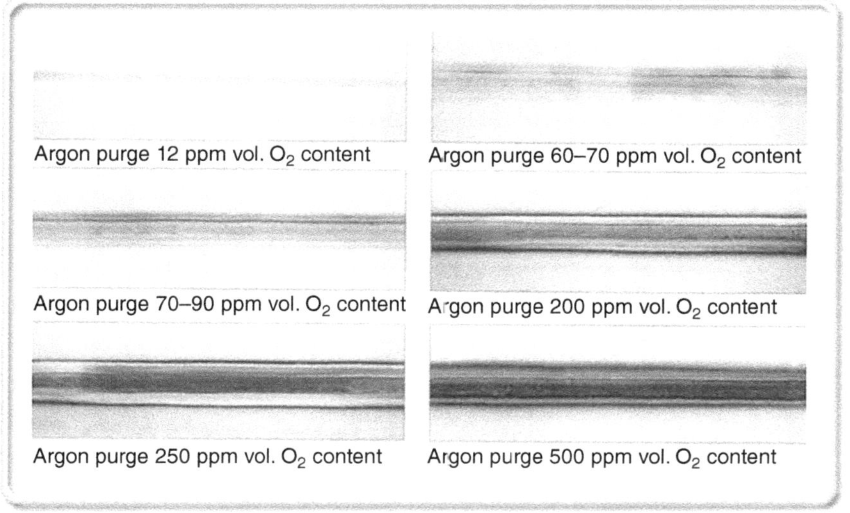

(b)

Figure 8.7 (*Continued*)

to various levels using argon purging, with the cleanest welds occurring with the lowest residual oxygen level.

8.4 Fillet Weld Faults

Fillet welds are used to join two pieces of metal (i.e. joint elements) together whether they are perpendicular or at an angle, and not in line, as in butt welds. They are most commonly found in L- or T-joints in which the two joint elements are at right angles to one another. They are also used to make lap joints in which two joint elements overlap and are welded at one or both edges. Figure 2.5 shows a tee and a lap joint.

As for any fusion weld, there are good and there are bad fillet welds. For the most part, what makes a fillet weld good or bad is whether it has what is considered a proper profile (or cross section). In worst cases, bad welds are unacceptable because they would fail to provide required structural integrity. In other cases, although the resulting fillet weld might be acceptable from a structural standpoint, it is, nevertheless, unattractive or unsightly. The upper portion of Figure 8.8 schematically illustrates some examples of "desirable fillet weld profiles" in (a) that would not only provide structural integrity but are also attractive. It also illustrates "acceptable fillet weld profiles" in (b) that are deemed would provide structural integrity but are less attractive. Most importantly, the figure illustrates several examples of "unacceptable fillet weld profiles" in (c) that would likely not provide suitable structural integrity, beyond being unattractive.

The keys to structural integrity for fillet weld profiles are (i) balanced leg lengths, known as "throats," in the two generally orthogonal directions,

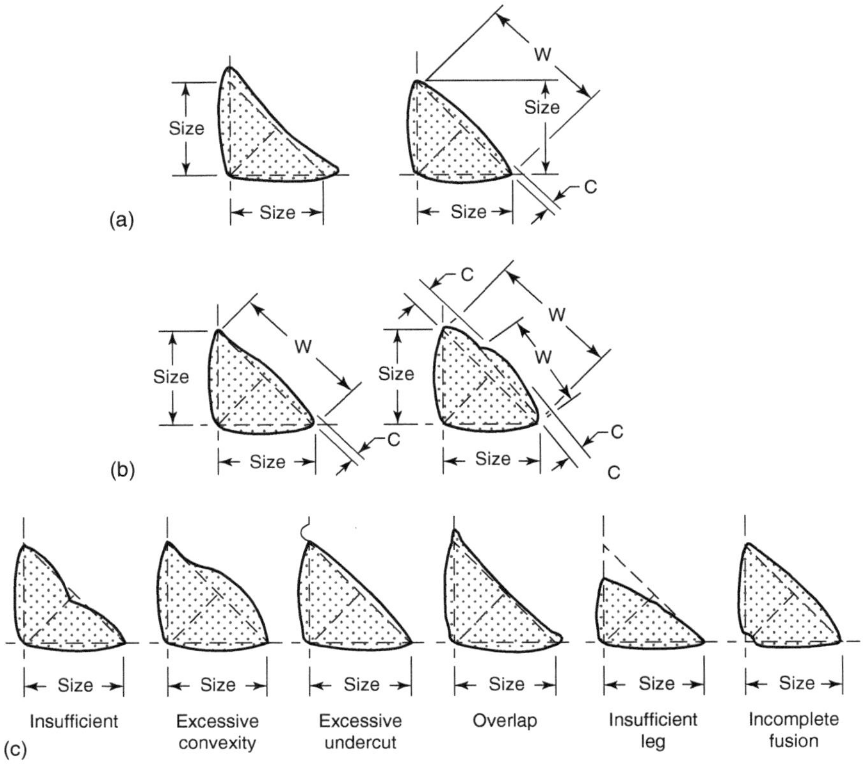

Figure 8.8 Schematic illustrations showing (a) desirable fillet weld profiles, (b) acceptable fillet weld profiles (that are less-than-attractive), and (c) unacceptable fillet weld profiles. Source: American Welding Society (2002). Used with permission of The American Welding Society, Miami, FL, USA.

(ii) sufficient penetration into each joint element, (iii) a face that extends beyond the hypotenuse of an imaginary triangle drawn from each toe of the fillet,[2] i.e. has convexity, (iv) good transition from the fillet weld to the joint elements at each toe, and (v) not pronounced shape irregularity. Figure 8.8 shows representative examples of the most common problems with fillet weld profile.

Like butt welds, fillet welds can exhibit other features that detract from visual appearance (e.g. weld spatter), as in the following sections.

8.5 Reading Weld Ripple Marks

Just as the wake left by a boat on the otherwise smooth surface of a lake hints at how fast the boat was traveling by how sharply V-shaped the wake is, so too do the ripple marks on the solidified crown surfaces of a fusion weld given an indication of the speed at which a welding source was moved (i.e. the welding speed). *Weld ripples* (or *weld ripple marks*) are known to form solely because

2 The "toe" of a fillet weld is where the fusion zone touches the surface of the joint elements.

of the oscillation of the weld pool, and no other mechanism. Such oscillation is caused by the pressure (or impingement force) of an electric arc or beam pushing molten metal in the weld pool out from under the arc or beam, only to return once the liquid returns after reflecting off the unmelted base material at the pool's edge. At the trailing end (or tail) of a moving weld puddle, where solidification of the weld has been completed, the periodic waves of moving molten metal freeze to form ripples.

For a stationary arc or beam source, the weld pool formed would be circular when viewed from above. As the welding heat source moves, the circular pool becomes progressively more elliptical as the welding speed is increased, and the ripples that are formed reflect the progressively sharper elliptical shape. At some critical welding speed,[3] the weld pool becomes tear-drop shaped, and the ripples that form take on a distinct V-shape. Faster and faster welding speeds produce sharper and sharper (and more and more narrow) elliptical weld pools, and sharper and sharper ripples. Eventually, welding speed can be so fast that portions of the tail end of the weld puddle separate from the main puddle, become isolated, and solidify to form shrinkage defects (i.e. cracks or pipes).

Figure 8.9 schematically illustrates how a weld pool elongates from a circle to an ellipse to a tear-drop shape as the travel speed of the welding heat source

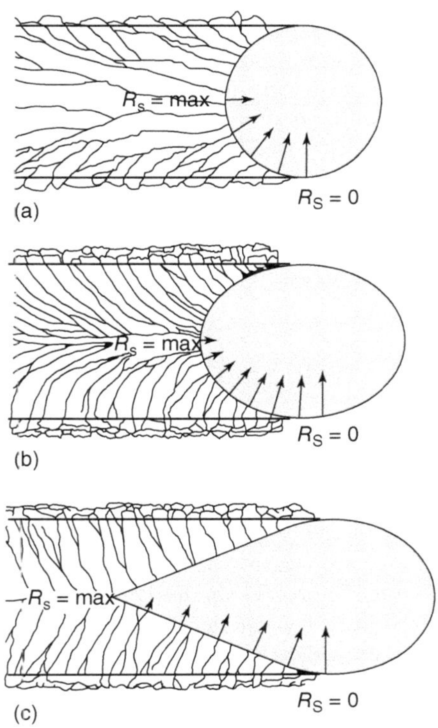

Figure 8.9 Schematic illustrations made from stills from a movie of weld pool shape versus welding speed showing the effect of weld pool shape on growth during solidification. Source: Triveldi et al. (2003). Used with permission of AIP Publishing.

3 The critical speed is reached when the rate at which latent heat of fusion (or, here, solidification) being released during solidification begins to exceed the rate at which that heat can be dissipated by lateral conductive heat flow.

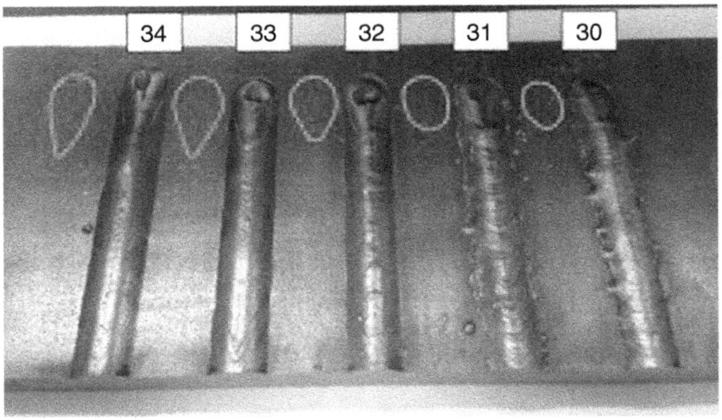

Figure 8.10 Photograph showing a topside view of fusion arc welds made at 7 ipm/2.98 mm/s, 11 ipm/4.66 mm/s, 15 ipm/6.35 mm/s, 19 ipm/8.04 mm/s, and 23 ipm/9.74 mm/s (from #30 to #34) with the effect on weld pool shape and resulting ripple marks on the weld crown. Source: Rausch (2015). Used with permission of EWI, Columbus, OH, USA.

is increased. The illustrations also show how solidification growth is altered by the shape of the trailing edge of a weld puddle, as discussed in more detail in Chapter 9. Figure 8.10 shows photographs of actual welds made using various speeds and the difference in ripple marks that result.

The upper limit for welding speed is not fixed. For one thing, it is different for different welding processes, being higher for processes with higher energy density heat sources, such as plasma arcs versus normal arc and electron beams versus plasma arcs. Quality limits the maximum welding speed. If penetration is lost or a bead gets too narrow or, without question, if cracks begin to appear along the weld centerline just behind a tear-drop-shaped weld puddle, speed is too high.

Like anything else in welding: so long as the resulting welds are of high quality, there is no reason to change anything.

8.6 Weld Spatter

Anyone who has watched the welding of steel has likely seen glowing streamers emanating from the weld beneath the arc (Figure 8.11).[4] They may even have had small molten globules of steel strike their pants or shoes. Provided they are wearing denim jeans and leather shoes, no damage will likely be done. However, if they are wearing polyester slacks or vinyl athletic shoes, small burn holes will result. To protect themselves, welders typically wear flame-retardant jackets or, better yet, leather sleeves or jackets, as well as leather gloves, leather skull caps or cotton dew rags, and, of course, helmets with dark (or self-darkening) glass lenses.

4 Although every metal or alloy is capable of producing spatter with consumable-electrode welding processes (SMAW, FCAW, and GMAW), not every metal's or alloy's spatter produces electromagnetic radiation (i.e. incandescence) in the visible spectrum. Mildly dull-red visible radiation begins to appear at 798 K/525 °C/977 °F, known as the Draper point.

Figure 8.11 Photograph showing sparks from globules of molten metal known as "spatter," typical of some welding process conditions. Here, gas–metal arc welding (GMAW) is being performed. Source: Photograph by William M. Plate, Jr., for the U.S. Air Force, as ID 040112-F-16633-001, 12 January 2004. Freely used from the public domain as the work of an employee of the U.S. government.

Weld spatter (or simply *spatter*) consists of small particles or droplets of molten metal expelled from consumable electrodes during the welding operation. Most fly away in all directions, landing on the ground. Some, however, land on the workpiece, get into gas nozzles (perhaps blocking the flow of shielding gas with GMAW or gas-shielded flux-cored arc welding [FCAW]), or at the tips of welding copper contact tubes (perhaps causing wire feeding problems in FCAW or GMAW). Although some spatter with consumable electrodes is inevitable, long arc lengths, high amperages, excessive wire electrode stick out (greater than about ¾ in./19 mm), severe electrode angles (off normal), or certain molten metal transfer modes with GMAW tend to exacerbate the problem.

When spatter lands on workpieces, it may be problematic. Besides always being unsightly, the roughness it creates when it adheres may be unacceptable for some applications where the surfaces of parts must be smooth (e.g. welds inside the drums of large, commercial washing machines or dryers, as spatter can snag delicate clothing). Of greater concern – and consequence – is when spatter adheres to the surface of steels with high hardenability (e.g. C_{eg} greater than 0.50 in C steels and greater than 0.40–0.45 in low- and medium-alloy steels). With such steels (especially the so-called quenched and tempered, Q&T, grades), the superheated

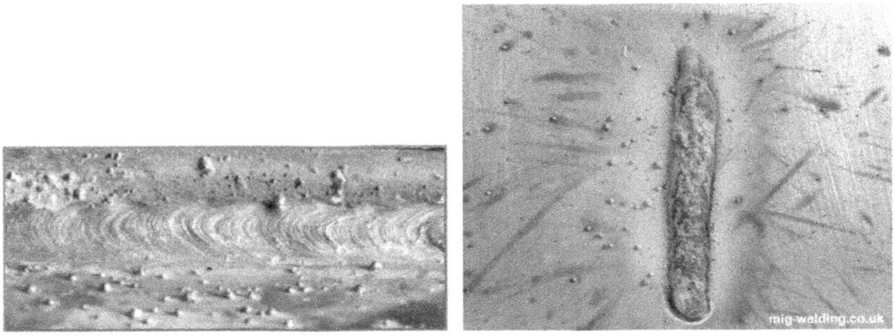

(a) (b)

Figure 8.12 Photographs showing spatter produced by shielded metal-arc welding (SMAW) (a) and by flux-cored arc welding (FCAW) without supplemental inert gas shielding (b), both on steel. Source: (a) Miller Electric Manufacturing Company. Used with permission of Miller Electric Manufacturing Company, Appleton, WI, USA. (b) Migwelding. Used with permission of www.migwelding.co.uk.

(molten) spatter particles can heat an extremely small volume and thin layer of the base material at the impact site to above the temperature where austenite is formed. When this extremely small volume cools very quickly as heat is extracted by the surrounding much cooler (perhaps, room temperature or below) large thermal mass, as-quenched, untempered martensite is formed. Unless the weldment is post-weld heat treated to temper any martensite that may have been produced by welding, these sites can initiate cracks from hydrogen or simply from fatigue.

Although sanding or grinding if often used to remove stubbornly adherent spatter, neither guarantees that such spots of untampered martensite will be removed. Besides correcting whatever is believed to cause excessive spatter in trouble-prone steels, for example, spatter-resisting, high-temperature paints and anti-spatter sprays can be used (following manufacturer's directions).

Figure 8.12 shows photographs of spatter on welds.

8.7 Arc Strikes

An *arc strike* occurs when an arc is initiated on the base material surface away from the weld zone. It may occur intentionally when a welder purposely (but inappropriately!) strikes the tip of a welding electrode to the workpiece to start the arc for welding, or accidentally when the welder inadvertently touches the tip of a welding electrode to a grounded workpiece. The arc strike is a critical discontinuity because there is a localized area where the base material has been melted and then rapidly cooled because of the massive heat sink created by the surrounding base metal. This *always* changes the structure of the base metal or alloy and often (as in the case of steels, as well as Ti and Ti alloys) causes it to become hard and brittle. This, in turn, can lead to cracking (perhaps immediately from shrinkage stresses, perhaps later from hydrogen, or perhaps by fatigue in service), which could lead to failure of the weldment.

Figure 8.13 shows a photograph of an arc strike on the surface of a workpiece, remote from the weld.

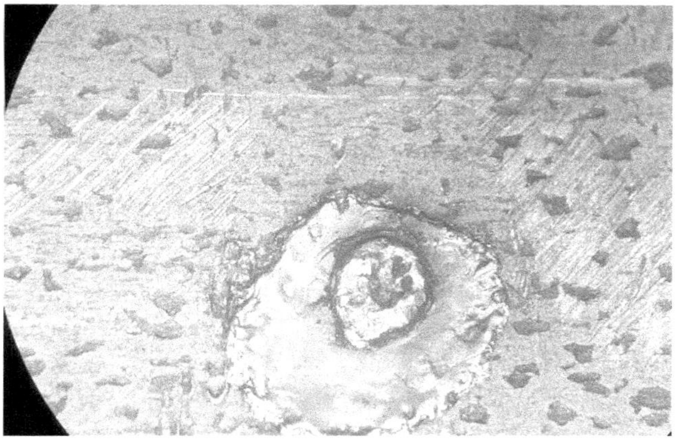

Figure 8.13 Photographs showing an arc strike, which may occur intentionally when a welder initiates an arc off the intended weld line or accidentally with an errant touch. Source: Miller Electric Manufacturing Company (2010). Used with permission of Miller Electric Manufacturing Company, Appleton, WI, USA.

The cause is always carelessness by the welder. Prevention of intentional strikes is simple, i.e. forbid welders from initiating arcs away from the intended weld line. Prevention of accidental strikes is, obviously, more difficult. However, regardless of whether an arc strike occurred by an intentional or accidental act, all arc strikes should be removed by grinding, typically to a depth of 1/8 in./3 mm below the original surface. All locations where arc strikes have been removed should be inspected using penetrant or magnetic particle, as well as hardness testing to assure freedom from cracks and any untampered martensite in steels or α case in Ti or Ti alloys. If repair is necessary following grinding, a qualified repair procedure should be used and the repair should be inspected.

8.8 Weld Heat Tint

Many metals and their alloys produce a very thin, adherent layer of oxide on their surfaces exposed to elevated temperatures in the atmosphere. Although these oxide layers are dull gray or black for most metals and their alloys, some exhibit colors indicative of the maximum temperature to which they were exposed.[5] Generically known as *heat tint*, they are sometimes desirable for their attractiveness and are developed intentionally (e.g. bluing of steels). Other times, they develop on their own, may or may not be considered attractive, and while often benign, in some cases are detrimental to the material's performance in service. Relevant to welding, heat tint occurs around fusion welds made in steels, stainless steels, and titanium and its alloys, predominantly in heat-affected zones, but, possibly, on base metal that was preheated to aid welding. Figure 8.14 shows heat tint around a fusion weld in steel.

---

5 A good example can be found with a Google search of "temper colors for steel," with examples of color versus tempering temperature found under "Images." Experienced heat treaters are able to tell what temperature tempering was done at simply by looking at the color of the heat-treated steel. A great example of the attractiveness heat tint can add to a weld can be seen in Figure 8.2a.

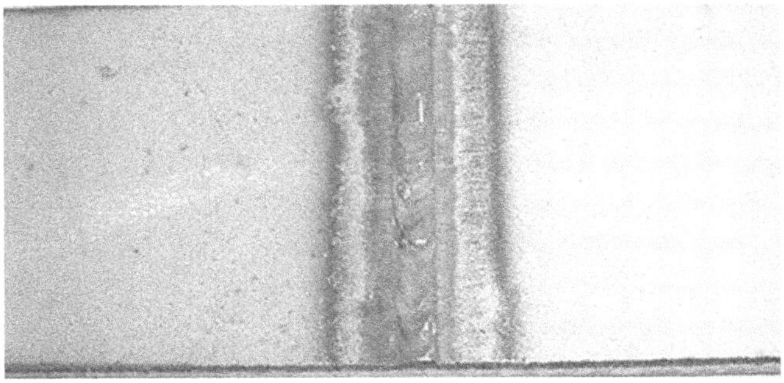

Figure 8.14 Photograph showing heat tinting on and adjacent to a fusion weld made in stainless steel. If not removed by proper cleaning, such as by a wire brushing, such oxidation can initiate pitting corrosion in some corrosive environments. Note: A color version can be seen online via Google image search under "weld heat tint in stainless steel." Source: From the website www.garcoamet.com. Used with permission of Garcoa Metallurgical Services, Mississauga, Ontario, Canada.

The objective of shielding by dry inert gas, air-excluding molten slag, or high vacuum during fusion welding by various processes (e.g. gas tungsten-arc welding [GTAW] and GMAW, SMAW and FCAW, and electron-beam welding [EBW], respectively) is to prevent unwanted oxidation of molten weld metal or newly formed welds, as such oxidation can adversely alter the composition and degrade the properties. The presence of heat tint on and near a weld indicates that shielding was less-than-perfect, and, depending on the material and the intended application, may imply compromised properties, most notably fatigue resistance, fracture toughness, and/or corrosion resistance. As but two examples, (i) the dark heat tint formed alongside welds during the welding of stainless steels is the result of a thicker chromium oxide scale that degrades corrosion resistance by reducing the chromium remaining in the solution and (ii) any heat tint formed on and around the welds in Ti and Ti alloys is the result of oxygen pickup and the formation of a thin layer of extremely brittle α case that degrades fatigue resistance.

Figure 8.15 shows the effect – and effectiveness – of lowering oxygen levels in the environment while fusion welding stainless steel, whereas Figure 8.16 shows how heat tint is used as an indicator of weld quality in titanium alloys.

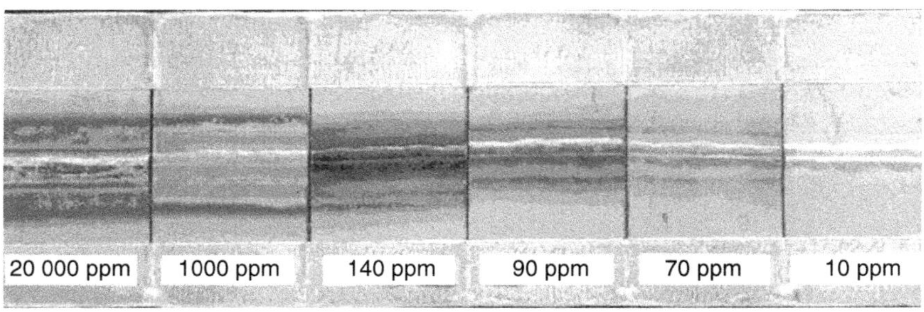

Figure 8.15 Reference photograph published in AWS D18.2, 1999, that shows the range of weld discoloration as a function of oxygen content in the internal argon purge gas of stainless steel tubing (check Google image for color version). Source: From work by Metarinka on 02 May 2009. Used with permission of The American Welding Society.

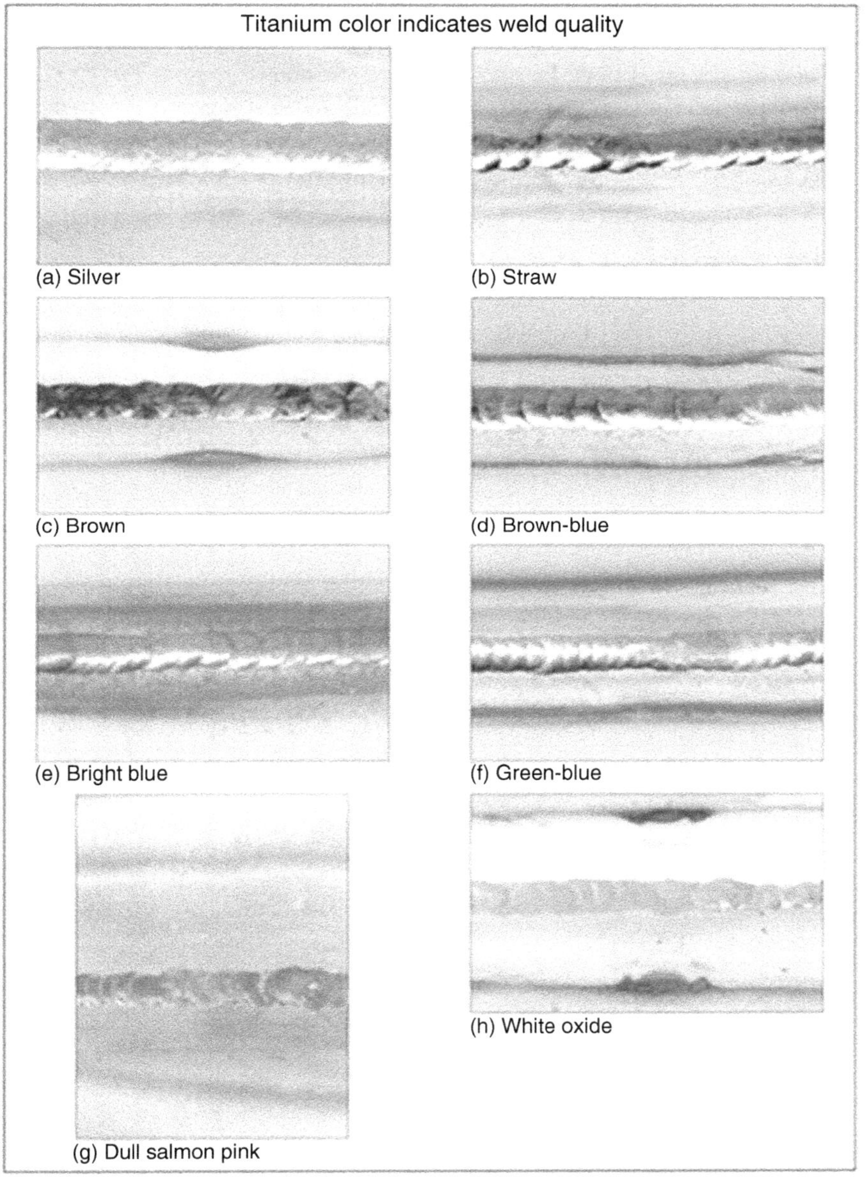

Figure 8.16 Photograph showing the heat tint color or titanium and titanium alloys when welded in the presence of oxygen at various levels, as shown. Because oxygen forms a brittle α case on Ti and Ti alloys, oxygen content must be kept very low in the welding (and heat treatment) environment or it must be thoroughly removed by acid pickling in aqua regia. Note: A color version can be seen online using a Google image search under "discoloration in titanium due to oxygen." Source: Miller Electric Manufacturing Company. Used with permission of Miller Electric Manufacturing Company, Appleton, WI, USA.

Removing heat tint may not be necessary for some materials, such as carbon or low- or medium-alloy steels, but it is essential for others, such as Ti and its alloys. Wire brushing is acceptable for steels and stainless steels, but Ti and Ti alloys must be acid-pickled in aqua regia to remove all traces of α case. Of course, the best practice is to prevent heat tint in the first place, by using proper shielding with problem-prone materials.

8.9 Troubleshooting Weld Appearance Problems

Table 8.1 lists the various problems with weld appearance that can arise during fusion welding.

Table 8.1 Troubleshooting problems with weld appearance.

Weld Crown Bead Faults:		
o Width	Too wide	Reduce amperage
	Too narrow	Increase amperage; slow down
o Profile contour	Excessive convexity; rollover	Increase amperage; slow down Increase arc length
	Excessive concavity	Add filler; slow down; check root gap
o Uniformity	Irregular	Check heat input is not marginal Check travel speed consistency
o Ripples	(see "Weld Ripple Marks")	
o Undercut		Shorten arc length; reduce V Speed up; adjust electrode angle
o Spatter	(see "Weld Spatter")	
Weld Root Bead Faults:		
o Lack of penetration	Intermittent	Marginal heat input; increase amperage or decrease speed
o Lack of fusion	Intermittent	(as above)
o Excessive penetration	"Icicles," "grapes"	Reduce amperage; reduce root gap
o "Sugaring" with SS	Crusty gray-black oxide	Improve root-side shielding
Weld Ripple Marks:		
o Excessive V-shape	Problem with cracks or voids	Slow down
o Irregularity		Improve manual technique or, with automation, check travel
Weld Spatter:		
	On torch body, nozzle, etc.	Reduce arc length; reduce I Reduce wire electrode stick out Reduce electrode angle off-normal

Table 8.1 (Continued)

	On and adjacent to weld	Check transfer mode with GMAW as above, plus, use anti-spatter spray or paint
Arc Strikes:		
	Intentional	Correct welder
	Accidental	Alert welder to exercise caution
Weld Heat Tint:		
o On stainless steel	Sensitivity to pitting	Remove by sanding or brushing
o On Ti or Ti alloys	Sensitivity to fatigue	Acid pickle in aqua regia to remove all α case

References

American Welding Society (2002). *Structural Welding Code – Steel*, AWS D1.1/D1.1M:2002. American Welding Society.

Collier, J. (2010a). MIG welding techniques by old timer. Welding Tips and Technique. www.weldingtipsandtricks.com.

Collier, J. (2010b). TIG welding aluminum. Welding Tips and Technique. www.weldingtipsandtricks.com.

Dunbar, D. (2000). Purging for pipe/tube welding – on corrosion-resistant materials. *Practical Welding Today*. www.tubenet.org.uk.

Khichade, A. (2015). *Welding Defects*, taken from Slide #22/22 in a presentation. Mumbai: KEC International Ltd.

Migwelding. How to weld with flux cored wire. www.migwelding.co.uk.

Miller Electric Manufacturing Company (2010). Stray arc strikes, temp fitments & stress points. Forum response on 09 June 2010, http://www.millerwelds.com.

Miller Electric Manufacturing Company. Welding tips for plant maintenance. http://www.millerwelds.com.

Miller Electric Manufacturing Company (2011). Semi-automatic TIG welding – weld positioner and pulse set-up question. Forum. http://www.miller-weld.com.

Miller Electric Manufacturing Company. Ten common TIG problems: a visual guide, figure 8. http://www.miller-weld.com.

Miller Electric Manufacturing Company. Five steps to improving your stick welding technique, figure 6. http://www.millerwelds.com.

Miller Electric Manufacturing Company. Titanium 101: best TIG (GTA) welding practice. http://www.millerwelds.com.

Rausch, J. (2015). Gas metal arc welding basics: travel speed and contact-to-work-distance, figure 2. 30 June 2015 on the website of the Edison Welding Institute.

Triveldi, P., David, S.A., Eshelman, M.A. et al. (2003). In situ observations of weld pool solidification using transparent metal-analog systems. *Journal of Applied Physics* 93 (8): 4885–4895.

Yusof, F. (2003). *Weld Inspection: Defects/Repairs*, from Slide #25/55 in the presentation. Course Reference WIS 5. Cambridge: TWI Ltd.

Further Reading

American Welding Society (AWS) (2001). *Welding Handbook*, Volume 1: Welding Science and Technology, 9e (ed. C.L. Jenney and A. O'Brien). Miami, FL: American Welding Society.

ASM International (1993). *ASM Metals Handbook*, Volume 6: Welding, Brazing, and Soldering, 9e (ed. D.L. Olson, T.A. Stewert, S. Liu and G.R. Edwards). Materials Park, OH: ASM International.

Part II

Location of Problems in Welds

A weld is a complex entity, consisting of several zones of microstructure from fusion zone and, for an alloy, partially melted zone, through heat-affected zone, to unaffected base metal for fusion welding processes. Multipass fusion welds have even more complex microstructures, whereas non-fusion welds, while free of fusion, have their own microstructural features. *Where* problems are located in a weld is the second clue to their root cause and, thus, solution.

9

Fusion Zone of Fusion Welds

9.1 A Refresher on Microstructural Zones in and Around Welds

As learned from Part I, many problems can – and do – arise during the process of welding. However, as was also learned, not every one of those problems manifests itself in the weld proper (see Chapters 3 and 4). Those that do can logically be divided by their location within the *weld zone*, which is distinctly different for welds made by fusion processes versus welds made by non-fusion processes.

The inextricable interrelationship among a material's structure, processing, and properties as these three influence performance in a design (see Figure 2.6) was introduced in Section 2.2. There too, the existence of various microstructural zones in and around a weld was described. In brief, for fusion welding, the heat of the welding process leads to changes in the microstructure of the base material, extending from complete melting and solidification in the hottest region through effects of highly nonequilibrium heat treatment (i.e. heat effects) to the unaffected base material far enough from the heat source at which point the temperature never rose. Precisely what occurs in a material depends on whether that material is pure or alloyed and on the mechanism(s) by which strengthening is achieved (i.e. on the strengthening mechanism or mechanisms employed). For non-fusion welding, where heating may or may not be involved, but, rather, some combination of pressure and solid-state diffusion leads to material continuity, the effects on the microstructure of the base material are always less severe and, as a consequence, the microstructural zones differ from welds made by fusion welding processes.

Figure 9.1 schematically illustrates the various, generic microstructural zones in a fusion weld made in an alloy[1] (in a) and in a pure metal (in b), here

1 Although the phase diagram for a generic, hypothetical eutectic alloy with limited solid solubility of solute B in solvent A is shown, the low-solute end of an alloy with complete solid solubility of B in A (i.e. an isomorphous system) would appear the same, going from high temperature in the liquid (L) region through a two-phase L + α region between the liquidus and the solidus temperatures, into a single-phase solid solution region (here, a) from below the liquidus to the solvus temperature, and, finally, into a two-phase α + β region below the solvus temperature. The key difference would be that the single-phase solid solution (here, α) region would persist in the isomorphous alloy below the solidus temperature.

A Practical Guide to Welding Solutions: Overcoming Technical and Material-Specific Issues,
First Edition. Robert W. Messler, Jr.

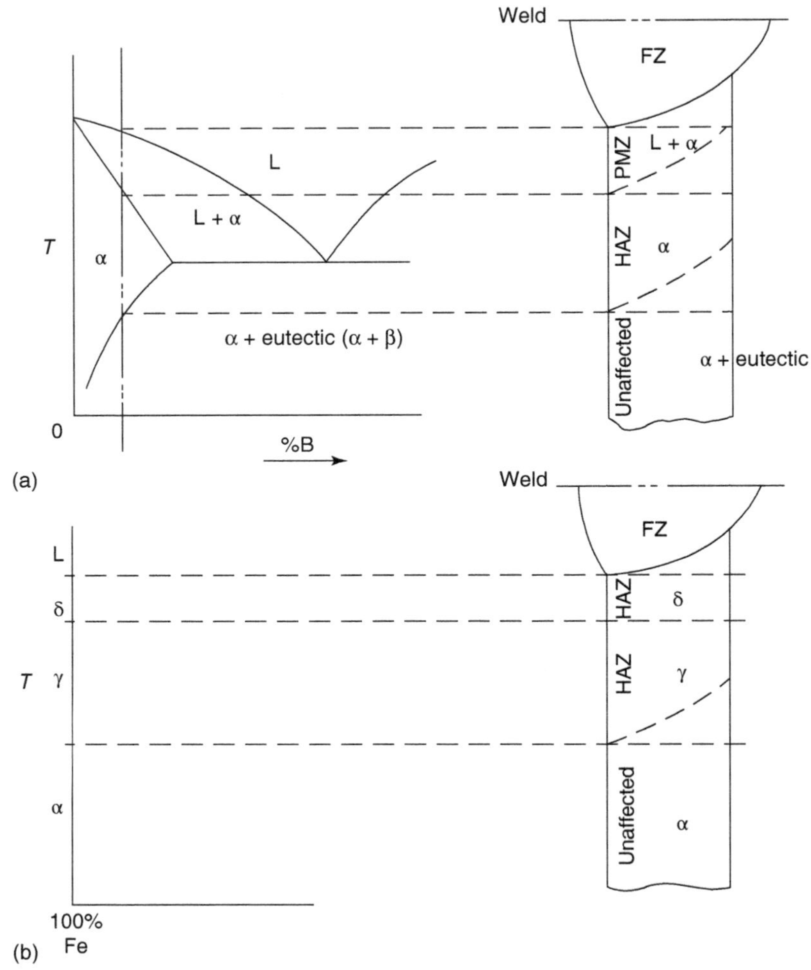

Figure 9.1 Schematic illustration showing the correlation between the various microstructural zones in a fusion weld in (a) an alloy and (b) a pure metal and the corresponding equilibrium phase diagram. Source: Messler (1993). Used under reciprocal agreement between Wiley and Elsevier for fair use.

pure iron.[2] From highest to lowest temperature in an alloy: a fusion zone (FZ) exists from the centerline of the weld (i.e. highest temperature in the weld pool) to the fusion boundary (or fusion line) fixed by the alloy's liquidus temperature;

---

2 Pure iron (Fe) exhibits allotropic transformations as temperature is increased, from BCC α-Fe below 910 °C/1670 °F to FCC γ-Fe from 910 °C/1670 °F to 1394 °C/2541 °F, to BCC δ-Fe from 1394 °C/2541 °F to melting at 1538 °C/2800 °F, to liquid Fe above 1538 °C/2800 °F. A fusion weld made in pure Fe would *not* result in an HAZ once the metal cooled back to ambient temperature. The phase diagram for the Fe–C (here, the metastable Fe–Fe_3C phase diagram) is shown in Figure 9.2. The HAZ γ and HAZ δ shown in Figure 9.1b would only exist immediately after weld cessation.

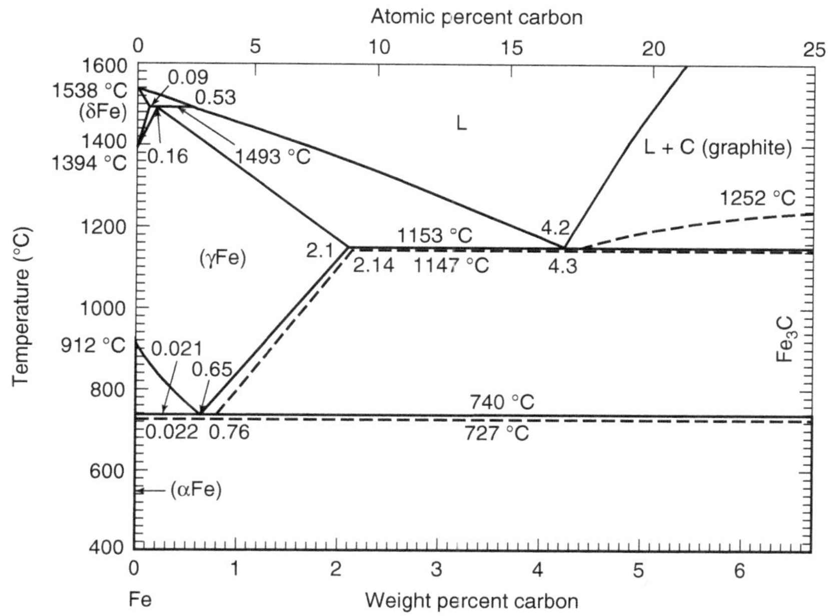

Figure 9.2 The metastable phase diagram for Fe–Fe_3C commonly used instead of the equilibrium Fe–C phase diagram. Source: ASM International (2016). Used with permission of the ASM International.

a partially melted zone (PMZ), here L + α, exists from the fusion boundary (at the liquidus temperature) to the point of the solidus temperature; a heat-affected zone (HAZ), here α, exists from the point of the liquidus temperature to the lowest temperature at which *any* change occurred to the base material by the time–temperature history created by welding; and, finally, the unaffected base material or metal (UBM) from the highest temperature at which *no* change occurred in the base material to the ambient temperature of the welding environment, here α + eutectic (α + β). For a pure metal, the only microstructural zones are, from highest to lowest (ambient) temperature, the FZ above the metal's discrete melting point (MP), a HAZ from just below the MP to the lowest temperature at which *any* change occurred to the base metal by the time–temperature history created by welding, and, finally, the unaffected base metal (UBM) from the highest temperature at which no change occurred in the base metal to the ambient temperature of the welding environment. What, if any, change occurs to a pure base metal by short-time exposure to elevated temperature around the FZ of a fusion weld depends on the starting grain size of the metal and any residual stored strain energy from prior cold work.

In non-fusion welds, there is, by definition, no FZ nor, for an alloy, any PMZ. Whether or not there is any HAZ, and what it consists of if there is, depends on how much heat is present, if any was used to facilitate solid-state welding, or whether any heat was generated by plastic deformation by pressure or by friction used to obtain material continuity. More details are provided in Chapter 14 on non-fusion welding and welds.

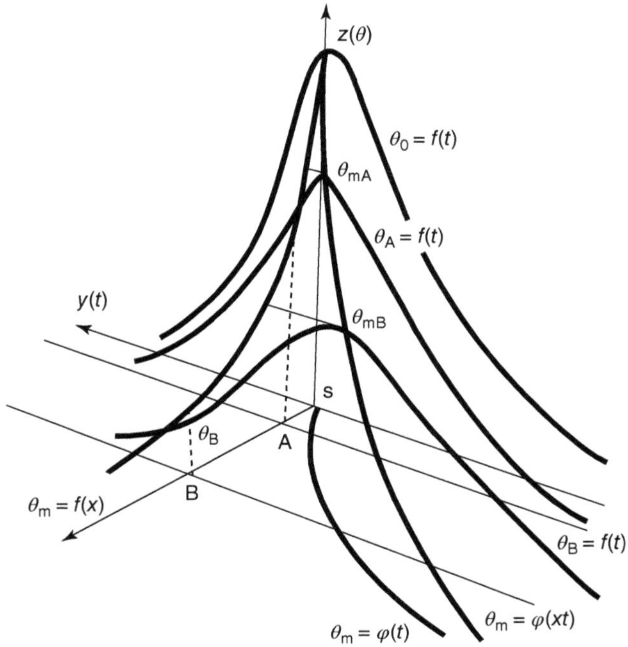

Figure 9.3 A schematic of the "heat solid" suggested by Portevin and Seferian (1934), showing the distribution of instantaneous temperatures around a heat source (moving under quasi-steady-state) at the moment it passes point S during movement along line S_x. Temperatures are plotted only along S_z for each distance *y* for the weld line. Source: Granjon (1991). Used under reciprocal agreement between Wiley and Elsevier for fair use.

Before leaving this refresher on the microstructural zones in and around a weld, a few additional factors are worth noting.

The width (or extent) of each of the just-described microstructural zones depends on the temperature profile that prevailed during welding. As the fusion welding process is highly dynamic, this profile changes as the weld pool moves, continuously dropping from a peak along a line perpendicular to the weld line through the center of the weld pool at time $t = 0$. This is shown schematically in Figure 9.3 and is known as a "heat solid," suggested by Portevin and Seferian (1934).

Because the processes that cause changes to the microstructure of a pure metal or alloy always involve nucleation and growth and generally rely on diffusion (in the liquid state for solidification and in the solid state for all other transformations), both the temperature *and* the time at temperature control all changes. Phase diagrams reflect what phases and constituents exist, and what transformations take place, under conditions of equilibrium (i.e. as much time as is required, without limit). In the real world of fusion welding, conditions are anything but equilibrium. In fact, for most fusion welding processes, the conditions are highly nonequilibrium, with peak temperatures reaching well above all heat treatment and most casting processes, and cooling rates being much faster than most casting processes and all but the most severe quenches during heat treatment. Not surprisingly, this all leads to microstructures that are often different than predicted by equilibrium phase diagrams.

The former notwithstanding, a few factors that affect the width of various zones are as follows:

- The width of an FZ is increased (for a given heat input) by lower melting point in a pure metal and lower liquidus temperature in an alloy
- The width of an FZ is increased in any pure metal or alloy by higher heat input
- The width of the PMZ of an alloy is increased (for a given heat input) by a wider freezing range (i.e. range between the alloy's liquidus and solidus temperatures)
- The width of the PMZ of a given alloy is increased by higher heat input
- The width of the HAZ for a given material is increased by higher heat input
- The width of all zones (FZ, PMZ, and HAZ) is increased (for a given heat input) by higher thermal conductivity in the material.

Higher heat input to a fusion weld produces a greater volume of melting, which, for a joint of a given thickness, results in a wider fusion zone and profile of elevated temperature from the centerline of the weld pool outward by heat flow. The latter results in a wider PMZ and HAZ. Obviously, the lower the melting point of a pure metal or solidus and liquidus temperature for an alloy, the wider the FZ for both, PMZ for the alloy and HAZ for both, as the microstructure of either would be affected by an elevated temperature.

Problems with welds, in the form of defects, can occur in any of the various microstructural zones or regions of fusion welds. As such, it is logical, as is done in Part II of this book, to describe the types of defects that can occur, along with the means for resolving such problems. The above factors can aggravate or alleviate certain problems with welds. Higher thermal conductivity of a pure metal or alloy causes deposited heat to spread farther, thus leading to a wider FZ and HAZ and, if relevant, PMZ.

9.2 Gas Porosity in the Fusion Zone of Welds

Gaseous elements and compounds (e.g. H_2S, CO_2, and SO_2) enter molten, liquid metals during smelting, refinement, electrolysis, alloy production, casting, and fusion welding and solid metals during acid cleaning, electroplating, or upon the introduction of metallic articles to a gaseous atmosphere, accelerated by elevated temperature, as during heat treatment or welding. Once they enter the metal, there are three types of interactions that can occur between gases and metals: (i) adsorption, (ii) dissolution, and (iii) the formation of chemical compounds.

In adsorption, the atoms or, in the case of gaseous compounds, molecules of gases react only with the surface of the metal, forming a film that is only one or two atoms or molecules thick. Bonding between the gas atoms or molecules and atoms of the metal is secondary van der Waals bonding enabled by weak dipole attraction. Adsorption decreases as the temperature of the metal is raised and as the pressure of the gas over the metal (i.e. the partial pressure) is lowered. Molecules of water vapor are especially adsorbed by already-existing layers of oxides or other tarnish layers on metals, as routinely is the case with rust on steels and aluminum oxide on aluminum or Al alloys or magnesium oxide on magnesium or Mg alloys.

Most gases, other than the noble or inert gaseous elements, form true solutions with both liquid and solid metals, the atoms going into the spaces among the metal atoms (i.e. crystal interstices in solid metals). For gases with molecules consisting of several atoms (e.g. diatomic H_2, O_2, and N_2; CO_2; H_2S; and SO_2), the molecules break up (dissociate) into atoms upon dissolution in liquid or solid metals.[3] The solubility of gases in molten metals is considerably higher than that in solid metals, as there are much larger spaces available among the metal atoms in the hot liquid. The concentration of a diatomic gas that can dissolve in a metal is given by Sievert's Law as

$$k = [\text{gas}]/P^{1/2}{}_{\text{gas2}}$$

where k is the equilibrium constant, [gas] is the concentration of a particular gas in the molten metal as weight percent (wt%), and P_{gas2} is the partial pressure of the particular gas in the diatomic form (see Figure 5.7). The actual concentration of diatomic gas found in a molten metal is higher than this, as some of the gas also enters into the solution in the nascent and/or ionic form. The solubility of gases in molten metal decreases as the temperature of the liquid decreases, with a significant and discontinuous decrease as the molten metal solidifies (see Figure 5.8 for the solubility of oxygen and nitrogen in pure Fe and in pure Ni and Figure 5.9 for the solubility of hydrogen in pure Al, pure Cu, pure Fe, and pure Ni). Solute additions can, and usually do, affect the solubility of a given gas in a given metal upward or downward (see Figures 5.10 and 5.11).

In some cases, gases form compounds with metals, such as hydrides, oxides, nitrides, and sulfides, as prime examples. As these compounds are not soluble in metals, they exist as independent phases, as so-called nonmetallic inclusions. The presence of such inclusions greatly decreases certain mechanical properties (notably, dynamic properties such as fatigue strength and resistance, fracture toughness, and impact toughness) and anticorrosion qualities by the effect of forming microscopic electrochemical cells.

All three forms of gas–metal reactions can lead to porosity formation in the fusion zones of fusion welds when the concentration of gases released from adsorbed surface layers, in solution from the original base metal or filler metal, or produced by the thermal decomposition of some nonmetallic inclusions (e.g. hydrides), exceeds the solubility limit of the solidifying melt. The most likely source, however, is gases drawn into the melt from a wide variety of external sources. Because Chapter 5 discusses the problem of porosity in welds in great detail, only certain highlights will be repeated here.

---

3 Besides facilitating penetration of gases into the metal by requiring small spaces for accommodation (i.e. requiring less energy to push metal atoms apart), some of the energy needed to do so is provided by the energy gained upon chemical interaction of the atoms of the gas and the metal.

Gas pores can be spherical (as globular) or elongated (as wormholes or pipes) and can occur as random, uniformly scattered porosity, clustered porosity, or linear porosity (see Sections 5.1 and 5.2). True to the mantra for producing quality welds – "Cleanliness. Cleanliness. Cleanliness" – gas porosity in fusion welds is always the result of contamination, with the major sources being the following:

- Contamination of the surfaces of joint elements by oxides, tarnish layers, or hydrocarbons
- Contamination of the welding environment by water in the atmosphere, on joint elements, or on or in welding consumables (e.g. coated electrodes, flux-cored or rusty bare wires, granular fluxes, and shielding gases)
- Contamination of the weld by air from the atmosphere (e.g. drafts, empty gas cylinders, aspiration into the arc)
- Contaminated or defective welding equipment (e.g. dirty wire feed cables, dirty contact tubes, dirty wire feeder rollers or gears, and loose or leaking hose fittings)
- Contamination from improper welding procedure or technique (e.g. improper root gaps or backing strips or shielding, improper electrode angle, and insufficient inert shielding gas flow rate)
- Contamination from within the base material (e.g. residual dissolved gases and certain nonmetallic inclusions).

Gas porosity in fusion welds is sometimes visible when it breaks the surface of the weld. Other times, surface-breaking porosity can be detected by using a fluorescent or dye penetrant or, for ferromagnetic materials (e.g. carbon, low- and medium-alloy steels, tool steels, and ferritic and martensitic stainless steels), magnetic particle inspection.[4] For subsurface or internal porosity, detection is by radiography or ultrasonic inspection, radiography normally being more effective for both detection and characterization. Detection of pores constituting less than 2% of the cross-sectional linear dimension can be difficult, so detecting porosity in thick sections is inherently difficult. Obviously, effective radiographic inspection requires a line of sight through the joint.

Figure 5.1 shows a photomacrograph of typical spherical porosity internal to the weld (a) and open to the surface of the weld's crown bead (b).

Section 5.4 provides details for each major source mentioned above. Section 5.5 describes means for preventing or minimizing problems, with an additional possibility being to use a filler with adequate levels of deoxidants (e.g. Si or Al) to deal with minor sources of gas contamination (e.g. from base metal residuals). Section 5.6 provides a troubleshooting guide.

9.3 Cracking in the Fusion Zone of Welds

As stated in the outset of Chapter 6, the most dreaded and unacceptable discontinuity of all in welding and welds is a crack, regardless of where or when it

4 The roughness of the surface of weld beads from ripples and other features can trap liquid penetrants, so inspection for porosity must be done properly and carefully.

is formed. Many metals and alloys are most susceptible to cracking during the process of solidification, when a metal or alloy has the lowest strength and least tolerance for strain.[5] Such solidification cracking, observed frequently in castings and cast ingots, can – and does – also occur in the fusion zone of fusion welds. Known generically as *hot cracking* (with the result being *hot cracks*), it occurs only in those zones where solidification is taking place, i.e. the FZ or, in alloys, the PMZ, as well. Several examples of hot cracking in fusion welds are shown in Figures 9.4 and 9.5.

Hot cracking can be characterized by the following: (i) it is *always* intergranular (or, more appropriately for fusion welds, interdendritic[6]), that is, it occurs along

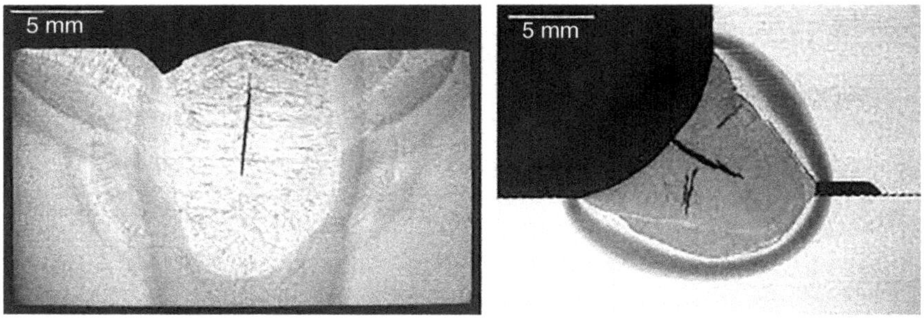

Figure 9.4 Photomacrographs showing centerline hot cracking in steel: (a) an internal crack in the FZ of a butt weld and (b) a surface-breaking crack in a fillet weld, with "flare cracks" along growth structure coming from opposite sides of the fillet weld. Source: The Welding Institute. Used with permission of TWI, Cambridge, UK.

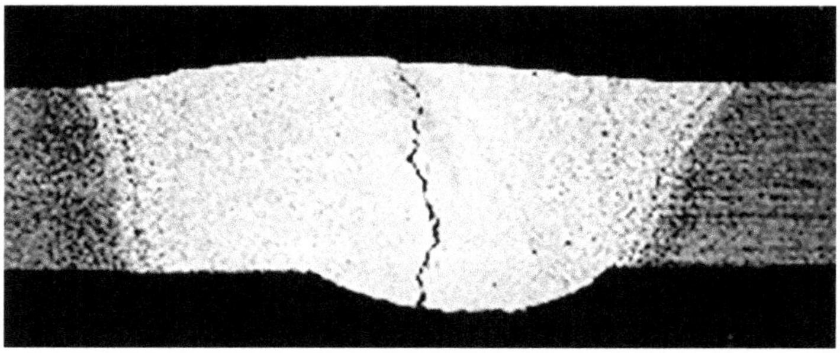

Figure 9.5 Photomacrograph showing a centerline hot crack all the way through a butt weld in an Al alloy. Source: The Welding Institute (1996). Used with permission of TWI, Cambridge, UK.

5 *Strength* is the measure of the ability of a solid material to resist deformation (whether elastic and recoverable or plastic and permanent) under stress (i.e. force or load per unit area), whereas *strain* is the amount by which a material undergoes a dimensional change as the result of stress. A low tolerance for strain indicates an inability to undergo strain without cracking and, ultimately, fracturing.

6 Molten metals and alloys solidify to form crystallites that grow in certain preferred (most closely packed) crystallographic (easy growth) directions and, as a result, form branched treelike structures known as *dendrites* (from the Greek *dendron*, meaning "tree"). Hence, the structure found in castings and in their small-scale cousins, the fusion zones of welds, is dendritic.

the grain or subgrain boundaries; (ii) it occurs at the terminal stages of solidification when shrinkage-induced tensile stresses from both weld metal solidification and thermal contraction become the highest; and (iii) severity increases with both the degree of constraint on the weld and the thickness of the weldment at the joint. Figure 9.6 shows several progressively higher magnification views of a hot crack along the centerline of a fusion weld that depict both its intergranular and terminal nature, i.e. along the line of impingement of oppositely growing dendrites in the weld. Although most commonly found in the fusion zone or FZ of fusion welds, hot cracking can also be found in the PMZ of some alloys (see Chapter 10) and in highly localized regions, the HAZs, of prior passes in multi-pass welds as the result of reheating and liquation cracking (see Chapters 13 and 17).

As described in detail in Section 6.3, hot cracks in the fusion zone of welds can appear in three forms, i.e. (i) *longitudinal* or *centerline* cracks (shown here in both Figures 9.4 and 9.5), (ii) *crater cracks* (not shown here, but shown in Figure 6.4), and (iii) *transverse cracks.* Crater cracks will not be discussed here, as they are discussed in Section 6.3, and as they are strictly the result of poor practice in terminating fusion welds on a workpiece. Transverse cracks will also not be discussed here, as they occur more rarely and almost always under conditions of severe restraint on the joint (i.e. severe constraint on a solidifying weld), and as they are sufficiently well described in Section 6.3. In addition, the underlying mechanism leading to the formation of both longitudinal and transverse hot cracks is identical, i.e. segregation of low-melting solutes, often as impurities.

The accepted theory of solidification cracking or, alternatively, hot cracking in welds (as well as in castings) is that a coherent, interlocking network of newly formed solid separated by an essentially continuous thin film of liquid is ruptured by tensile stresses arising from liquid-to-solid volumetric shrinkage and thermal contraction of the solid HAZ. If sufficient liquid phase is present near the cracks as they pull open (i.e. form), they can backfill and heal. If not, the cracks appear as open tears. Whether hot cracks form or not depends on how much liquid (versus newly formed solid) is present as lowering temperature (as the weld pool cools) approaches an alloy's solidus temperature (adjusted downward from the equilibrium value to account for the effects of nonequilibrium cooling).

Several factors favor hot cracking, some of which are metallurgical in origin and some of which are mechanical in origin. Aggravating metallurgical factors include (i) a wide (versus narrow) freezing range between liquidus and solidus temperatures, (ii) presence of low-melting eutectics (usually involving impurity elements such as S and P and elements in the same vertical groups as these two in the periodic table of elements), (iii) coarser growth structure or substructure in the solidified weld, and (iv) wetting of grains or subgrains by the grain boundary liquid phase (as the result of low surface tension). Aggravating mechanical factors include (i) large shrinkage stresses (based on the volumetric shrinkage of L-to-S for the alloy), (ii) large contraction stresses (based on the net heat input, as this affects HAZ width and the coefficient of thermal expansion [CTE] of the alloy), and (iii) the degree of restraint on the weld (based on the thickness of the joint and stiffness of the surrounding weldment structure).

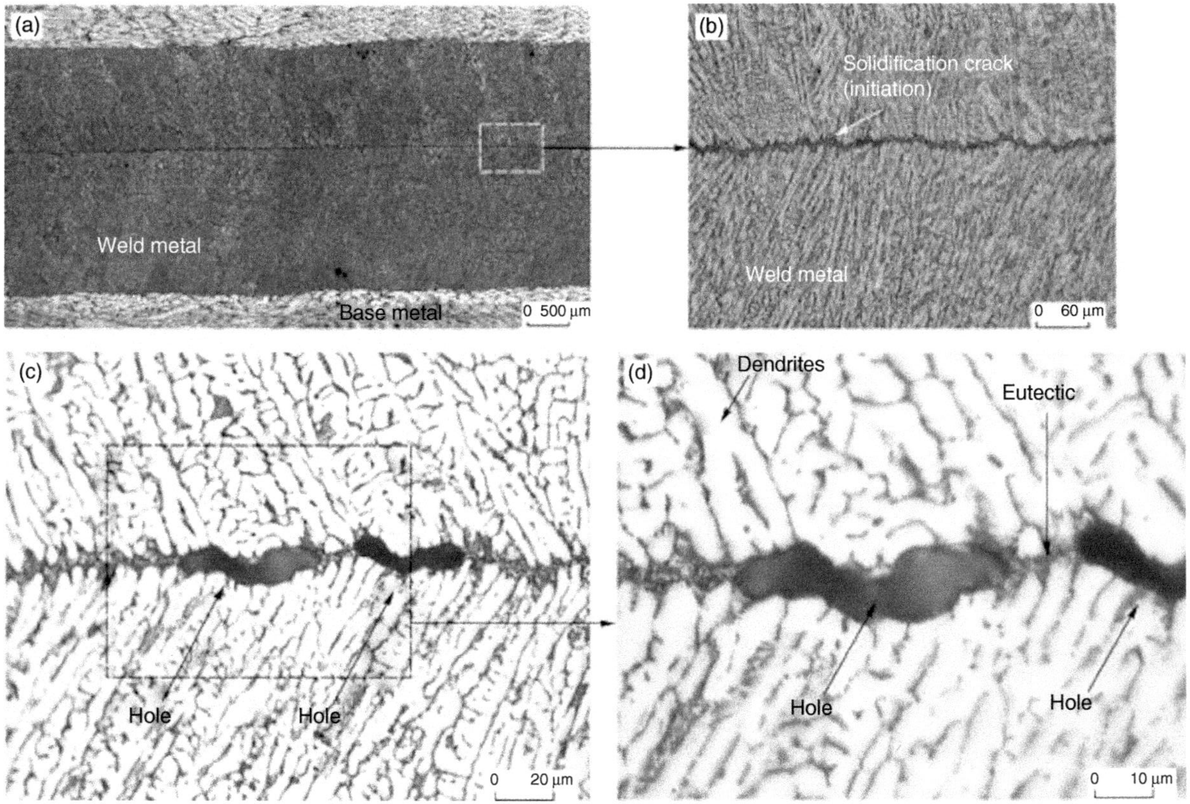

Figure 9.6 A series of progressively higher magnification photomacrographs and photomicrographs showing the path of a hot crack along the interdendritic boundary of impinging growth structure in a fusion weld, with initiation at a small void. Source: Ploshikhin et al. (2011). Freely used as Wiley is a STM signatory publisher with Springer Nature.

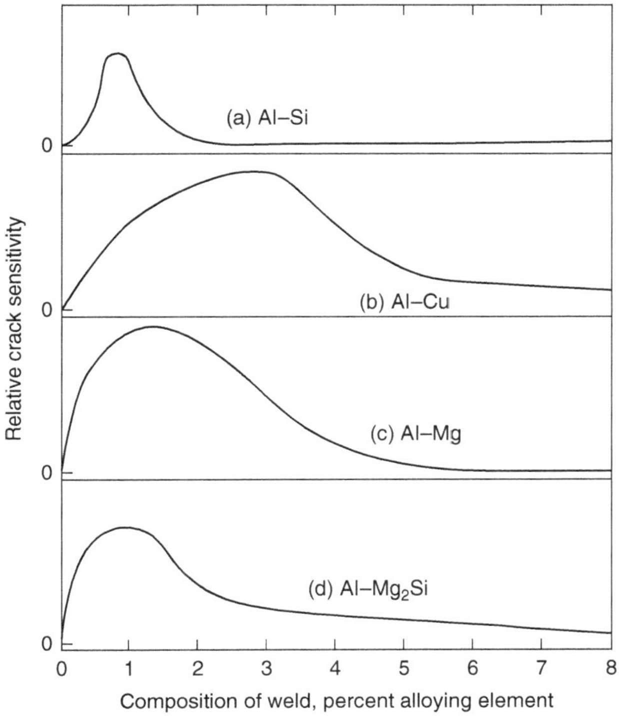

Figure 9.7 Plots showing the susceptibility of some aluminum alloys to solidification cracking as a function of major solute content. Source: Dudas and Collins (1966). Published by and used with permission of The American Welding Society.

Figure 9.7 shows the susceptibility of some Al alloys for hot cracking as a function of the alloy's major solute content using plots. Such plots are indicative of the alloy's freezing range and volume fraction of liquid phase near the termination of solidification. Similar crack sensitivity curves are available in various references for some alloy systems. Figure 9.8 schematically illustrates the effect of grain boundary wetting by a liquid phase as measured by the dihedral angle, with lower angles for liquids that wet better because of lower surface tension forming continuous films (versus isolated globules). Table 4.1 lists shrinkage factors and CTEs for major metals and alloys commonly welded for engineering applications. These data can be used (as described in Section 4.3) to assess a metal or alloy's susceptibility to problems of distortion or hot cracking.

The effect of weld fusion zone structure coarseness on susceptibility to hot cracking is that finer structure is more resistant hot cracking, all else being equal. The reasons are multifold and are as follows: (i) finer, more equiaxed grains can better deform to accommodate contraction stresses, thereby relieving stress across interdendritic boundaries; (ii) remaining liquid is more easily fed to forming (incipient) cracks to promote healing in finer grained structures; and (iii) finer grain size results in greater grain boundary area, so the concentration of any potentially harmful low-melting segregate is reduced, hopefully to below

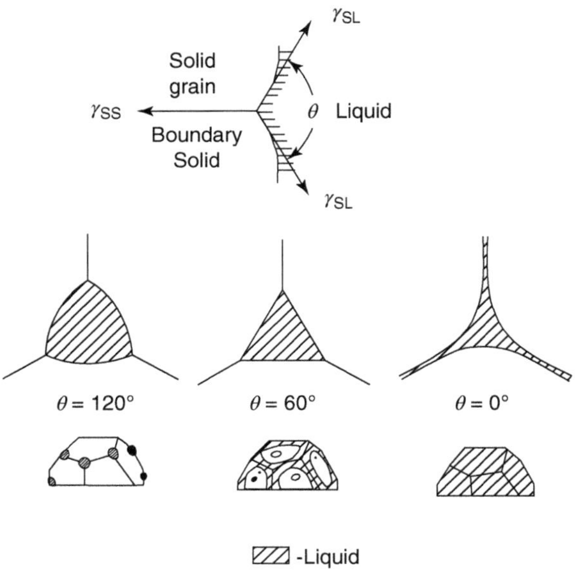

Figure 9.8 Schematic illustration showing the relationship between the surface tension of a liquid, the dihedral angle, and the distribution of the liquid in grain boundaries. Source: Smith (1948). Used with permission of the Minerals, Metals and Materials Society (TMS), Pittsburgh, PA, USA.

the threshold that causes hot cracking. A fourth factor is any condition prevailing during welding that promotes severe head-on impingement of dendrites growing from opposite sides of a weld along the centerline of a weld exacerbates centerline cracking susceptibility by the plowing effect on segregates.

Figure 9.9 schematically illustrates how the growth structure and substructure in a fusion weld can be made finer by employing welding parameters (e.g. lower welding current and/or higher welding speed) and conditions (e.g. faster cooling of the weld), and even process embodiments (e.g. higher energy density processes, such as plasma arc welding [PAW], laser-beam welding [LBW], and electron-beam welding [EBW]) move from columnar dendritic toward cellular or planar growth by increasing the ratio of G/R or by increasing the product of GxR. Detailed discussion of this G versus R plot can be found in suitable references on welding metallurgy.[7]

Although Section 7.3 describes ways to prevent or, at least, minimize susceptibility to hot cracking in the fusion zone of welds in detail, below is a summary of the major general approaches and techniques for addressing a problem with longitudinal or transverse cracking (but not crater cracking):

- For autogenous welding (where no filler is used), avoid base metal compositions that are prone to hot cracking
- When filler metals are used, filler metal composition, flux or shielding gas, process, operating parameters, and welding procedures must all be carefully considered and selected to control weld metal composition to minimize hot cracking, including

7 *Principles of Welding: Processes, Physics, Chemistry, and Metallurgy* by Robert W. Messler, Jr., 2005/1999, Wiley-VCH/John Wiley & Sons, Inc., Berlin, Germany/New York, NY, USA, is recommended; specifically, Chapter 13, Section 13.5.3.

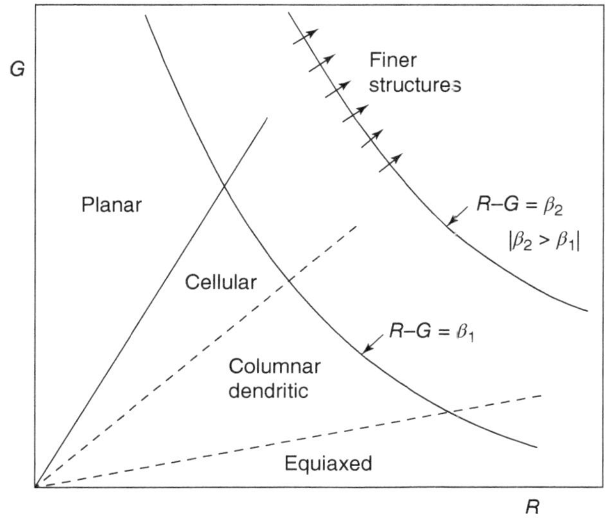

Figure 9.9 Schematic growth rate (*R*) versus temperature gradient (*G*) map showing transitions in solidification growth mode (sloped lines) as well as the refining effects of higher cooling rates (curved lines). Grain boundary area is greater for cellular than for columnar dendritic and for finer structures. Source: Giamei et al. (1976). Published by and used with permission of the ASM International.

- ✓ Minimizing Cu in Al alloys to narrow the freezing range and amount of low-melting eutectic liquid
- ✓ Minimizing residual element, especially S and P, in steels and, especially, Ni-based alloys
- ✓ Using higher Mn contents in steels to control the formation and distribution of low-melting segregates such as FeS
- ✓ Using alloy content of the filler to control fusion zone structure, such as having 5–10 vol.% ferrite in austenitic stainless steel filler metals
- ✓ Avoiding active additions, such as 1–2% oxygen, to inert the shielding gas with gas metal-arc (GMA) for steel welding.

o Control solidification structure of favor development of substructure modes with more boundary area and finer scale, including
- ✓ Selecting the appropriate process, favoring those that provide lower linear heat input via higher energy density, such as PAW, LBW, or EBW
- ✓ Adding grain-refining agents to fillers, such as Ti or Zr to Al alloys, Si to steels, etc.
- ✓ Employing transverse arc oscillation or circular arc movement (e.g. manual weaving) to increase the resolved welding velocity and, thus, growth rate R, also to produce a less demarcated centerline
- ✓ Stir or vibrate the weld pool during solidification to cause dendrite fragmentation to promote fine equiaxed structure.

o Use favorable welding conditions including
- ✓ Employing high-intensity, high energy density processes (such as PAW, LBW, or EBW)

✓ Minimize restraint across the weld joint
✓ Control weld profile (e.g. keep depth-to-width ratios between 1.0 : 1 and 1.4 : 1) and bead profile (e.g. convexity, smooth transition, and uniformity)
✓ Preheating to reduce contraction stresses and reduce cooling rates for materials that from inherently brittle phases and constituents athermally (e.g. ferrous martensite).

Although *cold cracking* from the formation of as-quenched martensite in the fusion zone of a weld can occur, it is easy to control by proper choice of filler metal. Martensite can form in the fusion zone of welds made in a carbon, low- or medium-alloy, or tool steel, as well as in martensitic stainless steel and some grades of cast iron if the C-equivalent in the fusion zone exceeds about 0.45.[8] One potential source of carbon pickup that raises C_{eq} enough to allow martensite to form in the FZ is weld dilution. This said, as discussed in Section 6.4, cold cracking will only occur if (i) hydrogen is present, (ii) tensile stress is present, *and*

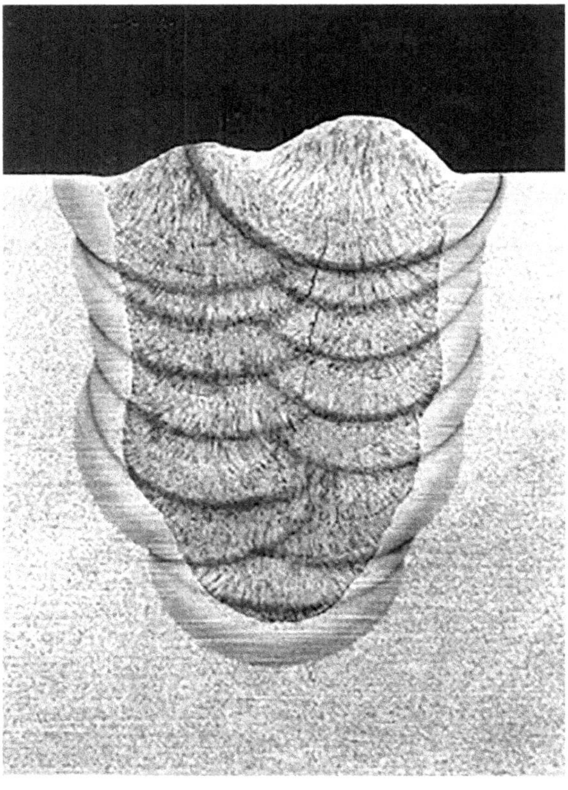

Figure 9.10 Photomacrograph showing cold cracks in a multi-pass weld joint made in HY100 steel. Source: Pargeter (2007). Used with permission of The American Welding Society.

8 *Weld dilution* occurs when the base metal melted during fusion welding mixes with the filler in a sufficient fraction to alter the composition of the filler. As it pertains to cold cracking in the FZ, problems can arise when a high-C base metal is welded to a low-C base metal (e.g. a 3.5 wt% C cast iron to a 0.30 wt% structural steel). With just a 15–20% dilution factor, the net C content of the diluted filler can lead to a C_{eq} high enough that martensite forms in the FZ. Techniques to avoid problems are described in Chapter 21.

(iii) the martensite is left untampered. Correcting any one of these situations will preclude the possibility of cold cracking, but prudent engineering suggests always seeking elimination of any and all sources of hydrogen (e.g. water and hydrocarbons) *and* preventing the formation of martensite by preheating *or* post-weld tempering a suspect weld *and* post-weld heat treating to relieve weld-induced stresses. Doing less might work, but is risky!

Figure 9.10 shows an example of cold cracking within the fusion zone of multi-pass welds in steel, whereas Figure 6.6a,b shows examples occurring in the HAZ at the root of single-pass and multi-pass welds.

9.4 Inclusions in the Fusion Zone of Welds

Inclusions are the result of foreign solid or, in one case, liquid material getting into a weld. Like gas porosity, inclusions can only occur in the fusion zone of a fusion weld, although they could conceivably occur at the faying surface (or interface) of welds made by some non-fusion welding processes (see Chapter 14). As described in Chapter 7, inclusions are of one or the other of two types; i.e. *nonmetallic inclusions* or *metallic inclusions*. Both are problematic for one or both of two principal reasons: (i) they represent a potential site for crack initiation by impact or fatigue or even sustained stress in service and (ii) they indicate improper welding technique. Perhaps equally problematic for some applications of the finished weldment, they indicate inadequate post-weld nondestructive evaluation for weld quality.

Nonmetallic inclusions are overwhelmingly the result of slag from a slag-protected welding process (e.g. shielded metal arc welding [SMAW], flux-cored arc welding [FCAW], submerged arc welding [SAW], and electro-slag welding [ESW]) becoming entrapped in the fusion zone of a single-pass weld or between passes during multi-pass welding. The usual location of a slag inclusion in a single-pass weld is along the edge of the FZ, near the fusion boundary, where it constitutes a local lack of fusion. For multi-pass welds, slag becomes trapped at the interface between passes, where, again, it constitutes a local lack of fusion.

Figures 7.1–7.3 show examples of slag inclusions in welds, including X-radiographs of these low-density defects.

Prevention of slag inclusions is largely about using a proper technique. Recommendations include the following:

- Employing joint designs and preparations that allow adequate access for consumable electrodes to be properly manipulated
- Using welding techniques that produce smooth weld beads and adequate inter-run fill to avoid forming pockets that can trap molten slag
- Using electrodes of the right size held at the proper angle
- Avoid operating with heat input that is too low, with current too low so that molten slag does not float or with travel speed that is too fast so that the weld pool solidifies too quickly
- Avoid using parameters that produce undercutting, which can trap slag and make removal difficult; do not allow heat input to get too high because of excessive amperage

- o Avoid the use of fluxes with high basicity indexes (i.e. basic fluxes) if possible, as these are often difficult to remove completely, increasing the chance of slag inclusions in multi-pass welds
- o Thoroughly remove all hardened slag before making another overlaying pass; grind out slag if wire brushing does remove.

Although foreign material (e.g. particles of dirt, particles of ceramic from nozzles, and particles of dry flux) can fall into a molten weld pool to become entrapped as nonmetallic inclusions, incidents are rare compared to slag entrapment. Likewise, autogenous welding of base materials with heavy scale that is not removed before welding or with excessive amounts of oxide or silicate inclusions can lead to nonmetallic inclusions in the fusion zone, but is relatively rare.

Metallic inclusions that occur in the fusion zone of welds arise from three possible sources, the first by far the most common, i.e.: (i) tungsten inclusions that come from the nonconsumable tungsten electrode used with gas tungsten-arc welding (GTAW), (ii) copper inclusions from gas metal-arc welding (GMAW) torch nozzles, or (iii) pieces of unmelted filler wire from the core of coated SMAW electrodes or from flux-cored or bare wire electrodes. Tungsten inclusions occur when the welder accidentally transfers some tungsten to either the joint with an accidental touch or into the weld pool with an accidental dip. In both cases, a small particle of solid tungsten becomes entrapped in the FZ. Another possibility is when the W electrode overheats to cause a sharp tip to melt and transfer a molten globule of tungsten into the weld pool to become entrapped. This is most likely to occur when GTAW is operated in DC reverse polarity (i.e. DCRP or DCEP) and the tip of the W electrode was not formed into a ball by melting before welding.

Figure 7.5 schematically illustrates the typical locations of W inclusions in a weld, as well as the appearance of these high-density defects in an X-radiograph.

Copper inclusions can result when the nozzle of a GMAW torch becomes overheated and some melting occurs and molten globules transfer into the weld pool. Inclusions of an unmelted filler metal can occur when consumable electrodes are dipped into the weld pool.

Because they act as stress risers, metallic inclusions must be removed by grinding and rewelding. To not remove metallic inclusions can seriously degrade the life of a structure operating under fatigue loading.

9.5 Macrosegregation in the Fusion Zone of Welds

Microsegregation refers to composition gradients that span a grain or dendrite width and occurs in virtually all fusion welds and castings as the result of solute redistribution occurring at the advancing solid–liquid interface under nonequilibrium conditions (i.e. real-world processing). Much more extensive segregation can also occur in large castings, where a much larger volume of liquid exists and much longer times are required for solidification. Such *macrosegregation* can also occur in fusion welds under certain circumstances. When it does, it can be problematic.

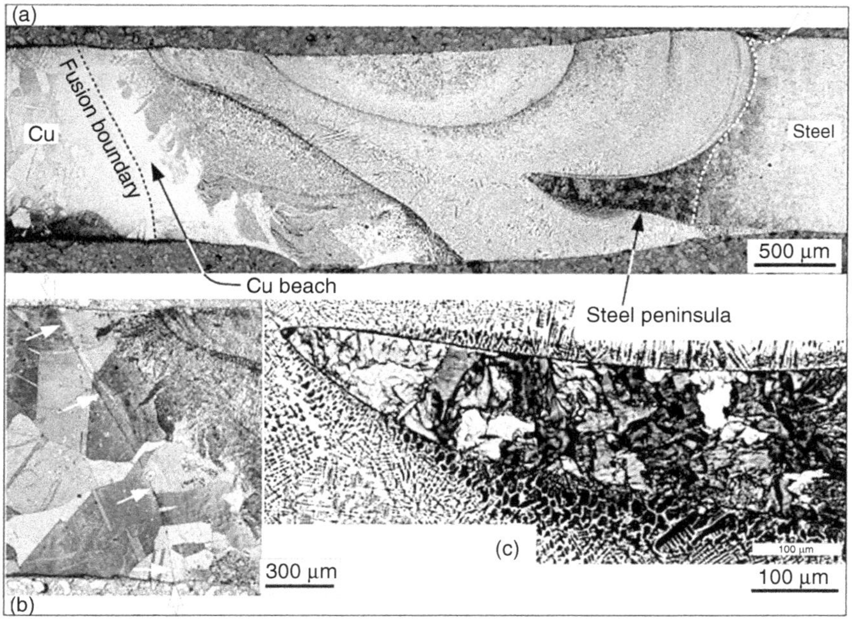

Figure 9.11 Transverse macrograph of a Cu-to-steel arc weld showing various features of macrosegregation for these two dissimilar metals. Source: Soysal et al. (2016). Used under reciprocal agreement between Wiley and Elsevier for fair use.

The most probable and pronounced source of macrosegregation in the fusion zone of welds is when fusion welds are made between base metals of significantly dissimilar composition. As molten metal obtained from each substrate tries to mix, with or without a filler, a marbling effect typically results, as shown in Figure 9.11 for a weld between steel and pure copper. The problem caused by such gross chemical inhomogeneity is the unpredictability of mechanical and, especially, corrosion properties, both of which are inevitably degraded.

Preventing macrosegregation may be a simple as-increasing weld pool stirring via normal convection forces (e.g. Lorentz force or impingement force) or using current pulsing (using appropriate power supplies). In more extreme cases, magnetic stirring of the weld pool during welding using an externally applied magnetic field might be attempted. In the most severe cases, any of a variety of techniques described in Chapter 21 might be tried.

9.6 Troubleshooting Problems in the Fusion Zone of Welds

Table 9.1 lists the various ways problems can arise in the fusion zone of fusion welds.

Table 9.1 Troubleshooting problems in a weld's fusion zone.

Gas Porosity:	
o From contaminated joint element surfaces:	
✓ Because of mill or heat treat scale	Grit blast + solvent clean
✓ Because of oxide or tarnish layer	Acid pickle or alkali clean
✓ Because of hydrocarbons (e.g. oil)	Solvent clean
o From contamination by water from the environment	
✓ Because of dew, rain, frost, etc.	Dry and keep dry; use heat
✓ Because of excessive humidity	Preheat joint area; use low-H electrodes
✓ Because of moisture in consumables	Bake electrodes and fluxes, store in dry boxes
	Use drying filters in gas lines
o From contamination by air from the atmosphere:	
✓ Because of drafts or fans	Eliminate drafts; keep fans >20 ft./6 m.
✓ Because of excessive arc gap	Shorten arc gap
✓ Because of improper electrode angle	Readjust angle
✓ Because of aspiration at fittings	Check tightness of fittings
✓ Because of porous hoses	Check hoses
o From contaminated or defective welding equipment:	
✓ Because of dirty cable liners	Check; replace cable liners
✓ Because of dirty wire feeder rollers	Check; clean or replace rollers
✓ Because of aspiration at fittings	Check fittings; tighten or replace
✓ Because of porous hoses	Check hoses; replace hoses
✓ Because of emptying gas cylinder	Check gas level before welding
o From improper welding technique	Check for proper gas flow rate
	Check, readjust electrode angle
	Check root gap and back shielding
o From excessive gases in base material	Check certification chemistry
	Vacuum degas to lower hydrogen
	Consider fillers with deoxidizers (e.g. Al)
Cracking:	
o Interdendritic hot cracking	Check certification for S, P, Se, and As
	Check and reduce alloy-freezing range, if possible
	Check and reduce weld structure coarseness
	Consider fillers to eliminate boundary wetting
o Transgranular cold cracking	Reduce all sources of hydrogen
	Prevent martensite formation via filler choice
	Post-weld temper if martensite may have formed
	Consider post-weld thermal stress relief

(Continued)

Table 9.1 (Continued)

Inclusions:	
o Tungsten inclusions	Avoid dipping GTAW electrode tip into pool
	Avoid overheating electrode in DCRP (reduce I)
o Copper inclusions	Avoid overheating nozzle in GMAW
o Inclusions from wire in electrode	Avoid dipping
Macrosegregation:	
o Dissimilar metal joints	Stir weld pool better
	Consider dilution by base metal
	Employ techniques in Chapter 21

References

ASM International (2016). *ASM Handbook,* Volume 3: Alloy Phase Diagrams, 10e. Materials Park, OH: ASM International.

Dudas, J.H. and Collins, F.R. (1966, 1966). Preventing weld cracks in high-strength aluminum alloys. *Welding Journal Research Supplement* 45 (6): 241s–249s.

Giamei, A.F., Kraft, E.H., and Lemkey, F.D. (1976). *New Trends in Materials Processing*. Materials Park, OH: ASM International.

Granjon, H. (1991). *Fundamentals of Welding Metallurgy*. Cambridge, UK: Abington Publishing, Woodhead Publishing.

Messler, R.W. Jr. (1993). *Joining of Advanced Materials,* figure 13.1, p. 376. Stoneham, MA: Butterworth-Heinemann.

Pargeter, R. (2007). Evaluation of necessary delay before inspecting for hydrogen cracks, originally from a figure in the paper. *Welding Journal Research Supplement* 82 (11): 321s–329s.

Ploshikhin, V., Prihodovsky, A., and Ilin, A. (2011). Experimental investigation of the hot cracking mechanism in welds on the microscopic scale. *Frontiers of Materials Science,* Higher Education Press 5 (2): 133–145.

Portevin, A. and Seferian, D. (1934). Heat study of torch and arc welding. XIth International Congress of Acetylene and Autogenous Welding, Rome, Italy, June 1934, vol. III, pp. 269–297.

Smith, C.S. (1948). Grains, phases, and interfaces: an interpretation of microstructure. *Transactions of the AIME* 175: 15–51.

Soysal, T., Kou, S., Tat, D., and Pasang, T. (2016). Macrosegregation in dissimilar-metal fusion welding. *Elsevier's Acta Materialia* 110: 149–160.

The Welding Institute. *What is hot cracking (solidification cracking)?* figure 3 (in a) and figure 1 (in b). Cambridge, UK: TWI www.twi-global.com.

The Welding Institute (1996). Job Knowledge 21. In: *Aluminum alloys/Weldability of Materials*. Cambridge, UK: TWI www.twi-global.com.

Further Reading

Kou, S. (2002). *Welding Metallurgy*, 2e. New York, NY: Wiley-Interscience, Wiley.

Lippold, J.C. (2014). *Welding Metallurgy and Weldability*, 1e. New York, NY: Wiley.

Messler, R.W. Jr. (2004/1999). *Principles of Welding: Processes, Physics, Chemistry, and Metallurgy*, Chapter 13. Weinheim, Germany/New York, NY: Wiley-VCH/Wiley-Interscience, Wiley.

Phillips, D.H. (2016). *Welding Engineering: An Introduction*, 1e. New York, NY: Wiley.

10

Partially Melted Zone of Fusion Welds

10.1 Origin and Location of the PMZ in Fusion Welds

Like all materials that form crystalline solids upon solidification from their liquid state, pure metals solidify at a single, discrete temperature, known as the material's *melting temperature* or *melting point*. Metal alloys, on the other hand, like all impure materials that form crystalline solids upon solidification, progressively solidify as temperature is lowered through a so-called *freezing range*, beginning at the material's *liquidus temperature* and ending at the material's *solidus temperature*.[1] This is the case for all alloys, whether exhibiting complete solid solubility or partial solid solubility of a solute B in a solvent A in a two-component binary system of A and B. For low concentrations of solute, the isomorphous phase diagram for the former situation and the eutectic phase diagram for the latter situation each exhibit liquidus and solidus temperatures that bound the alloy's freezing range. Different solutes in a given solvent can cause the freezing range to be narrow or wide.

Figure 10.1 schematically illustrates the characteristic solidification (or cooling) curves for a pure metal and for an alloy, thereby depicting the single, discrete melting temperature or melting point for the former and liquidus and solidus temperatures for the latter.

When alloys are fusion welded, a *partially melted zone*,[2] *PMZ*, forms from thermal cycles associated with welding that heat an alloy to peak temperatures between the liquidus and solidus (Figure 10.2). For alloys that underwent prior nonequilibrium solidification that resulted in pronounced interdendritic microsegregation of solute, a PMZ forms above that alloy's effective solidus temperature (T_{Seff}) and below its effective liquidus temperature (T_{Leff}). Pure metals do not exhibit a PMZ because melting occurs at a single, discrete temperature (i.e. melting point) and, thus, along a line (i.e. fusion line or fusion boundary) not over a region. Figure 10.2 shows the thermal cycles that lead to partial melting having peak temperatures lower than those that cause complete

1 The true liquidus temperature and the true solidus temperature only occur for the equilibrium condition. For real-world nonequilibrium conditions, both temperatures are depressed on cooling more and more as deviation from equilibrium increases (that is, as cooling occurs faster and faster).
2 Many references use partial melted zone, while this author prefers using partially melted zone to denote the action that leads to the zone's formation during welding.

A Practical Guide to Welding Solutions: Overcoming Technical and Material-Specific Issues,
First Edition. Robert W. Messler, Jr.

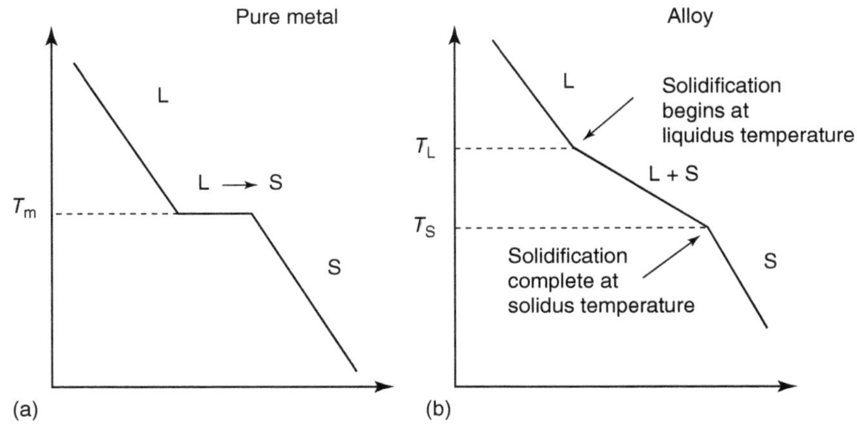

Figure 10.1 Schematic illustration of characteristic solidification or cooling curves for a pure metal (a) and for an alloy (b). An arrest in cooling of a pure metal is the result of solidification occurring at a single, discrete temperature (melting point) until complete. For an alloy, liquid progressively transforms to solid as temperature drops more and more over a freezing range from a liquidus temperature (here, T_L) to a solidus temperature (here T_S), under equilibrium conditions. The result is the existence of a partially melted zone (PMZ) on heating (e.g. by fusion welding) and a mushy zone on cooling.

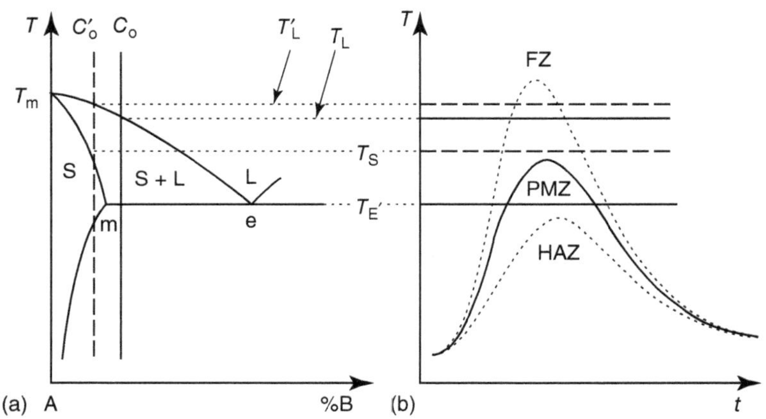

Figure 10.2 Schematic illustration showing the relationship between (a) a portion of an eutectic phase diagram and (b) welding thermal cycles in the fusion zone (curve FZ), partially melted zone (curve PMZ), and heat-affected zone (curve HAZ). Source: Kou (1987). Used with permission of John Wiley & Sons.

melting in the fusion zone (FZ), but above those that do not cause melting but affect the microstructure of the solid outside this region, that is, the *heat-affected zone (HAZ)*.

The region of partial melting that exists all the way around a stationary molten weld pool in an alloy, where the temperature rose to between the effective solidus T_{Seff} (or $T_{eutectic}$ in the eutectic region of a eutectic phase diagram) and the effective (usually normal[3]) liquidus T_{Leff}, only persists along the lateral edges of the pool once it has moved and solidification has occurred behind it. At the leading edge (front) of a moving weld pool, everything is fully melted as the pool advances. At the trailing edge, solidification has occurred; therefore, here a "mushy zone" of mixed liquid and solid existed as temperature dropped from T_{Leff} to T_{Seff}. Only once temperature drops below T_{Leff}, is weld solidification complete. Of course, for a partial penetration weld, remnants of a PMZ persist around what was the underside of the weld pool.

Figure 10.3 schematically illustrates the location of the PMZ around a weld pool in plan view, both instantaneously ahead of the pool and permanently at the end of weld solidification.

The extent (i.e. width) of a PMZ depends on several factors, including (i) the liquidus and solidus temperatures (i.e. melting or freezing range) of the alloy being welded, (ii) the thermophysical properties of the alloy being welded (as these affect heat flow[4]), (iii) the total heat content and intensity (or energy density) of the welding process employed, (iv) the operating parameters for the process (as these affect the net linear heat input), and (v) the operating conditions (e.g. preheat, source manipulation, as by weaving or arc oscillation). PMZ width is increased by

- wider freezing range in an alloy
- higher thermal conductivity in an alloy (because of greater heat spreading)

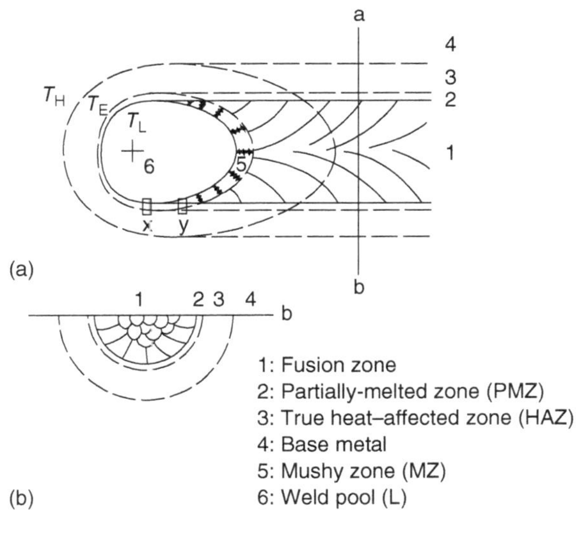

Figure 10.3 Schematic illustration showing the location of the partially melted zone around a fusion weld pool (a) in plan view both instantaneously ahead of the pool (where the PMZ is destroyed as the pool advances) and permanently after solidification along both sides of the weld pool path, as well as in (b) cross section. Source: Kou (1987). Used with permission of John Wiley & Sons.

3 Phase transformation temperatures are altered from their equilibrium value by nonequilibrium heating and/or cooling in a direction that causes the transformation to lag the equilibrium case. Heating, however, has less of an effect than does cooling, with highly nonequilibrium cooling cause depression of transformation temperatures (like T_{solidus}) as much as 100 °C/180 °F or so.

4 Thermophysical properties that affect heat flow include thermal conductivity, specific heat, and density, with thermal conductivity generally having the most pronounced effect.

- o lower energy density in a fusion welding process (because of lower melting efficiency)
- o higher net linear heat input via higher welding current or lower welding speed
- o slower effective welding speed from weaving technique
- o preheat (because of more shallow temperature gradients).

Figure 10.4 schematically illustrates the effect of base material thermal conductivity on the temperature gradient into the base material for a high and for a low thermal conductivity alloy.

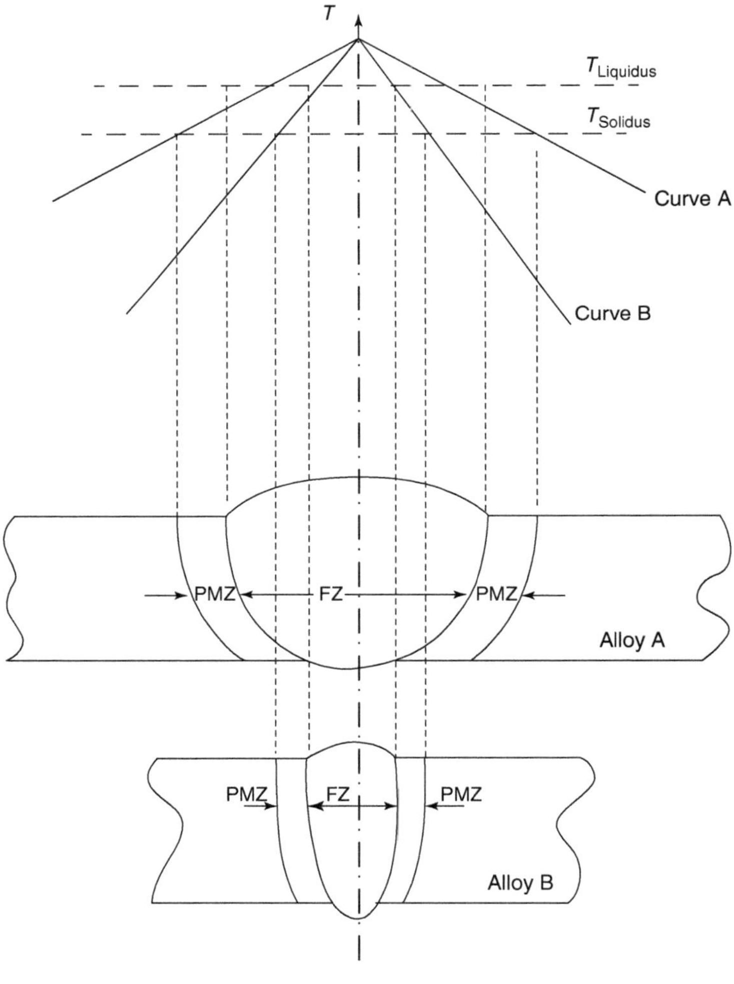

Figure 10.4 Schematic illustration showing the effect of base alloy thermal conductivity on the temperature gradient into the unaffected base material for a high thermal conductivity (curve A) and low thermal conductivity (curve B), with the effect on FZ and PMZ widths also being shown. Source: Messler (1999). Used with permission of John Wiley & Sons.

10.2 Conventional Hot Cracking in the PMZ

Because a PMZ exists where the peak temperature reached during fusion welding exceeded the alloy's effective solidus temperature (T_{Seff})[5] but did not exceed the alloy's liquidus temperature ($T_{\text{Leff}} \sim T_{\text{Lequilibrium}}$ or T_{Lnormal}), a mixture of liquid phase (L) and solid phase (usually designated S_1 or α) was present. The relative proportion (or fraction or percent) of each phase present for an alloy of a given composition (i.e. wt% solute) depends on the peak temperature reached between T_{Leff} and T_{Seff}. The closer to T_{Leff}, the greater the proportion of L. For most alloys, liquid phase dominates in the upper third or fourth of the freezing range. Below this, unmelted solid dominates in the form of an evermore interconnected network as the proportion of liquid decreases; making transport of the remaining liquid phase more and more difficult.

If an alloy is susceptible to conventional hot cracking (because of normal solute redistribution under any cooling condition *and* microsegregation under nonequilibrium cooling leaving a low-melting, solute-enriched liquid phase at structure and/or substructure boundaries), such hot cracking can – and does – occur in the PMZ. Such cracking typically occurs in the hotter regions of the PMZ, nearest to the FZ boundary, where more liquid phase is present. Characteristically, such hot cracking will occur along grain or dendrite boundaries (i.e. be intergranular or, more correctly, interdendritic).

Although remedies for such conventional hot cracking are similar to those used for hot cracking in the fusion zone, success is often harder to achieve.

10.3 Constitutional Liquation Cracking in the PMZ

There is no doubt the principal reason for the existence of a PMZ alongside and beneath the fusion zone in an alloy is that alloys melt over a range of temperatures rather than at a single temperature like pure metals do.[6] There is, however, another reason for a PMZ. Under the conditions of rapid, highly nonequilibrium heating associated with all fusion welding processes, there are occasions when partially melted grain boundary (GB) films are found to have occurred significantly below the bulk equilibrium solidus temperature of an alloy. Known as *constitutional liquation*, his phenomenon extends the width of a PMZ to lower-than-expected temperatures and can have profound adverse effects on weldability because it often results in a special form of hot cracking generically as "subsolidus cracking" but properly known as constitutional liquation cracking or, simply, as liquation cracking.[7]

---

5 Recall, conditions of nonequilibrium cooling cause the equilibrium value of an alloy's solidus temperature ($T_{\text{Sequilibrium}}$ or T_{Snormal}) to be depressed to a lower value (T_{Seff}).

6 As made clear earlier, alloys also solidify over a range of temperatures between effective liquidus and effective solidus temperatures (under nonequilibrium cooling). However, to be precise, this leads to a mushy zone. The PMZ of welds made in alloys is the result of heating, not cooling.

7 There are actually a couple of variants of subsolidus liquation cracking, but only constitutional liquation cracking is discussed herein, as the other variant is much more rarely encountered. Some

The phenomenon is found in alloys that have a second-phase compound of the generic form A_xB_y. These include a number of important engineering alloys that obtain additional strength (beyond solid solution strengthening) by precipitation-strengthening or age-hardening (notably heat-treatable Al alloys and Ni–base alloys). However, such compounds are also found in a number of elevated temperature alloys in which secondary hardening is developed by heat treatment to cause compound formation by reaction (e.g. carbides or nitrides) rather than by precipitation from supersaturated solid solution. Let us look at the mechanism involved in constitutional liquation.

Figure 10.5 schematically illustrates a portion of a phase diagram for a hypothetical alloy system exhibiting the necessary characteristics for constitutional liquation to occur. In this diagram, A_xB_y is a second-phase precipitate (e.g. ceramic carbide, nitride, carbo-nitride, phosphide, sulfide, or silicide, or intermetallic, such as an aluminide) distributed in the α solid solution phase, which serves as a matrix phase. When an alloy of nominal composition C_0 is heated very slowly (so that equilibrium is essentially maintained), the solubility of solute B in the α matrix increases as the solvus temperature T_2 is reached (at point n). The particles of A_xB_y shrink as atoms of B leave the B-rich phase to go into solution in the α matrix phase via solid-state diffusion. At the solvus temperature T_2, the last remnants of A_xB_y particles disappear, converting the alloy to a homogeneous single-phase solid solution consisting entirely of α of

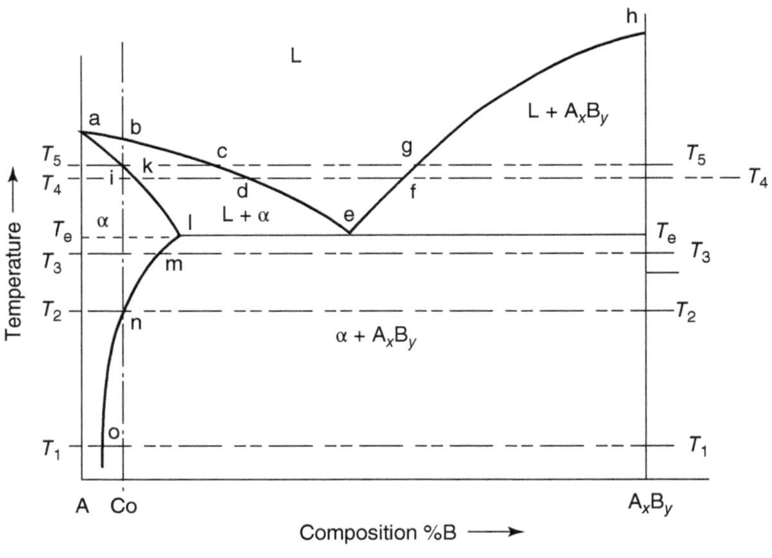

Figure 10.5 Schematic illustration showing a portion of a phase diagram for a hypothetical alloy system exhibiting the necessary characteristics for constitutional liquation to occur in the PMZ of a fusion weld, namely, a primary solid solution phase (here, α), a second-phase compound with the general stoichiometric formula A_xB_y, and a eutectic reaction $L \leftrightarrow \alpha + A_xB_y$. Source: Pepe and Savage (1967). Used with permission of The American Welding Society.

sources refer to "HAZ liquation cracking," when they are actually talking about cracking originating from constitutional liquation, as the PMZ is bounded at its outer (lowest temperature) extreme by the effective solidus as depressed by nonequilibrium.

composition C_0. Diffusion under conditions of equilibrium has allowed complete dissolution and homogenization.

If heating is rapid, as occurs in fusion welding, A_xB_y particles do not have enough time to dissolve by having B atoms enter solid solution in the α phase by solid-state diffusion and nonequilibrium prevails. As a result, some remnants of the original larger particles remain in the α matrix even after the temperature rises above the equilibrium solvus temperature T_2, say to T_3. With further heating to the eutectic temperature, T_{eutectic} or T_e, liquid phase of the eutectic composition begins to form at the interface between dissolving A_xB_y particles and the α matrix, as somewhere in the region near the particle–matrix interface, the concentration in the solute profile that develops in the matrix reaches the eutectic composition. Further heating to T_4 would provide additional time for more of the A_xB_y particles to dissolve, with the formation of more liquid phase forming around the remaining particles. In fact, at T_4, every remnant of A_xB_y particles still undissolved should be completely surrounded by a liquid film of varying compositions from f to d as distance from the particle–matrix interface increases.

Figure 10.6 schematically illustrates the sequence of constitutional liquation in a susceptible alloy during rapid heating during fusion welding and the extension

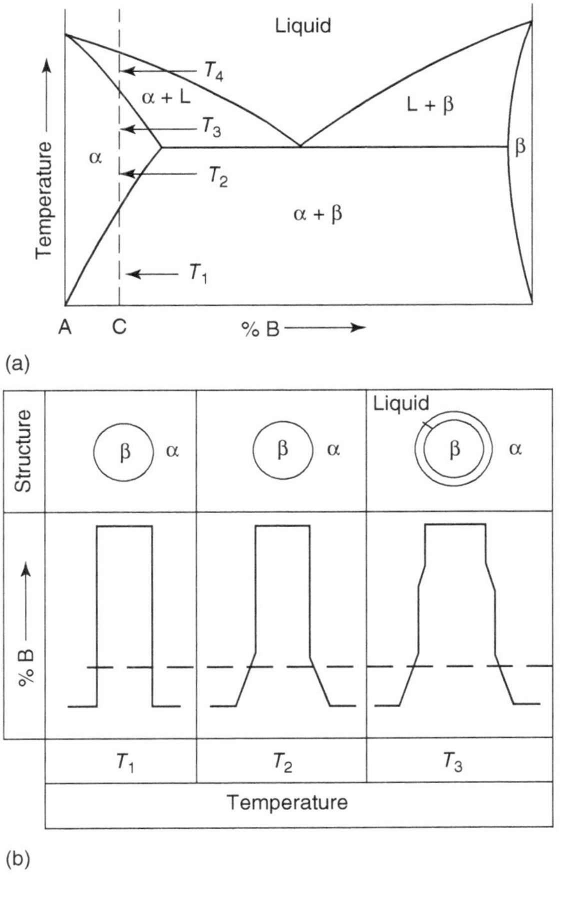

Figure 10.6 Schematic illustrations showing extension of the partially melted zone by constitutional liquation of particles of β phase as these progressively dissolve (at the bottom in b) at three progressively higher temperatures shown in the simplified phase diagram for a hypothetical binary alloy (at the top in a). Source: Yeniscavitch (1987). Published by and used with permission of John Wiley & Sons.

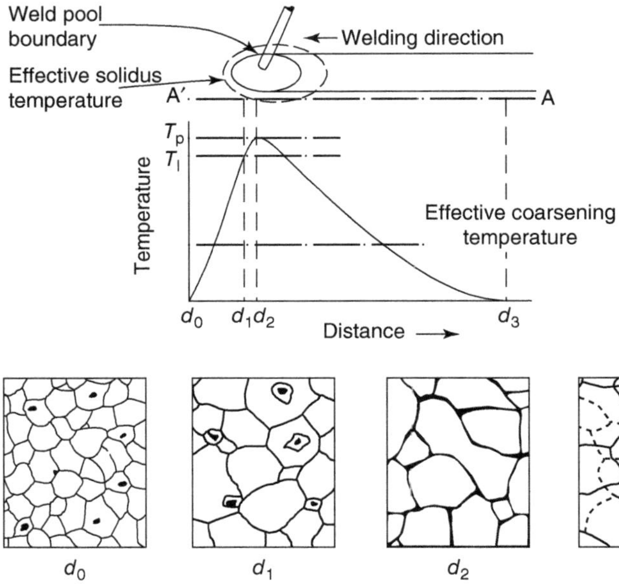

Figure 10.7 Schematic representation of the formation of a ghost boundary network formed by constitutional liquation and solute segregate left behind following grain boundary migration. Source: Pepe and Savage (1970). Published by and used with permission of The American Welding Society.

of the PMZ that results. As for all hot cracking, cracks form as thermally induced shrinkage and contraction stresses pull open boundaries containing liquid films. The occurrence of constitutional liquation is sometimes indicated by the appearance of solute-rich ghost grain boundaries or grain boundary networks in some alloys (Figure 10.7). These are created by pinning of migrating grain boundaries during nonequilibrium heating. Such ghost boundaries have been found to be associated with hot ductility dip in some alloy systems known to be susceptible to constitutional liquation.

Table 10.1 lists some important engineering alloys susceptible to liquation cracking in their PMZ, whereas Figure 10.8 schematically illustrates the suggested sequence for the development of microfissuring in Nb-containing Inconel 718 Ni–base alloy and 347 Nb-stabilized austenitic stainless steel.

10.4 Cold Cracking in the PMZ

Hydrogen-induced cold cracking is observed in the PMZ, as well as in the adjacent FZ, of certain high yield, HY steels in which mixing between the molten filler and the melted base metal is incomplete. The operative mechanism has been attributed to the formation of liquated films along the grain boundaries of the PMZ, serving as easy paths for hydrogen picked up in weld metal to diffuse across the fusion boundary. Because these films contain molten iron that can dissolve three to four times as much atomic (nascent) hydrogen as the surrounding solid, they become supersaturated with hydrogen upon solidification.

Table 10.1 Major engineering alloys prone to PMZ constitutional liquation or hydrogen-induced cracking.

Constitutional Liquation Cracking Prone Alloys:

- Al alloys
  - ✓ Al–Cu and Al–Cu–Mg 2xxx alloys
  - ✓ Al–Mg + Si 6xxx alloys
  - ✓ Al–Zn–Mg 7xxx alloys
  - ✓ Al–Li 8xxx alloys
- Fe alloys
  - ✓ 316 LN (nitrides) and 347 (Nb compounds) austenitic SS
  - ✓ Some duplex stainless steels, such as Alloy 2205 and Ferralium 255
  - ✓ Precipitation hardenable Fe–base alloys, such as A286, Discoloy, Haynes 556, Incoloy 903 and 909
- Precipitation-hardenable Ni alloys
  - ✓ Alloy 800
  - ✓ Hastelloy X
  - ✓ Haynes 214
  - ✓ Incoloy 702, 706, cast 738, 825, and 901
  - ✓ Inconel 718, 722, and X-750
  - ✓ M252
  - ✓ Rene 41 and Rene 80
  - ✓ Udimet 700
  - ✓ Waspalloy

Hydrogen-Induced Cracking Prone Alloys:

- HY80, HY100, and HY130 steels
- 18Ni-maraging steel

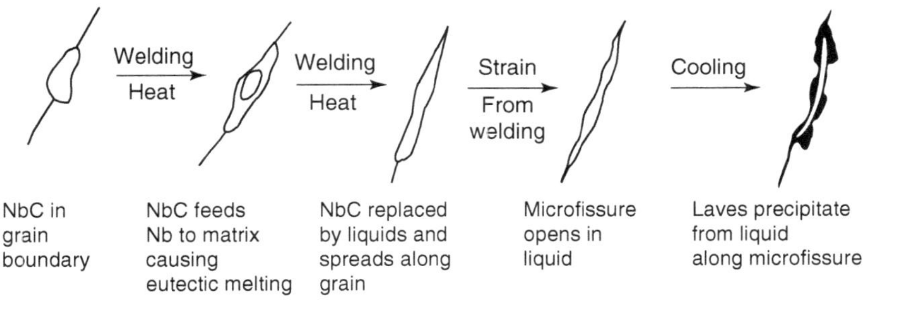

Figure 10.8 Schematic illustration showing the suggested sequence for the development of microfissuring in Nb-containing Inconel 718 and other austenitic alloys. Source: Thompson et al. (1985). Published by and used with permission of The American Welding Society.

As these films are also enriched with solute as the result of solute rejection at the S–L interface, they are considerably more hardenable and, therefore, more prone to the formation of martensite than either the fusion zone or the HAZ base metal. The result is that as-quenched, untampered martensite forms right where hydrogen content is the highest and in the presence of some of the highest thermally induced residual tensile stresses found anywhere around the weld. Thus, the three essential criteria for hydrogen-induced cold cracking – hydrogen, tensile stress, and untampered martensite – could not be better met.

10.5 Overcoming Cracking Problems in the PMZ

Knowing the sources of cracking that tend to prevail in the PMZs of some alloys is a necessary but, regrettably, not always sufficient condition for remediation of problems. General recommendations, however, include the following:

1. Minimizing the width of the PMZ by
   a. Selecting a welding process offering inherently low net linear heat input via high energy density (e.g. plasma arc welding (PAW), laser-beam welding (LBW), electron-beam welding (EBW), or even gas tungsten-arc welding (GTAW))
   b. Keeping welding current as low, and welding speed as high as possible, without sacrificing weld penetration and bead profile
   c. Using multi-pass welding techniques to minimize the heat input for each pass (e.g. stringer beads)
2. Decreasing the level of restraint as much as possible (for both liquation cracking and hydrogen-induced cracking), using one or more of the following:
   a. Presetting or prebending during joint setup
   b. Balancing heat input using double grooves and/or sequencing
   c. Employing intermittent welding
   d. Preheating
3. Minimizing problems from base metal by
   a. Avoiding alloys prone to PMZ defects (e.g. with wide freezing ranges)
   b. Avoiding base metals with pre-existing segregation (e.g. castings, prior welds needing rewelding)
   c. Keeping base metal grain size as small as possible (to maximize grain boundary area) by grain-refining heat treatments before welding (e.g. normalized steels), as shown to be effective in Figure 10.9

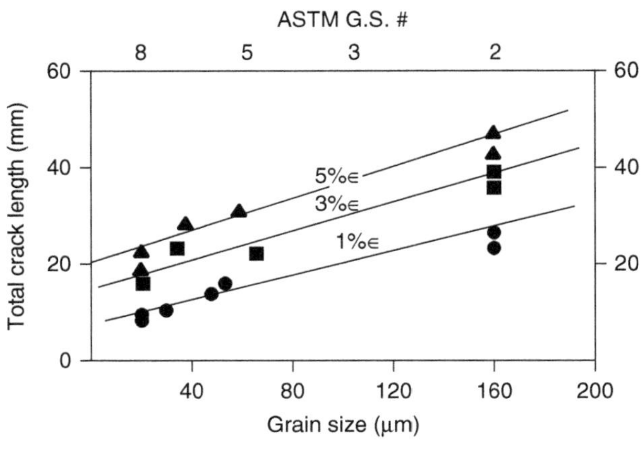

Figure 10.9 Plot showing the favorable effect of reducing grain size on the tendency to liquation cracking in the PMZ. Source: Thompson et al. (1985). Published by and used with permission of The American Welding Society.

4. Choosing a proper filler metal (i.e. one whose melting range assures that solidification throughout the PMZ will be completed *before* weld metal in the FZ solidifies). (In employing this tactic, care must be taken to be sure that the filler does not render the weld metal more susceptible to solidification cracking!)

10.6 Troubleshooting Problems in the PMZ

Table 10.2 lists the various ways problems can arise in the fusion zone of fusion welds.

Table 10.2 Troubleshooting problems in a fusion weld's PMZ.

For all problems, minimize the width	Minimize net linear heat input by reducing I and increasing TS; consider use of multi-pass welding using stringer beads
	Attempt to minimize joint restraint
Conventional Hot Cracking:	
Similar susceptibility as FZ	
Interdendritic morphology	Reduce low-melting solutes and impurities
	Reduce W/D ratio to 1 : 1 to 1.4 : 1
	Produce convex bead contour
Constitutional Liquation Cracking:	
Rapid, nonequilibrium heating	Avoid susceptible alloys (per Table 10.1)
Melting around partially dissolved A_xB_y particles	Choose filler that solidifies in PMZ before FZ
Hydrogen Cold Cracking:	
Easy diffusion of H from FZ to PMZ via liquated GBs and solute enrichment in GBs promotes martensite formation	Eliminate all potential sources of hydrogen (per Section 6.4)

References

Kou, S. (1987). *Welding Metallurgy*. New York, NY: Wiley.

Messler, R.W. Jr. (1999). *Principles of Welding: Processes, Physics, Chemistry, and Metallurgy*, figure 15.3 on p. 504. New York, NY: Wiley.

Pepe, J.J. and Savage, W.F. (1967). Effects of constitutional liquation in 18Ni maraging steels. *Welding Journal (Research Supplement)* 46 (9): 411s–422s.

Pepe, J.J. and Savage, W.F. (1970). The heat-affected zone in 18Ni maraging steels. *Welding Journal (Research Supplement)* 49 (12): 545s–553s.

Thompson, R.G., Dobbs, J.R., Mayo, D.E., and Cassimus, J.J. (1985). The relationship between grain size and microfissuring in alloy 718. *Welding Journal (Research Supplement)* 64 (4): 91s–96s.

Yeniscavitch, W. (1987). Joining. In: *Superalloys II* (ed. C.T. Sims, N.S. Stoloff and W.C. Hagel). New York, NY: Wiley.

Further Reading

Kou, S. (2002). *Welding Metallurgy*, 2e. New York, NY: Wiley-Interscience, Wiley.

Lippold, J.C. (2014). *Welding Metallurgy and Weldability*, 1e. New York, NY: Wiley.

Messler, R.W. Jr. (2004/1999). *Principles of Welding: Processes, Physics, Chemistry, and Metallurgy*, Chapter 15. Weinheim, Germany/New York, NY: Wiley-VCH/Wiley-Interscience, Wiley.

11

Heat-Affected Zone of Fusion Welds

11.1 Origin and Location of the HAZ for Fusion Welds

Any welding process that employs heat to achieve continuity between the atoms of abutting metallic joint element can be expected to have some effect on the microstructure and, thus, properties of the base materials, even if there is no melting of those base materials.[1] The region affected by the heat of welding (whether intentional or not) is called the *heat-affected zone*, commonly abbreviated as HAZ. Not surprisingly, fusion welding processes create more pronounced heat-affected zones (HAZs) than other welding or weld-related processes, such as non-fusion welding processes, brazing, and soldering that use or generate heat. The HAZ lies immediately outside the fusion zone (FZ) of fusion welds in pure metals and immediately outside the partially melted zone (PMZ) of fusion welds in alloys (see Figure 9.1).

As distance from the centerline of a weld increases, the HAZ begins where the peak temperature of the welding cycle is just below the effective solidus temperature (T_{Seff}) of an alloy and extends to a point where the peak temperature was high enough for long enough (under conditions of nonequilibrium) to cause some measurable change in the microstructure of the base material(s). Such a measurable change would alter one or more of the properties of the base material(s). The criterion that the peak temperature must be high enough and persist for long enough is terribly important because a high peak temperature in and of itself is not a problem if the structure, and thus any property, is not altered.[2]

1 Some non-fusion welding processes employ heat to facilitate the plastic flow of metal to achieve material continuity, albeit without requiring or causing melting (e.g. hot pressure welding, HPW). Other non-fusion welding processes generate heat in the course of achieving material continuity, albeit without causing melting (e.g. friction welding, FRW).
2 Because virtually all phase transformations take time for the new phase to nucleate and for the transformation to progress, time at temperature is important, where deviation from the ideal conditions of equilibrium alters the structure and properties in real materials processed in the real world.

A Practical Guide to Welding Solutions: Overcoming Technical and Material-Specific Issues,
First Edition. Robert W. Messler, Jr.

11.2 Manifestation of Problems in the HAZ of Fusion Welds

Table 3.2 lists the various weld joint discontinuities that can arise during or as a result of welding. The majority of these discontinuities are defects in the weld (e.g. gas porosity, cracks, inclusions, defective profile, or unsightly appearance) or weldment (e.g. dimensional shrinkage, geometric distortion, and structural instability). However, obvious or even more subtle defects are not the only problems that can be encountered. There is also the possibility of problems with the properties of a weld, as these fail to meet design requirements and compromise performance of the weldment in service. Nowhere are problems with properties more troublesome – and troubling – than in the HAZ of fusion welds. Property problems can include hardening/embrittlement, softening/weakening, sensitization to corrosion, and loss of ductility from excessive grain growth. In fact, although defects in the form of cracks can occur in the HAZ, problems with properties are far more common.

The precise effect of exposure to elevated temperature on the microstructure and properties of a metallic base material depends on what the base material is and, most importantly, on how it obtains its properties, that is, on the operative mechanism or mechanisms of strengthening. Although there are seven strengthening mechanisms in physical metallurgy that can be used singly or in combinations,[3] only two lead to problems of great enough significance with great enough frequency following fusion welding that they merit discussion in this chapter. These two are (i) precipitation strengthening (or age hardening) and (ii) transformation hardening or strengthening (or quench hardening). The former includes major engineering materials among heat-treatable Al alloys, heat-treatable Mg alloys, heat-treatable Ni alloys, and heat-treatable stainless steels. The latter includes quench-hardenable steels, including carbon steels, low- and medium-alloy steels (including high-strength, low-alloy [HSLA] and high yield [HY] steels), some tool steels, and martensitic stainless steels. Interestingly, for each of these two strengthening mechanisms, problems can appear as either cracks or strictly degraded or compromised properties.

Let us consider the defects and property problems associated with the HAZs of fusion welds, first addressing problems found in the HAZ of precipitation-hardening alloys, then a problem found in the HAZ of some austenitic stainless steels because of sensitization to intergranular corrosion, then problems found in the HAZ of transformation-hardening steels, and, finally, a cracking problem found in both types of alloys, i.e. reheat cracking.

---

3 The seven mechanisms of strengthening enumerated in *The Essence of Materials for Engineers* by R.W. Messler, Jr., 1st ed., 2011, Jones & Bartlett Learning, Burlington, MA, are (i) grain refinement (as per the Hall–Petch relationship); (ii) strain hardening; (iii) solid solution strengthening/alloying (substitutional or interstitial); second-phase strengthening via (iv) coherent precipitates (from aging), (v) solid solution phase boundaries, or (vi) dispersed phases not inherent in the alloy system; and (vii) transformation hardening (e.g. the martensite transformation in steels).

11.3 Precipitation-Hardening Alloy HAZ Problems

In many alloys, adding a solute beyond its solubility limit in the solvent leads to the formation of a second phase. That second phase can be a solid solution (virtually always having a different crystal structure than the primary solid solution phase) or an ordered ceramic or intermetallic compound. Either way, the second phase is virtually always much richer in solute than the primary solid solution phase. When the second phase forms from the parent phase as the result of precipitation of an ordered compound, the alloy is said to be *precipitation hardened* or *precipitation strengthened*. Because the precipitate phase requires time – often considerable time – to form (by diffusion-controlled nucleation and growth), such alloys are also said to have *age hardened*. Such aging can occur at room temperature as natural aging or, more often, to dramatically accelerate the process, at an elevated temperature as artificial aging. In either case, the second phase always increases the strength of an otherwise single-phase alloy by hindering the motion of dislocations that lead to plastic deformation by slip because of the strain energy they introduce into the lattice of the matrix phase at a coherent interface with the precipitate.[4]

As shown by the hypothetical binary phase diagram for an alloy with solute B and solvent A in Figure 11.1, three conditions must coexist for an alloy to be

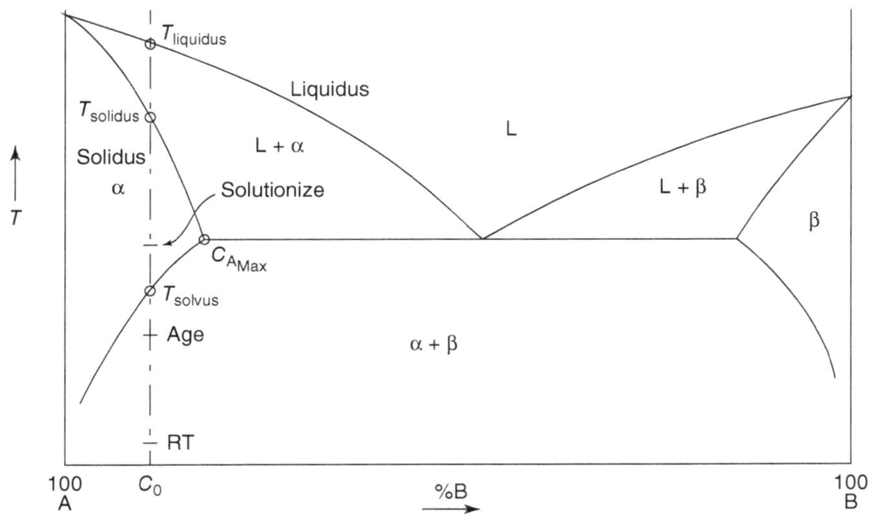

Figure 11.1 Schematic illustration showing a hypothetical binary phase diagram for an alloy (with nominal composition $C_0 < C_{A_{Max}}$) that can be precipitation or age hardened. Source: Messler (1999), figure 16.11, p. 531. Used with permission of John Wiley & Sons.

4 Coherency means that certain (usually closest-packed) planes of atoms in the precipitate phase attempt to align with certain (usually closest-packed) planes of the matrix phase, but because of the differences in lattice spacing between the two, distortional strain gives rise to stresses that hinder movement of dislocations more effectively than if there is no coherency. Maximum strain results in optimum strengthening with optimal aging.

precipitation or age hardenable: (i) The solute must exhibit limited solid solubility in the solvent; (ii) this solubility must decrease with decreasing temperature; and (iii) the alloy of interest must have a nominal solute concentration (i.e. composition) below the maximum limit of solubility.

Development of the second-phase precipitate requires the following steps (also shown in Figure 11.1 for alloy $C_0 < C_{A_{Max}}$) during a heat treatment process:

1. Alloy of nominal composition $C_0 < C_{A_{Max}}$ must be heated above the solvus temperature (T_{solvus}) into the single-phase α solid-solution region and held there to solutionize as it equilibrates. This is called a solution heat treatment, solution annealing, or solutionizing step.
2. The solutionized alloy must be cooled rapidly enough to not allow sufficient time for the solute-rich β second phase to form by diffusion of B as the temperature falls below T_{solvus}, thereby retaining more solute in the solution than permitted under equilibrium. The result is a supersaturated solid solution of solute B in solvent A. This is called quenching treatment or quenching step.
3. The supersaturated alloy must be given sufficient time to allow second-phase β to nucleate and grow as a precipitate. This is called aging heat treatment or aging step and is accelerated by heating the supersaturated solution to nearer but below the solvus temperature, thereby speeding up the solid-state diffusion of B in A.[5]

Without going into unnecessary details here, the formation of a precipitate phase occurs in stages involving (i) diffusion-controlled clustering of B atoms above their nominal concentration in the supersaturated solid solution phase (i.e. formation of Guinier–Preston [GP] zones in one or two stages) until their concentration approaches the level found in the precipitate (typically 25–67 at.%), (ii) usually forming a semicoherent precursor to the ideal, fully coherent precipitate, (iii) formation of the ideal, fully coherent precipitate with optimum aging time at temperature, followed by (iv) loss of coherency (and loss of optimum strengthening effectiveness) as the precipitate grows too large in what is known as overaging. Figure 11.2 schematically illustrates the generic correlation of hardness or strength of a precipitation-hardening alloy with the time of aging for various aging heat treatment temperatures. Note that the peak hardness or strength (i.e. optimal aging) occurs in shorter and shorter times as the aging temperature is increased. Also note how longer time–lower temperature aging cycles (or heat treatments) result in higher peak hardness or strength values. Finally, note how overaging occurs after the point of optimal aging, with very slow, albeit inevitable loss of hardness or strength (on the logarithmic time axis).

During fusion welding (or, to a lesser extent, with non-fusion welding involving intentional or incidental heating), an age-hardened base material joint element loses hardness and strength in the heat-affected zone. Figure 11.3 schematically illustrates how the β-phase precipitate first overages and then redissolves into solid solution[6] as the peak temperature rises as the fusion boundary is

5 The rate of solid-phase diffusion follows an exponential relationship, which, as a rule-of-thumb, results in a doubling of rate with every 30 °C/50 °F temperature rise above room temperature.
6 The process of solid-phase dissolution of precipitate in an age-hardened alloy is known as *reversion*, and the affected region is called the *reversion zone*.

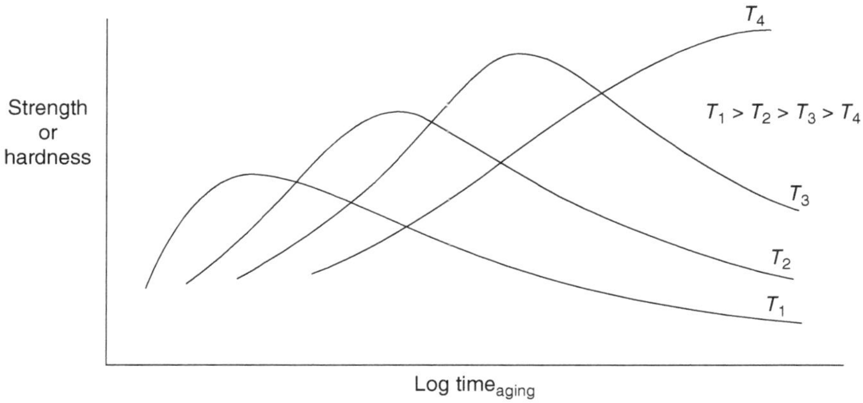

Figure 11.2 Schematic illustration showing the generic correlation between the strength or hardness of a precipitation-hardening alloy as a function of aging time at various aging temperatures. Source: Messler (1999), figure 16.13, p. 534. Used with permission of John Wiley & Sons.

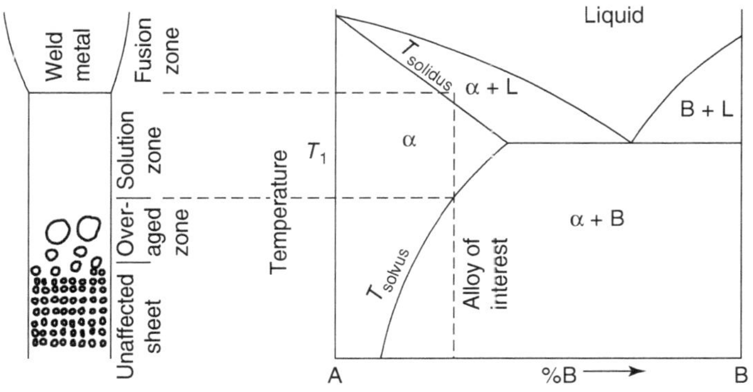

Figure 11.3 Schematic illustrations showing the effect of the heat of welding on the microstructure in the heat-affected zone of an optimally aged precipitation-hardening alloy. HAZ microstructure is shown as it correlates with various temperature ranges on a hypothetical equilibrium phase diagram. Note: As fusion welding leads to nonequilibrium heating and cooling, such a diagram must be used with caution. Source: Connor (1987). Used with permission of The American Welding Society.

approached. Figure 11.4 schematically illustrates how hardness and strength are lowered more and more through the overaged zone of the HAZ, reaching a minimum in the solution zone.

A loss of strength or, for some applications, hardness, caused by the heat effects of fusion welding a structure builtup from precipitation-hardening alloys poses a problem. As always, there are three options, the first of which is probably not acceptable: (i) accept the property loss,[7] (ii) if possible, prevent (or, at least,

7 Loss of a key design property is often referred to as a "knock-down," which a designer may be able to accept, especially during repair welding as opposed to original equipment manufacture (OEM).

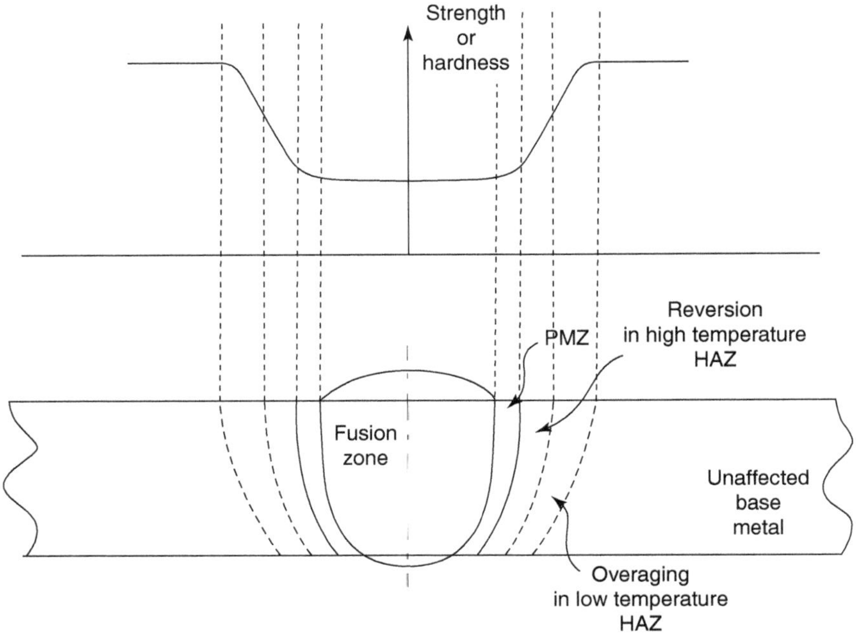

Figure 11.4 Schematic illustration showing a plot of the strength or hardness from the centerline of a fusion weld outward through the partially melted zone and high-to-low heat-affected zone to the unaffected base metal of an optimally aged precipitation-hardening alloy. Source: Messler (1999), figure 16.16, p. 541. Used with permission of John Wiley & Sons.

minimize) the property loss, and, if necessary, (iii) attempt to recover properties as much as possible. Prevention, or at least minimizing adverse effects, would likely entail joining aged joint elements using a non-fusion welding process. The most successful has been friction stir welding (FSW) used with aged Al-alloys. Other options might be to use very high-speed welding using a high energy density process such as laser-beam welding (LBW) or electron-beam welding (EBW). These might be better options for aged Ni alloys or aged precipitation-hardening (PH) stainless steels.

When it comes to fusion welding age-hardening alloys, there are three general approaches: one for minimizing adverse property loss in joint elements that start in the optimally aged condition, one for recovering properties following welding, albeit with some compromise, and one that offers full properties but is often impractical. These three, in the order just described, are as follows:

- Optimally aged, weld, and post-weld age: A + W + PWA
- Solution treat, weld, post-weld age: ST + W + PWA
- Weld plus post-weld solution treat plus age: W + PWST + PWA.

Figure 11.5 schematically illustrates the effect of each approach, as described in detail below.

Fusion welding optimally aged joint elements will result in an HAZ ranging from an overaged region, in which there is some loss of strength or hardness obtained from aging, to a reversion region, in which there is complete loss of strength or hardness obtained from aging. By applying the proper post-weld aging cycle of temperature and time, there would be little further loss of properties in the overaged region (as further aging would take much more time if the post-weld aging temperature was selected to be lower than the original aging temperature, as should be the case) and fairly complete recovery of properties in the reversion region. What happens in the fusion zone depends on whether welding was done autogenously (without filler) or with a filler metal. Without filler, post-weld aging would develop properties comparable to the unaffected base metal (UBM). With a filler, final properties depend on the choice of filler metal (e.g. whether and how it ages). A commonly used "trick" – really, a strategy – is to overage the joint elements before welding, accepting the lower properties compared to optimal aging, and post-weld aging the weldment. The result would be lower but more uniform properties from base metal to base metal.

By solution-treating joint elements before welding and then welding, only modest aging occurs in even the hottest regions of the HAZ, as time at temperature

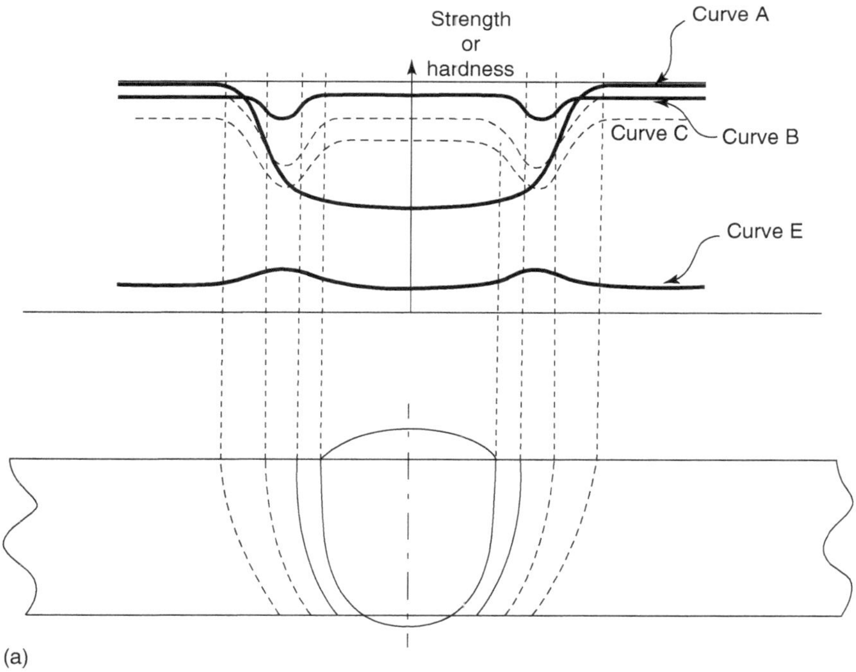

Figure 11.5 Schematic illustration showing the effect of several approaches to attempt to recover or maximize strength or hardness across the HAZ of a fusion weld made autogenously (without filler) in a generic age-hardening alloy. The uppermost and middle figures are schematic plots of strength or hardness, as these correlate with the weld zones shown in the lowermost figure. In the uppermost figure, Curve A is for the case of Aged + W, whereas Curve B is for the case of Aged + W + PWA. Curve E is for the case of ST + W, with Curve A essentially representing the case of ST + W + PWA. In the middle figure, Curve C is for the case of overaged (OA) + W, whereas Curve D is for the case of OA + W + PWA. Curve F essentially represents the case of W + PWST + PWA.

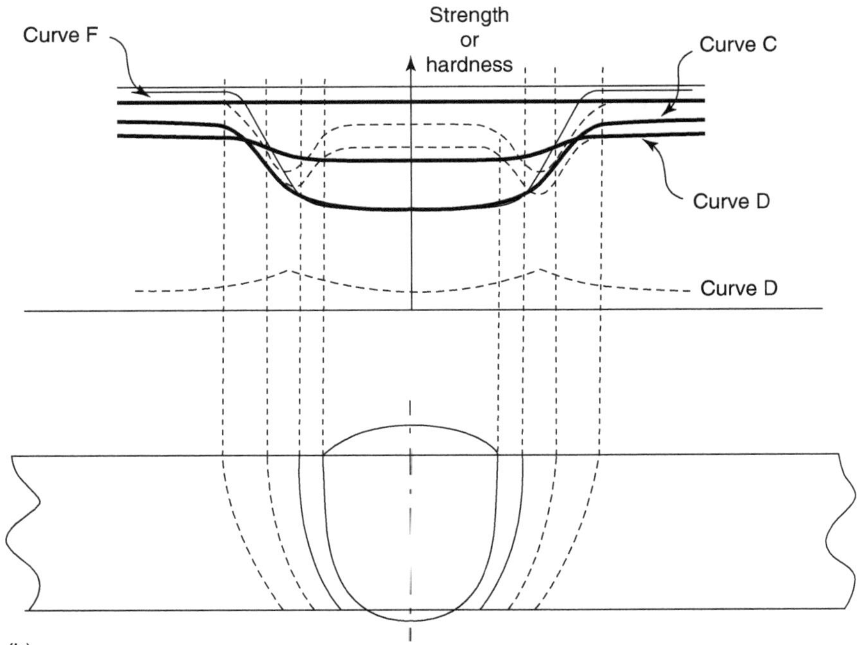

(b)

Figure 11.5 (*Continued*)

is very short (allowing little time for precipitate formation by solid-stated diffusion). Post-weld aging the entire weldment tends to develop properties to the level desired. With both this approach and the prior approach (above), by choosing a low-temperature/longtime aging cycle, there tends to be minimal risk of distortion for weldments during heat treatment. Not so for the next approach!

Although it sounds great to postpone any heat treatment of age-hardening alloys until *after* detailed parts have been welded into a final configuration (i.e. weldment), performing the solution treatment and aging treatment on the final weldment could result in optimal aging everywhere or, more likely, could destroy the geometric integrity of the weldment because of the high-temperature/longtime solution treatment and quench. Whether this approach is viable for any particular case must be determined case by case.

When fusion welding age-hardening alloys, doing as well as possible should be the goal. Not perfection! One should take solace in the fact that uniformity of properties across a joint is often more important to a designer and a design than optimum properties.

11.4 Sensitization in the HAZ of Austenitic Stainless Steels

A problem caused by unintended and unwanted precipitation of carbides at grain boundaries in the HAZ of fusion welds can occur in austenitic stainless steels

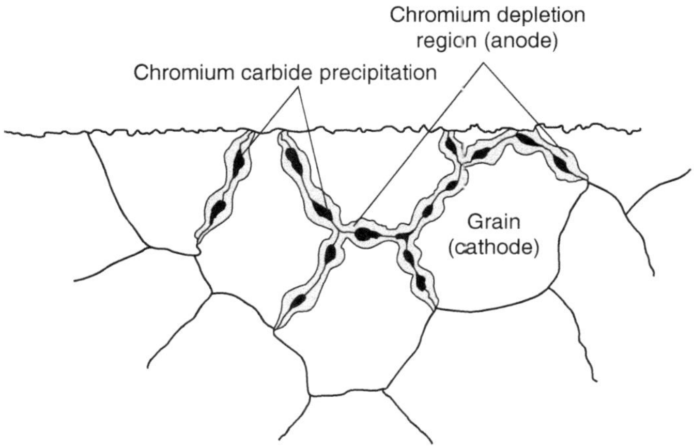

Figure 11.6 Schematic representation of the precipitation of chromium carbide $(Fe,Cr)_{23}C_6$ at the grain boundaries during sensitization in austenitic stainless steel. Source: Almubarak et al. (2013). Used with permission of Hindawi Publishing Corporation.

containing more than about 0.08 wt% C, rendering the affected areas susceptible to intergranular corrosion, cracking, and failure. The general phenomenon is known *as sensitization,* whereas when it occurs because of welding, it is known as *weld sensitization.*

The phenomenon is caused by the unintended precipitation of $M_{23}C_6$ Cr,Fe carbides preferentially along grain boundaries,[8] thereby depleting or "denuding" surrounding austenite of Cr remaining in solid solution to below the concentration required to provide resistance to corrosion (i.e. greater than about 13 wt%), as shown schematically in Figure 11.6. With the Cr-depleted region along the grain boundaries being more anodic compared to the more Cr-rich bulk of the grains, susceptibility to corrosive attack by an intergranular mode occurs (shown in Figure 11.7). Corrosive attack eventually causes grains to separate (i.e. leads to extensive cracking) and can result in the failure of the weldment by either leaking or structural failure. For these reasons, weld sensitization is alternatively known as *weld decay.*

Figure 11.8 shows a photograph of a severe crack in the HAZ of the sensitized HAZ of an austenitic stainless steel weldment.

As the problem of sensitization is treated in much more detail in Chapter 16 in Part 3 on Material-specific Weld-related Problems, it is suffice to say, the following:

- The rate of carbide formation is greatest in the temperature range between 600 and 800 °C (about 1100–1550 °F), i.e. the *sensitization range*

8 Precipitate formation is favored along grain boundaries for two reasons: (i) because the resulting increased internal energy from strain at the interface between the mismatched crystal lattices of the carbide and the austenite solid solution matrix phase is partially offset by the pre-existing strain in the grain boundary (i.e. by a synergistic effect) and (ii) because the rate of solid-state diffusion in grain boundaries is higher than in the bulk structure of the grain (i.e. because of what is known as "pipeline diffusion"), favoring both nucleation and growth of the new precipitate phase.

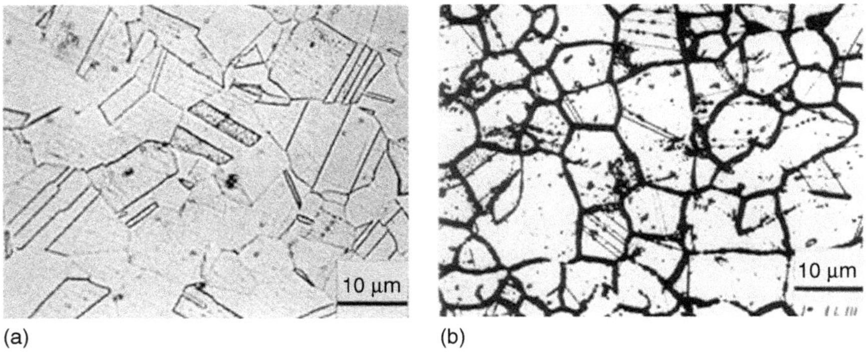

Figure 11.7 Photomicrographs showing (a) the appearance of the normal grain structure of an austenitic stainless steel (with characteristic annealing twins) and (b) the appearance of the grain structure of an identical alloy sensitized by heating into a critical range, often during fusion welding. Source: From two photomicrographs on www.wikipedia.com in article on "Intergranular corrosion," images by Webcorr on 4 November 2011. Freely used under Creative Commons ShareAlike CC BY-SA 3.0.

Figure 11.8 Photograph showing a severe stress corrosion crack running along the HAZ in a cast CF8M (type 316) flange welded to a pipe formed from sheet 316. Cracking was due to weld sensitization and subsequent corrosive attack. Source: Chakrapan (1992). Used with permission of ASM International.

- Susceptibility to weld sensitization is greatest when (i) the austenitic stainless steel contains higher amounts of carbon (typically over 0.06–0.07 wt%, which is rare by modern practice, despite 0.08 wt% maximum still being listed in American Iron & Steel Institute (AISI) and related specifications by UNS, AMS, ASME, and ASTM), (ii) weld heat input is high, (iii) the base material element was deformed by cold work before welding, and (iv) there are no intentional alloy additions with greater affinity for C than Cr (i.e. the alloy is not a stabilized grade[9]).

- o With the mechanism responsible for weld sensitization or weld decay known, three ways of avoiding the problem are
  - ✓ reducing the carbon content of the alloy to decrease the amount of Cr,Fe-carbide that can form, thereby reducing the amount of Cr that can be removed from the matrix (i.e. use a low-carbon grade, such as 304L or 316L, both with C limited to 0.035 wt% maximum)[10]
  - ✓ adding a small amount of a solute element having greater affinity for C than Cr, so carbides of these elements form preferentially, leaving Cr vital to corrosion resistance in solution (i.e. the so-called stabilized grades 321 with added Ti and 347 with added Nb)
  - ✓ keeping net linear heat input as low as possible and favor fusion processes with high energy density (e.g. plasma-arc welding [PAW], LBW, EBW, and even gas tungsten-arc welding [GTAW]) to dramatically shorten time at peak temperatures in the sensitization range

  In addition, as a measure of last resort for its general impracticality,
  - ✓ Once sensitization has occurred following welding, subject the weldment to a post-weld solution heat treatment at 1000–1100 °C (1800–2000 °F) for 1-in./25-mm° per inch/25-mm of thickness) to redissolve the Cr,Fe carbides, followed by rapid quenching (usually in water, possibly in air) to retain the Cr in solution. (This approach has limited practicality because of the likelihood of distortion of large and/or geometrically complex weldments exposed to such high, softening temperatures and gradient-producing distortion stresses from such rapid quenches, necessitating weldments be heat treated in shape-holding fixtures which, themselves, can tolerate such high temperatures and severe quench rates.)

Therefore, although not a problem in the HAZ of a precipitating-hardening alloy, weld sensitization or weld decay is, nevertheless, a problem because of weld-induced precipitation.

11.5 Transformation-Hardening Steel HAZ Problems

Alloys that form a particularly hard, high-strength metastable phase[11] during highly nonequilibrium (i.e. very rapid) cooling exhibit transformation hardening and are said to be hardenable. Almost inevitably, such phases are the result of diffusionless, athermal massive shear transformations, typified by the martensite transformation found in steels. Materials capable of such transformations under

9 The two most popular stabilized grades of austenitic stainless steels are 321 with Ti and 347 with Nb additions.

10 Recall that modern practice of the most reputable producers of austenitic stainless steel routinely keep carbon content around 0.035–0.040, without compromising the strength, as in the past with L-grades.

11 A metastable phase is in relative equilibrium.

controlled conditions during heat treatment are prone to similar, but uncontrolled transformations during cooling following welding. Uncontrolled transformations, like anything uncontrolled in engineering, are rarely good and often cause problems. Thus, transformation-hardening alloys are prone to problems in the HAZs of welds.

Important steels prone to transformation hardening include (i) carbon steels, including medium- (0.3–0.6 wt%) and high- (0.6–1.0 wt%) carbon steels, (ii) the so-called Q&T (quenched and tempered) steels, (iii) low-alloy (less than 4 wt% total alloy content, including C) steels, including the so-called HSLA steels, (iv) high-alloy (typically greater than 10 wt% total alloy content) steels, including heat-resisting grades and tool steels, and (v) martensitic stainless steels, as this group probably represents more than half of all the metals and alloys welded each year, around the world. Problems that arise have serious economic impact and, if unresolved, can have even more serious consequences.

Fusion welding exposes steels to much more severe conditions in the HAZ than normal heat treatment in several important ways, as follows:

- Peak temperatures encountered in the HAZ during fusion welding are much higher than those encountered during heat treatments (around 1400 °C/~2550 °F near the FZ versus 950–1000 °C/~1750–1850 °F maximum during homogenization and spheroidization annealing), with possible consequences of (i) partial melting by constitutional liquation around carbides of Cr, Mo, V, or W; (ii) severe grain coarsening; and (iii) reaustenization; the latter two facilitating formation of brittle, untampered martensite upon cooling.
- Much faster rates of heating and much shorter retention times at peak temperatures with fusion welding accentuate nonequilibrium effects, increasing the effective transformation temperatures (i.e. lower critical temperature A_{C1} and upper critical temperature A_{C3} on the $Fe{-}Fe_3C$ metastable phase diagram) and hindering diffusion-controlled transformation of ferrite + pearlite to austenite with possible consequences of nonhomogeneous austenite leading to localized high-carbon martensite and widely scattered hardness in the HAZ.
- Post-weld cooling rates without preheat that exceed the steel's critical cooling rate (CCR) and, so, often lead to the formation of as-quenched, untempered brittle martensite, with the possibility of hydrogen-induced cold cracking as a consequence.

Figure 11.9 schematically illustrates the various microstructural zones found in a typical hypoeutectoid steel (here, an AISI 1018 with nominal 0.18 wt% C). When the carbon equivalent (C_{eq})[12] of a steel exceeds around 0.45 for a plain carbon grade or 0.35 for a low-alloy grade, the rate of cooling under ambient (near- or below room temperature of 20 °C/68 °F) can result in the formation of martensite, with progressive ease in the intercritical HAZ (between 3 and 2 in the figure), fine-grain HAZ (between 2 and 1 in the figure), and coarse-grain HAZ (from 1 to the FZ in the figure). Two reasons for martensite forming most easily in the coarse-grained region are (i) cooling rate being fastest where the temperature gradient is steepest (see the plot of peak temperature at the left in Figure 11.9) and (ii) massive shear transformation of austenite to martensite being easier the

12 See Footnote 11 in Chapter 6 for empirical equations for calculating C_{eq} for plain C and low-alloy steels.

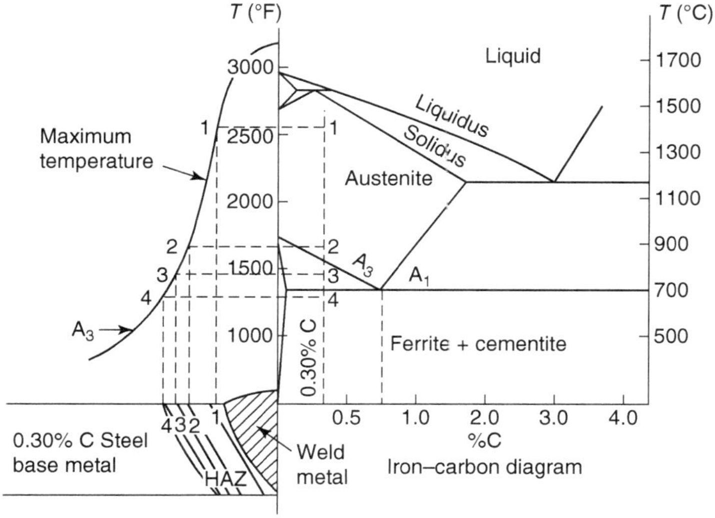

Figure 11.9 Schematic illustration showing the various microstructural regions of a typical mild steel, say AISI 1018 (at the left in the figure), showing correlation with the hypoeutectoid portion of the Fe–Fe_3C phase diagram (at the right in the figure). Source: Connor (1987). Used with permission of The American Welding Society.

coarser its grain size. As described in Section 10.4, martensite can also form in the PMZ (analogous to Zone 2 in the figure), and even in the fusion zone (Zone 1 in the figure), if the filler metal also has a suitably high value of C_{eq}.

Figure 11.10 shows a cold crack propagating through the CGHAZ of a hardenable steel, after initiating at the toe of the weld crown.

Figure 11.10 Photomicrograph showing a hydrogen-induced cold crack through the coarse-grained region of the HAZ of a fusion weld made in steel. Source: Lucas (2000). Used with permission of TWI.

Although martensite can crack upon formation because of prevailing thermal stresses, the ∼+11% volumetric change as austenite transforms to martensite, and the inherent brittleness of as-quenched martensite in what is known as *quench cracking*, it is more likely to crack in the presence of hydrogen by what is known as *hydrogen-induced cold cracking*. The keys to preventing the latter are described in Section 6.4.

The *best* strategy when welding steels susceptible to martensite formation due to their inherent hardenability, as assessed using an appropriate equation for calculating the alloy's carbon equivalent C_{eq} (as per Footnote 12 in this chapter) is to employ preheat during welding. The goal of preheating the region around a joint to be welded is to slow post-weld cooling to below the steel's CCR above which 100% martensite will form. Typical preheat temperatures versus values of C_{eq} for a carbon steel are as follows:

Value of C_{eq}	Suggested preheat temperature
Up to 0.45	Optional (if no history of cracking)
0.45–0.60	200–400 °F/95–205 °C)
Above 0.60	400–700 °F/205–385 °C)

Techniques for preheating include (i) resistance-heated blankets, (ii) induction-heating using coils, (iii) high-intensity quartz lamp heating, and (iv) localized oxy-fuel gas flame heating with burners or torch(es). Although thermocouples attached to the joint area can be used to monitor temperature, a more practical method generally uses temperature-indicating markers (e.g. Tempilstik®).

Figure 11.11 illustrates the effectiveness of preheating in preventing martensite formation in the HAZ of a steel, here an AISI 1040 steel.

Although no one – certainly no manual welder – would disagree that welding on joint elements that are preheated is quite literally a pain, beyond being an inconvenience, preheating is often mandated by structural codes (e.g. ASME Boiler & Pressure Vessel Code). Whether mandated by code or not, if problems with cold cracking have been encountered before with a particular material or weldment, preheating is, by far, the most effective method of preventing

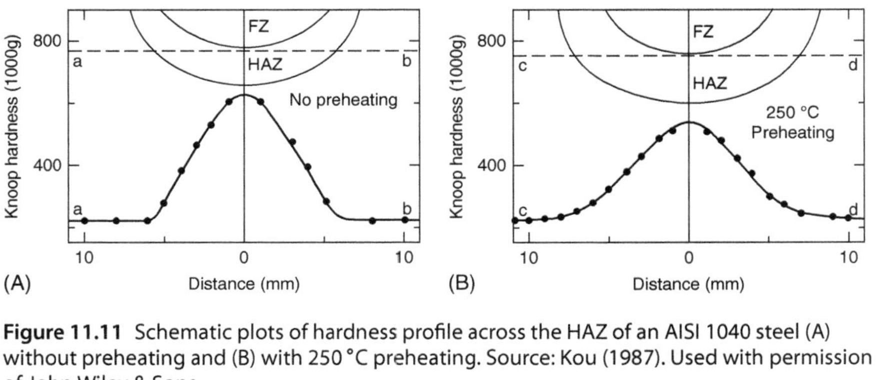

Figure 11.11 Schematic plots of hardness profile across the HAZ of an AISI 1040 steel (A) without preheating and (B) with 250 °C preheating. Source: Kou (1987). Used with permission of John Wiley & Sons.

the problem. However, if, for whatever rationalization (as opposed to reason), use of preheating is shunned, when there is *even a chance* that martensite formed in the HAZ, PMZ, or FZ during fusion welding, the weldment should be post-weld tempered. Tempering is intended to take the edge off the brittleness of as-quenched martensite[13] in order to prevent hydrogen-induced cold cracking. Specific temperatures and times, usually slow and low favored over high and fast, vary for different classes of steels (e.g. carbon steels, low-alloy steels, Q&T steels, and tool steels). This said, some types of steel (e.g. Q&T grades and low-alloy steels containing additions of Cr, Mo, V, and/or W) are subject to embrittlement during tempering (e.g. temper embrittlement, tempered martensite embrittlement, and blue brittleness) in certain mid-temperature ranges to the best practice temper around 1000–1100 °C/~1800–2000 °F. Hence, precaution is needed. Appropriate references (e.g. *American Welding Society's Welding Handbook*, 9th edition, Volume 4 Materials and Applications, Part 1 or *ASM International's Metals Handbook*, Volume 6, Welding, Brazing and Soldering) should be consulted for specific recommendations.

In summary, making excuses for why preheating during welding and/or heat treatment to temper (or, in some cases, to stress relieve) immediately following welding (i.e. post-weld heat treating) cannot be done are just that, i.e. excuses! Although each adds to process complexity, adds to production time, and costs money, one or the other or both are often prudent, if not essential, for avoiding more catastrophic losses that can cost lives beyond money and out-of-production times that cost orders of magnitudes more than fabrication delays.

11.6 Reheat Cracking

Cracking sometimes occurs during the post-weld heat treatment (PWHT) of weldments. As such, this particular *discontinuity* (among the types listed in Table 3.2) is unlike any that has been discussed so far in this book, other than it being a form of cracking associated with fusion welding, as discussed in Chapter 6. Unlike hot cracking, including solidification cracking (discussed in Section 6.3) and liquation cracking (discussed in Section 10.3), which both occur during the process of welding, or cold cracking, which may occur as soon as welding has been completed or after some delay (as discussed in Section 6.4), this form of cracking only occurs after welding if the weldment is subjected to PWHT. Then, when it occurs, it always occurs by or with an intergranular mode. As always, an intergranular fracture mode indicates an abnormality for metallic materials in which the grain boundary area is weaker than the bulk of the grain, with "weaker" indicating either less strong or less tolerant of strain (i.e. more brittle). The phenomenon is known as *reheat cracking*, in obvious reference to its occurrence during the reheating of a weldment comprised from a susceptible

13 Most references state the purpose of tempering as being to soften the steel. That is, usually *not* the goal. Why would anyone want to soften a hardened steel knife or sword blad, when it is hardness that keeps the blade sharp? The usual goal is actually to relieve brittleness enough to prevent fracture under flexing or impact, as per a knife or sword.

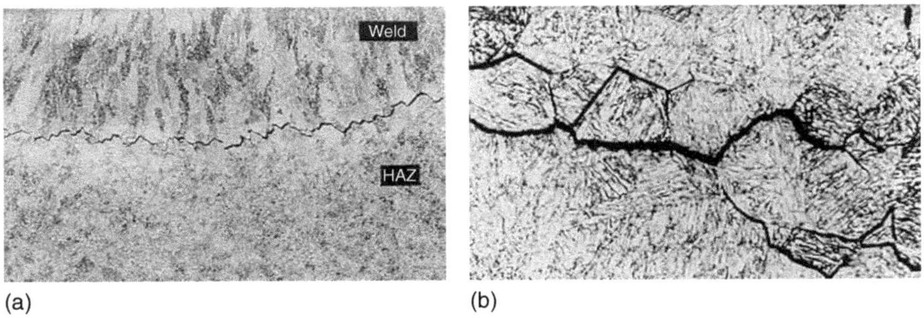

Figure 11.12 Photomicrographs showing reheat cracking associated with the coarse-grained heat-affected zone (CGHAZ) of a fusion weld in susceptible low-alloy steel (a) and close-up micrograph showing the intergranular mode of propagation and branching along prior austenite grains in the martensitic structure of the zone (b). Source: TWI (2000). Used with permission of TWI.

alloy. It is alternatively known as *strain-age cracking*, which is actually a tip-off to the operative mechanism for the discontinuity.

The principal cause of reheat cracking or strain-age cracking is that when post-weld heat treating susceptible alloys, the grain interiors (i.e. bulk) becomes strengthened by distributed (i.e. general) precipitates, forcing relaxation of residual stresses by creep deformation at grain boundaries. The presence of any impurities that segregate to grain boundaries or carbides that preferentially form in grain boundaries and promote embrittlement increases the susceptibility to this form of cracking. Joint design can also have an influence, with joints likely to contain stress concentration (e.g. partial-penetration welds) being more likely to initiate cracks. Last, but by no means least, the welding procedure has an influence, with large weld beads being undesirable, as they produce coarse columnar growth structure within the weld metal and a more pronounced CGHAZ, which is more susceptible to reheat or strain-age cracking.

Cracking usually, but not always, initiates in the CGHAZ and propagates by an intergranular mode into the regions unaffected by welding heat (i.e. the UBM).[14] Cracking may be in the form of coarse macrocracks or colonies of microcracks. A macrocrack appears rough, often with branching, following the CGHAZ and always parallel to the direction of welding.

Figure 11.12 shows photomicrographs of reheat cracking in the CGHAZ of a fusion weld in steel (a) along with propagation and branching by an intergranular mode along prior austenitic grain boundaries of the martensitic structure (b).

Reheat/strain-age cracking is known to occur a wide variety of alloys, including

- Low-alloy steels containing Cr–Mo (e.g. AISI/SAE 41xx) and, more so, Cr–Mo–V, including many construction grades (e.g. A508 Class 2, A514 Grade F, A517 Grade F, A533 Grade B, P23 pipe steel) … with PWHT at 350–550 °C/662–1022 °F
- HSLA-80 and HSLA-100 … with PWHT as for above

14 In steels susceptible to reheat cracking, cracks will propagate along prior austenite grain boundaries in the coarse-grained region, should martensite have formed on post-weld cooling.

- Steels containing Mo–V or Mo–B … as for PWHT as above
- Ferritic creep-resisting steels … as for PWHT as above
- 17-4 PH stainless steel (due to Nb content)
- Maraging steels (200, 250, 300, and 350 grades) because of Ti and Al
- Austenitic stainless steels 316L(N) and stabilized grades 321 (with Ti) and 347 (with Nb) … with PWHT at 500–700 °C/932–1292 °F
- Ni-based age-hardening superalloys containing Al + Ti exceeding about 6 wt% (e.g. Inconel 713, Inconel 718/Alloy 718, Incoloy 800/Alloy 800, IN-100, Udimet 500, 600 and 700 … with PWHT at 590–980 °C/1100–1800 °F.

Although the susceptibility to strain-age cracking in heat-treatable Ni-based superalloys is well known to progressively increase as the Al + Ti content (needed to form precipitates of γ' Ni_3Al) increasingly exceeds 6 wt%, the risk of reheat cracking in steels cannot be reliably predicted from chemical composition, although various indices have been suggested. The three most popular are as follows:

1. $\Delta G1 = 10C + Cr + 3.3Mo + 8.1\,V - 2$
2. $P_{SR} = Cr + Cu + 2Mo + 10\,V + 7Nb + 5Ti - 2$
3. $R_S = 0.12Cu + 0.19S + 0.10As + P + 1.18S + 1.49Sb$

Steels that have a value of $\Delta G1 < 2$, $P_{SR} < 0$, or $R_S < 0.03$ are less susceptible to reheat cracking.

The risk of reheat cracking in steels, ferritic heat-resisting steels, and austenitic stainless steels can be minimized by the following:

- Producing the maximum refinement of the CGHAZ by keeping net linear heat input as low as possible or, for the low-alloy steels and ferritic heat-resisting steels, applying a thin buttering layer using a small-diameter (3.2 mm/1/8-in.) electrode, then completing welding using a larger diameter (4–4.8 mm/5/32 3/16-in.) electrode to refine any remaining CGHAZ under the buttering layer
- Reducing or eliminating the CGHAZ by employing a temper-bead technique
- Limiting the degree of austenite grain growth by using a low heat input, while considering preheat to avoid hydrogen-induced cold cracking from martensite formation
- Employing joint design and welding technique that minimize localized stress concentrations from notches apt to occur with the use of a backing bar, partial penetration welding that can leave a root imperfection, lack of sidewall fusion, or a weld with a sharp toe
- Employing special two-stage PWHTs (as recommended by alloy manufacturers).

For age-hardening Ni-based superalloys susceptible to strain-age cracking, risk can be minimized by

- Selecting a base alloy with Al + Ti content below 6 wt%
- Preweld overaging via multistep overaging processes or by slow cooling (versus quenching) from the solutionizing temperature

- o Use of carefully selected – and coordinated – solutionizing and aging treatment cycles for particular alloys (by alloy manufacturer recommendations)
- o Using vacuum or inert gas atmosphere during PWHT to prevent oxygen pickup and embrittlement along grain boundaries
- o Use of low-restraint joint designs.

In summary, reheat cracking in low-alloy and ferritic heat-resisting steels or strain-age cracking in Ni-based alloys and stabilized grades of austenitic stainless steel (i.e. 321 and 347) is insidious. Best practice for avoiding problems from the residual stresses that are inevitably introduced in and around fusion welds in weldments and welded structures, especially as joint thickness increases over 1½-in./38 mm, recommends PWHT for stress relief. Yet, certain low-alloy and ferritic heat-resisting steels are susceptible to reheat cracking during such treatments. Even more insidious, Ni-based superalloys widely used for heat engines (e.g. steam and, especially, land-based and aircraft gas turbines) require post-weld aging to attain optimum properties for service, yet are prone to strain-age cracking during PWHT of alloys with Al + Ti exceeding about 6 wt%. Finally, when stabilized grades of austenitic stainless steels (i.e. 321 and 347) are called for to avoid sensitization during welding and retain resistance to corrosion in service (see Chapter 16), strain-age cracking again becomes a problem. The answer is be aware of the risks and act accordingly.

11.7 Troubleshooting Problems in the HAZ of Fusion Welds

Table 11.1 lists the various ways problems can arise in the fusion zone of fusion welds.

Table 11.1 Troubleshooting problems with HAZs.

HAZ Softening in Precipitation-Hardening Alloys:	
Loss of strength/hardness	Prevent in Al alloys with Friction Stir Welding (FSW)
	Prevent in Ni alloys and PH SS by using high-energy-density EBW, LBW, PAW
	Recover by employing
	A + W + PWA sequence, second best
	ST + W + PWA, best
Sensitization in HAZ of Austenitic Stainless Steels:	
Form grain boundary $(Cr,Fe)_{23}C_6$	
Deplete Cr in matric near gb's	
Render susceptible to IG corrosion	
	Reduce C content with L grades
	Use stabilized 321 or 347
	Keep net linear heat input low
	Favor higher energy-density processes

(Continued)

Table 11.1 (Continued)

If possible	ST@1000–1100 °C/1800–2000 °F 1 hr. per inch or 25 mm thickness + quench in water or air
HAZ Embrittlement in Transformation-Hardening Steels:	
As-quenched martensite forms in CGHAZ, then FGHAZ, then intercritical region	Preheat to prevent
	PWHT to temper, if unsure
Reheat Cracking in HAZ of Some Steels (Per Text):	
	Check values of $\Delta G1$, P_{SR}, R_S Refine CGHAZ by – keeping heat input low – using temper beads – using notch-free joints – employ two-step PWHT
Strain-Age Cracking in HAZ of Age-Hardening Alloys:	
Heat-treatable Ni alloys	Select Ni alloys with Al + Ti < 6 wt%
Maraging steels	
17-4 PH stainless steel	
316L(N), 321, 347 stainless	
	Preweld overage (OA) Coordinate ST and aging cycles Use vacuum or inert gas for HT Use low-restraint joint designs

References

Almubarak, A., Abuhaimed, W., and Almazrouee, A. (2013). Corrosion behavior of the stressed sensitized austenitic stainless steels of high nitrogen content in seawater, figure 10. *International Journal of Electrochemistry* 2013: 970835, 7 pages.

Chakrapan, D.G. (1992). Intergranular corrosion/cracking of a stainless steel pipe reducer section in bleached pulp stock service, figure 1. In: *Handbook of Case Histories of Failure Analysis* (ed. K.A. Esakul), 164–167. Materials Park, OH: ASM International.

Connor, L.P. (1987). *Welding Handbook*, Volume 1: Welding Technology, 8e. Miami, FL: American Welding Society.

Kou, S. (1987). *Welding Metallurgy*. New York, NY: Wiley.

Lucas, B. (with help from Gene Mathers and David Abson) (2000). Defects – hydrogen cracks in steels – identification, figure 2. Job Knowledge 45. Altered from the original in *Connect*, January/February 2000. www.twi-global.com.

Messler, R.W. Jr. (1999). *Principles of Welding: Processes, Physics, Chemistry, and Metallurgy*. New York, NY: Wiley.

TWI (2000). Defects/imperfections in welds – reheat cracking, figure 1a,b. Job Knowledge 48. Altered from the original in *Connect*, July 2000. www.twi-global.com.

Further Reading

Kou, S. (2002). *Welding Metallurgy*, 2e. New York, NY: Wiley-Interscience, Wiley.

Lippold, J.C. (2014). *Welding Metallurgy and Weldability*, 1e. New York, NY: Wiley.

Messler, R.W. Jr., (2004/1999). *Principles of Welding: Processes, Physics, Chemistry, and Metallurgy* (Chapter 16). Weinheim/New York, NY: Wiley-VCH/Wiley-Interscience, Wiley.

12

Unaffected Base Metal Cracking Associated with Welding

12.1 Weld-Related Problems in the Unaffected Base Metal

With the usual descriptor "unaffected base metal," one would not expect there to be a chapter about problems due to welding in such a region of a weldment. However, there are actually some, the three most common being (i) lamellar tearing, (ii) stress corrosion cracking (SCC), and (iii) fatigue cracking. Each of the three is, properly, weld related, as each occurs under the influence of weld-induced stresses. If lamellar tearing is to occur, it will occur as soon as the weld has been made and thermally induced volumetric shrinkage and coefficient of expansion (CTE) contraction stresses develop to a level sufficient to pull the base material apart in its short transverse, thickness direction. SCC and fatigue cracking, on the other hand, each takes time; the former requiring enough time for the prevailing corrosive service environment to act to cause material degradation, and the latter requiring enough time for fluctuating service loads to cause enough microscopic damage to accumulate that a crack is initiated and then grows, perhaps leading to an ultimate fracture by overload.

Lamellar tearing only occurs in base metals, overwhelmingly a thick-section, hot-rolled steel plate containing defects. SCC only occurs in materials susceptible to this form of corrosive attack in certain corrosive environments. Weld-related fatigue cracking, on the other hand, can – and does – occur in any material, if conditions in the weldment and loading environment are right.

Let us consider each in the following sections.

12.2 Lamellar Tearing in Thick Steel Weldments

Some metals and alloys contain excessive amounts of nonmetallic inclusions aligned in the longitudinal direction by rolling or other hot deformation processes (e.g. as stringers). The most notable example is wrought iron, which routinely consists of around 25 by weight of fibrous silicate stringer inclusions in very low carbon (less than 0.08 wt%) ferrite (Figure 12.1a) and was used in the construction of the world-famous Eiffel Tower in Paris. However, the condition can also be found in poorly made mild and low-alloy steels, especially in thick,

A Practical Guide to Welding Solutions: Overcoming Technical and Material-Specific Issues,
First Edition. Robert W. Messler, Jr.

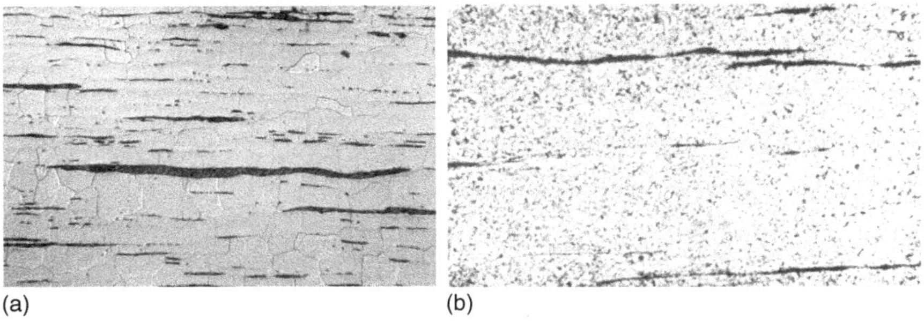

Figure 12.1 (a) An optical micrograph of wrought iron showing typical numerous, large fibrous nonmetallic silicate inclusions that impart toughness to the soft, low (<0.08 wt%) carbon ferrite matrix. Source: (a) Meier (2008). Published by and used with permission of ASM International. (b) An optical micrograph showing lamellar defects consisting of Mn–sulfide stringer inclusions in steel pipe, both at 50× magification. Source: Hutchings and Unterweiser (1981a). Used with permission of the ASM International.

hot-rolled plate product (Figure 12.1b), with stringers of either Mn silicates or sulfides or both. Such materials can exhibit *lamellar tearing* in welded structures as the result of the combination of high localized stresses due to weld contraction and low ductility of the base metal in its through-thickness (i.e. short transverse) direction parallel to the plate joint element's rolling direction because of the presence of elongated stringers of inherently weak and brittle nonmetallic inclusions. Tearing (or cracking) is triggered by decohesion of these inclusions, which causes tears or cracks to follow grain boundaries (i.e. to occur with an intergranular morphology). The problem always occurs outside the fusion zone (FZ) and often outside the heat-affected zone (HAZ), depending on at what location thermally induced shrinkage and contraction (i.e. tensile) stresses peak.

Figure 12.2 shows a photomicrograph of lamellar tearing beneath a fusion weld. Figure 12.3 shows a photograph of the BP Forties offshore drilling platform in

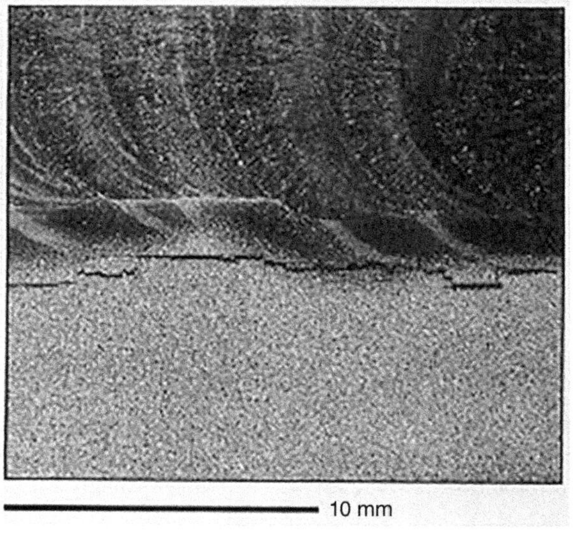

Figure 12.2 Photomacrograph showing lamellar tearing beneath a fusion weld, near but just outside the HAZ in steel. Source: Farrar and Dolby (1971). Used with permission of TWI.

Figure 12.3 Photograph showing a portion of the steel support structure of an offshore platform in the BP Forties field in the North Sea in which lamellar tearing was encountered during attempts to repair weld caused because of a lack of root penetration defects in brace welds. Source: TWI (2000). Used with permission of TWI.

which lamellar tearing occurred during repair welding because of the lack of penetration in a brace weld.

Two practical ways of avoiding lamellar tearing during fusion welding of thick-section susceptible base metals (given the general impracticality of changing to a less "dirty" material) are to (i) employ joint designs that cause contraction stresses to act along rather than across the short transverse direction or (ii) grinding or machining away the volume of the joint element where tearing is anticipated and "buttering"[1] the cutaway portion with a softer filler metal better able to accommodate strain from contraction stresses. Figure 12.4 schematically illustrates several examples of how susceptible joints can be reconfigured to be less susceptible to lamellar tearing.

There are other defects (e.g. shrinkage cavities or pipes in ingots flattened but not healed by rolling or forging, forging laps, etc.) that can lead to delamination of thick-section plate or other rolled forged or extruded shapes under through-thickness tensile stresses, but these tend to be rare.

Although far less likely to lead to tearing analogous to classic lamellar tearing, so-called *banded microstructures* of ferrite and pearlite in mild steels and even low-alloy steels (Figure 12.5) can also lead to problems in unaffected base metal, whether related to welding-induced stresses or not. Excessive banding can greatly reduce through-thickness, short-transverse strength. If nothing else, such a microstructure indicates that thermomechanical processing at the mill was not well controlled and executed.

---

1 *Buttering* is the addition of material, by welding, on one or both faces of a joint, before the preparation of the joint for final welding, for the purpose of providing a suitable transition weld deposit for subsequent completion of the weld. Often used to limit the pickup of base metal into the weld metal in the joint by the dilution of the latter by melted volumes of the former, it is also used to provide strain accommodation better than provided by either the base material or the weld metal. Check "Buttering" under Google "All" and "Image" searches.

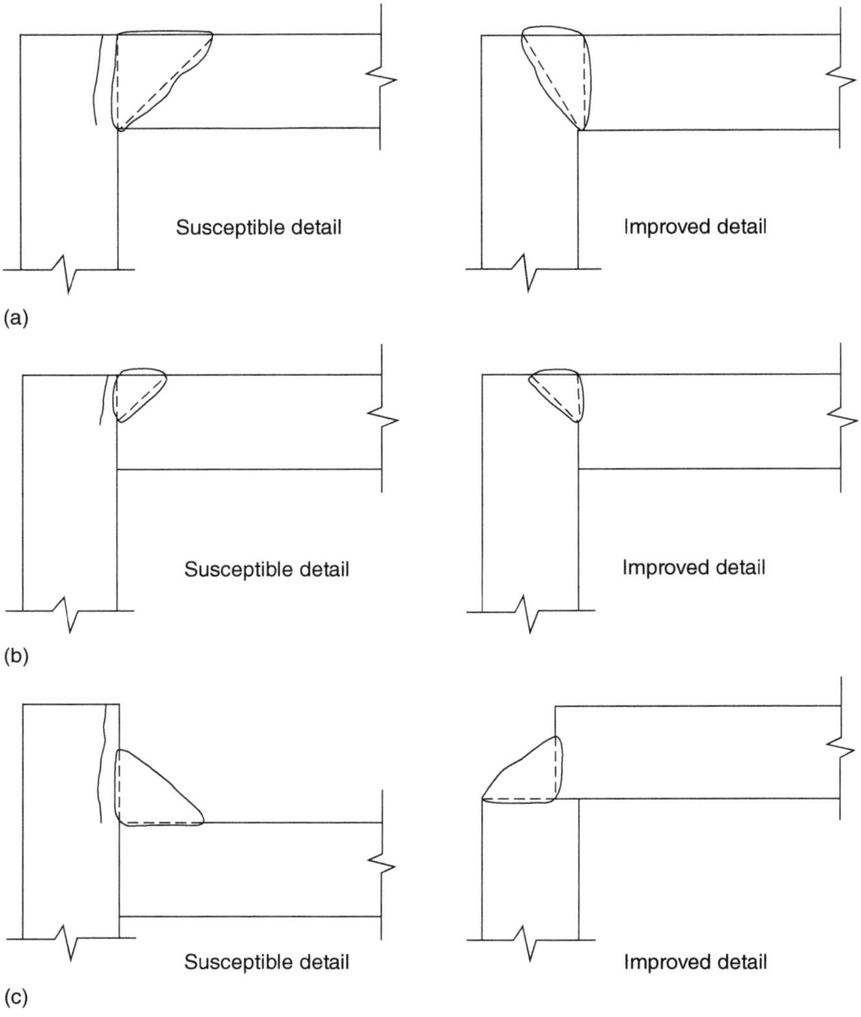

Figure 12.4 Schematic illustrations showing several weld joint details susceptible to lamellar tearing in one of the joint elements, along with suggested improved details for avoiding such defects. Source: Moyer (2012). Used with permission of the American Institute of Steel Construction.

12.3 Corrosion Cracking Caused by Fusion Welding

Corrosion is the natural process by which metals achieve a more stable state[2] by the chemical process of oxidation in which they give up (or lose) one or more

2 All elements in the periodic table of elements seek to achieve a stable electron configuration (like the inert gases) in which their outermost occupied level or shell of electrons is filled as per the rules of quantum physics. As metallic elements have one, two, or three electrons in the outermost occupied shell of their atoms, they seek to lose these to form +1, +2, or +3 ions during the chemical process of oxidation.

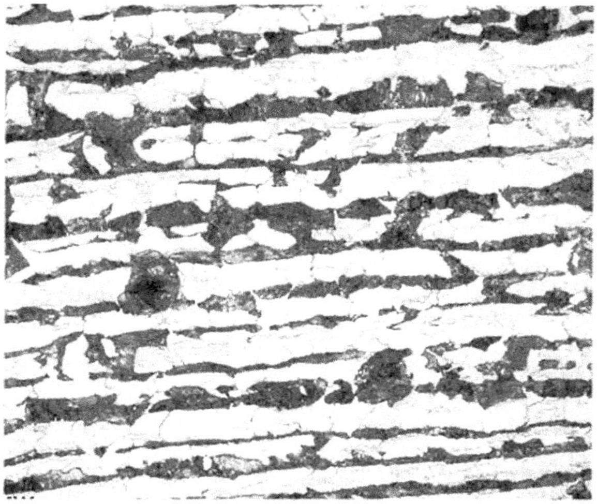

Figure 12.5 Photomicrograph showing banding of soft, low-carbon (<0.02 wt%), and high-carbon (0.77 wt%) pearlite in a 0.25 wt% C, 1.5 wt% Mn hot-rolled mild steel plate subjected to improper thermomechanical processing at the mill and susceptible to delamination failure with through-thickness loading. Source: Grange (1971). Used with permission of SpringerLink.

electrons from their atomic form to become positive ions in compounds (e.g. oxides, sulfides, carbonates, or other corrosion products). The gradual destruction of materials, usually metals, by chemical or electrochemical reaction with the environment is corrosion, with classification into eight forms: (i) uniform (or general) corrosion, (ii) galvanic (or bi-metal) corrosion, (iii) crevice corrosion, (iv) pitting corrosion, (v) intergranular corrosion, (vi) selective leaching, (vii) erosion corrosion, and (viii) stress corrosion or SCC.

All eight forms of corrosion are accelerated by two conditions that add energy to the chemical energy needed to cause the oxidation process: (i) elevated temperature (which provides thermal energy) and (ii) stress (which provides mechanical energy). However, one form only occurs in the presence of stress, i.e. SCC.

SCC involves the formation and intergranular growth of cracks in a corrosive environment, when no such formation or growth would normally occur. The process can lead to unexpected sudden (i.e. catastrophic) failure of normally ductile metals subjected to a tensile stress, especially when temperature is elevated. Initiation is always delayed, with crack growth progressing slowly (10^{-9}–10^{-6} m/s) until the crack reaches a critical size and could become unstable and result in catastrophic fracture. The stress required for the process to occur can be external in origin (i.e. directly applied) or internal (i.e. residual).

Figure 12.6 provides photomicrographs that show the typical torturous branching (in a) and intergranular morphology (in b) of SCC. Figure 12.7 shows areas of a formed and welded pipe subject to SCC.

SCC is highly chemically specific, occurring only in certain alloys (more than in pure metals) only when exposed to a limited number of chemical environments containing only very small concentrations of certain highly active chemicals. With the environment that causes SCC for a given alloy often being only mildly corrosive, SCC can be severe, with the metal or alloy being filled with microscopic cracks but can still appear bright and shiny, making it common for the condition to go undetected before failure.

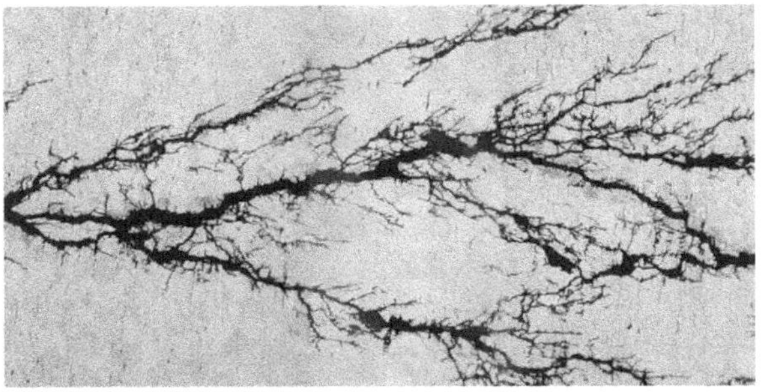

(a)

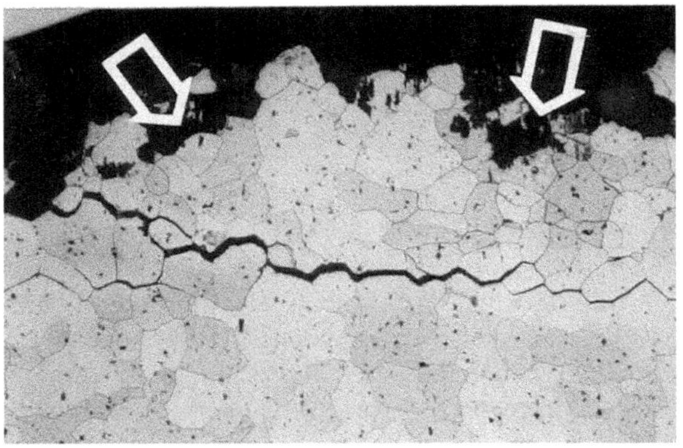

(b)

Figure 12.6 (a) SEM micrograph showing the unique river branching that allows identification of SCC, here in an austenitic stainless steel and (b) an optical micrograph (at 200× magification) showing the typical intergranular morphology of SCC, here in an Al alloy containing residual stresses and exposed to salt water. Source: McDaniels (1997). Used with permission of NASA.

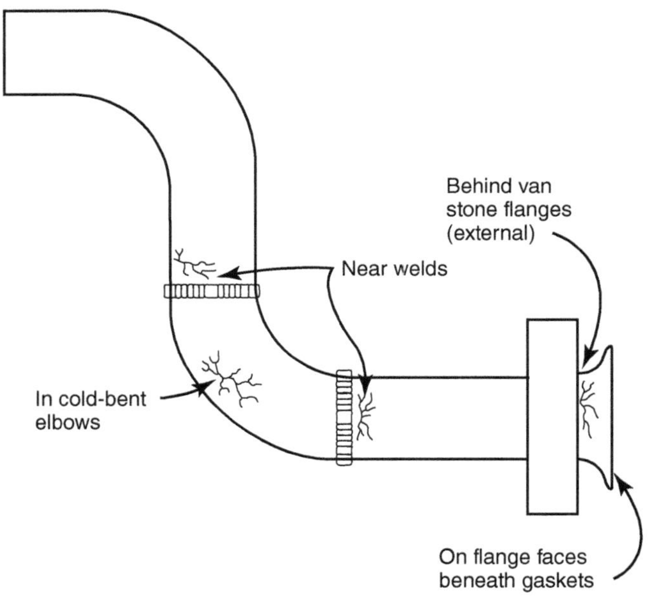

Figure 12.7 Schematic illustration showing typical locations vulnerable to stress corrosion cracking due to residual stresses from cold forming of bends or flanges or welding. Source: From the website www.flowgeeks.com in an article by Dr. Hira Ahluwalia, President of Material Selection Resources, Inc. Used with permission of Hira Ahluwalia.

Table 12.1 Alloy/environment systems that exhibit stress corrosion cracking.

Alloy	Environment
α-brass	Ammoniacal solutions (i.e. with HN_4^-)
Aluminum alloys	Aqueous halide solutions (Cl^-, Br^-, and I^-)
Austenitic stainless steels	Hot, concentrated chloride solutions; chloride-containing steam
Carbon steel	Hot nitrate, hydroxide, and carbonate/bicarbonate solutions
Copper alloys (some)	Natural atmospheres with SO_2, oxides of nitrogen, and NH_3 or HN_4^-
High-nickel alloys	High-purity steam
High-strength steels	Aqueous electrolytes, particularly when containing H_2S
Magnesium alloys	Aqueous Cl^- solutions
Martensitic stainless steels	Some alloys to some environments
PH stainless steels	Some alloys to some environments
Titanium alloys	Aqueous halide solutions (Cl^-, Br^-, and I^-); chlorinated solvents and fluids
Zirconium alloys	Aqueous Cl^- solutions, organic liquids, I_2 at 350 °C/600 °F

Table 12.1 lists alloy/environment systems known to exhibit SCC.

There is only one sure option for precluding SCC; i.e. avoiding susceptible alloy/environment combinations. Short of this, if a potentially susceptible alloy is to be fusion welded, thoroughly post-weld stress relieve all the welds. This said, there is still the chance for SCC from externally applied loads and direct tensile stresses.

12.4 Fatigue Cracking Outside Fusion Welds

No loading leads to more failures than fatigue loading. Many reputable sources state that it may account for 90% of all service failures from mechanical (versus thermal, wear, or corrosion) sources. Precisely accurate or not, the point is that fatigue is the cause of most mechanical failures, many of which are catastrophic as they injure or kill as final rupture occurs suddenly, with little or no warning. Fatigue has become progressively more prevalent as technology has developed a greater variety and amount of equipment and large structures, such as automobiles, aircraft, railroads, steel and lightweight aluminum ships, compressors and pumps, engines and turbines, bridges, etc., all of which are subjected to repeated loading and vibration.

Fatigue failures are characterized by the following:

- Repetitive fluctuating loading gives rise to stresses that cause microscopic damage to accumulate over time
- Initiation of fatigue damage overwhelming occurs at a surface,[3] typically at some stress-rising geometric or material notch, with Stage 1 occurring along

3 Although much rarer, fatigue cracks can initiate at the interior of a material, almost always at some stress-rising sharp flaw (e.g. crack and brittle inclusion).

slip planes that intersect the surface at an acute angle, producing steps by offset, and forming a crack in shear at stress levels only a fraction of the allowable yield stress for the material
- 80–85% of the time it takes for fatigue to lead to total failure by fracture or rupture (i.e. the part's fatigue life) is spent initiating a crack
- Once initiated, a crack grows or cracks grow perpendicular to the surface by a transgranular mode during Stage 2 with each opening cycle of loading, with propagation occurring in steps demarcated by striations that may or may not persist on the final fracture surface, depending on whether the material was too hard to allow plastic deformation to form striations or too soft, so that striations were rubbed off by wear
- In Stage 3, ultimate failure occurs by overload rupture once a crack has or cracks have propagated to the point that the remaining intact material cannot tolerate the next cycle of loading; with smaller overload areas indicating lower stresses during fatigue.

Fatigue tends to be divided into two regimes: (i) low cycle/high stress when the frequency of loading (i.e. cycles per unit time) is low, stresses are high, and failure occurs in less than 10^3 cycles and (ii) high cycle/low stress when the frequency of loading is high (typically more than several cycles per second), stresses are low, and failure occurs after more than 10^6 cycles. For high-cycle/low-stress fatigue in particular, for which crack growth rate tends to be slow, regular periodic inspection for evidence of cracking at surfaces is beyond advisable. It can literally be a lifesaver. The collapse of the I-35 W bridge across the Mississippi River into downtown Minneapolis, MN, on 1 August 2007 during 6 p.m. rush hour traffic killed 13 and injured 145 people. Evidence of fatigue cracking in the 40-year-old bridge was first detected just 10 years after it opened in November 1967. Unfortunately, no authority took these early warning signs of trouble as serious enough to investigate thoroughly enough to find the root cause and implement any remedial actions. Tell-tale signs were there!

Figure 12.8 schematically illustrates the path of a fatigue crack through a metal during Stage 1 and Stage 2. Figure 12.9 shows the characteristics of a fatigue fracture surface as a schematic representation (in a) and as an actual fractograph (in b). Figure 12.10 shows the striations that give evidence of fatigue when material properties are suitable.

Fatigue failures are not unusual with welded structures for two reasons. First, welds typically have geometric features that act as stress risers at the surfaces of parts (e.g. toes of fillet welds and edges of crown beads). Second, the various microstructural zones of fusion welds, with their typically widely differing mechanical properties, act as metallurgical notches. As stated in Section 6.5, fatigue cracks can also form in regions near a weld because of left-in residual stresses that assist in the formation and growth of such cracks in the presence of otherwise acceptable cyclic (fatigue) loading. Such cracks often initiate at the boundary between the FZ and HAZ or in a softened or embrittled region of the HAZ. However, as also stated, and as shown in Figure 6.9, they have been observed in locations beyond the HAZ where residual surface tensile stresses were produced by welding and not removed by post-weld stress relief.

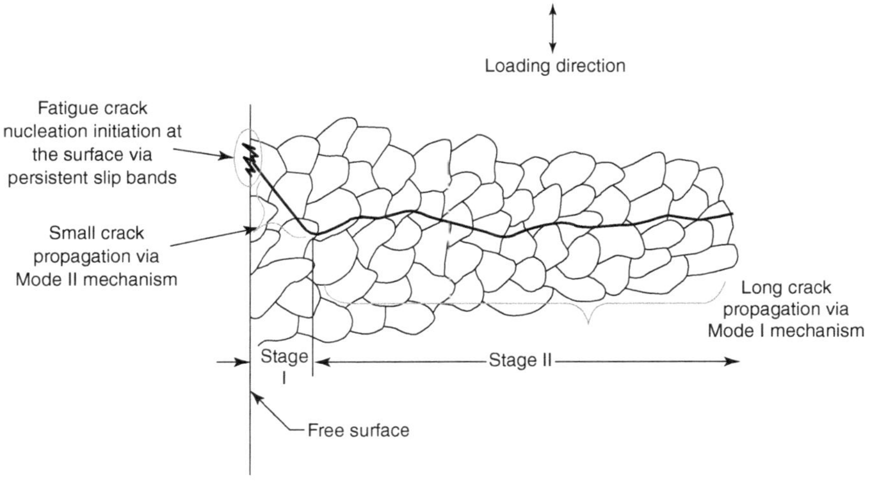

Figure 12.8 Schematic illustration showing the path of a fatigue crack during Stage 1 initiation via Mode II shear mechanism and Stage 2 propagation via Mode I transgranular mechanism. Source: Stephens et al. (2000). Used with permission of John Wiley & Sons.

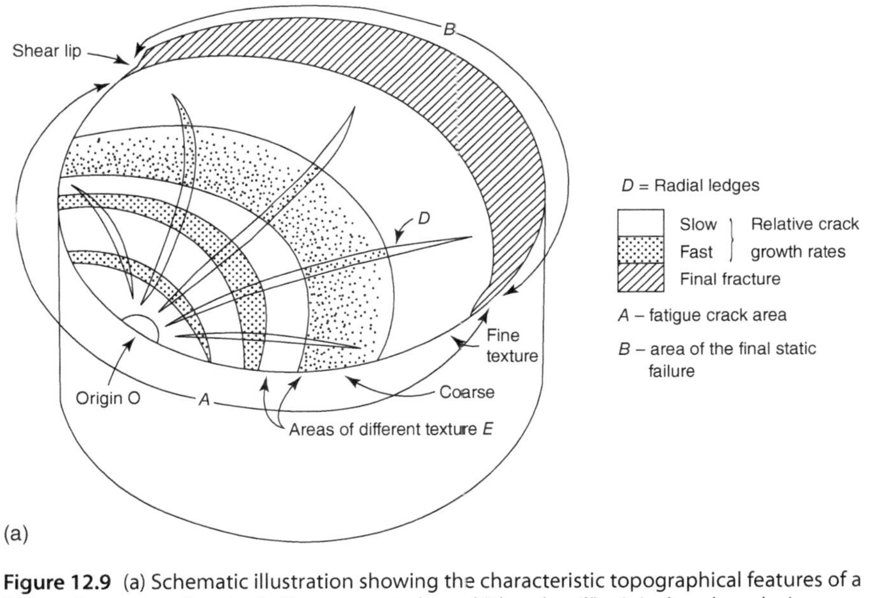

Figure 12.9 (a) Schematic illustration showing the characteristic topographical features of a fatigue fracture surface, including crescent-shaped "thumbnail" origin, beachmarks (as alternating concentric white and shaded regions), radial ledges, and terminal overload area of rupture (cross-hatched). Source: Collins (1981). Used with permission of John Wiley & Sons. (b) A fractograph showing comparable features in an actual fatigue fracture surface. Source: Hutchings and Unterweiser (1981b). Used with permission of the ASM International.

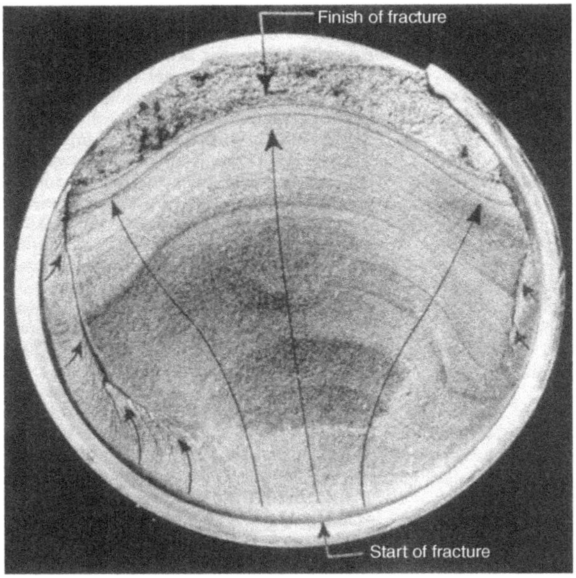

(b)

Figure 12.9 (*Continued*)

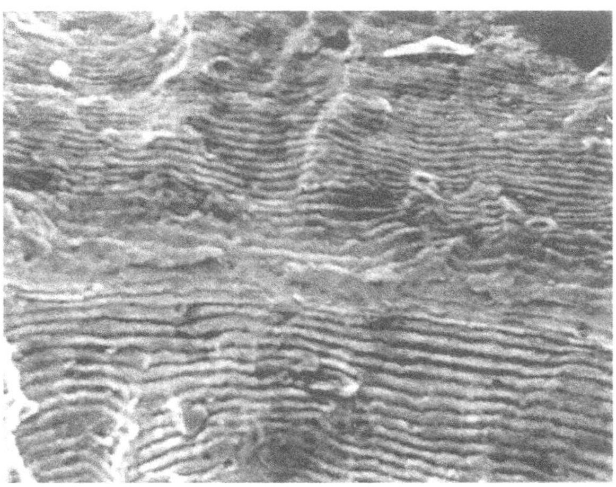

Figure 12.10 SEM micrograph showing typical fatigue striations demarcating individual crack advances under the opening cycle of loading. Source: Esakul (1992). Used with permission of ASM International.

Care should be given to developing welding parameters and procedures that produce proper crown bead and root bead profiles and contours. For the most demanding fatigue environments and performance, machine (or grind) crown and root beads produce smooth surfaces.

The best and safest option for lowering risk of fatigue around welds is post-weld stress relief heat treatment, but, short of this, localized hammer or shot peening

(a) (b)

Figure 12.11 (a) Photograph showing devastation after The Great Hanshin Earthquake in Kobe, Japan, on 17 January 1995, in which inherently stiffer structures suffered greater damage than more flexible structures, here collapse of a section of the Hanshin Expressway. Source: Photograph by Pan-Asia Newspaper Alliance, Singapore (no longer in existence). (b) A photograph showing fatigue cracking damage at welds at exterior fin beam-to-column connections of a three-story restaurant building damaged in the Tohuko Earthquake on 3 March 2011. Source: From Tohuko Japan Earthquake and Tsunami Clearinghouse, Tokyo (no longer in existence). Used with permission of the Earthquake Engineering Research Institute (EERI).

or planishing (all of which introduce some compressive stress to offset tensile stresses) is better than doing nothing and is often prudent.

Figure 12.11 shows how buildings and other large structures constructed using welding rather than bolting or riveting for assembly experienced worse fatigue damage during earthquakes in Japan, such as those occurred in Kobe on 17 January 1995 and in Tohuko on 11 March 2011. Similar results were seen in other earthquakes; the reason being more rigid welded joints cannot dissipate vibrational energy by friction.

12.5 Troubleshooting Weld-Related Problems in the Unaffected Base Metal

Table 12.2 lists the various ways problems can arise in the unaffected base metal of fusion welds.

Table 12.2 Troubleshooting problems in the unaffected base metal around fusion welds.

Lamellar Tearing in Thick-Section Plates or Extrusions:	
Wrought iron	
Steels with excessive nonmetallic inclusions	Limit inclusions
Severely banded steels	Restrict banding For all of the above, modify joints to avoid through-thickness tension
Stress Corrosion Cracking (SCC):	
Susceptible alloys/environments (per Table 12.1)	Avoid or Perform post-weld (PW) stress relief heat treatment (HT)
Fatigue Cracks and Failures:	
Related to notches around weld	Grind to smooth transitions at beads
Related to HAZ	Employ PW thermal stress relief (SR) and/or Consider mechanical SR treatments
Related to unaffected base metal (UBM)	Employ PW thermal stress relief and/or Consider PW mechanical SR treatments

References

Collins, J.A. (1981). *Failure of Materials in Mechanical Design: Analysis, Prediction, Prevention*. New York, NY: Wiley-Interscience Publishers.

Esakul, K.A. (1992). Failure of a flexible pump connector. In: *Handbook of Case Histories of Failure Analysis*, vol. 1 (ed. C. D'Antonio). Materials Park, OH: ASM International.

Farrar, J.C.M. and Dolby, R.E. (1971). *Lamellar Tearing in Welded Steel Construction*, 4. Cambridge, UK: The Welding Institute www.twi-global.com.

Grange, R.A. (1971). Effect of microstructural banding in steel. *Metallurgical Transactions* 2 (2): 417–426.

Hutchings, F.R. and Unterweiser, P.M. (1981a). Explosion in section of a 10-in. Steam main due to internal laminar defects. In: *Failure Analysis, The British Technical Report*. Materials Park, OH: ASM International www.asminternational.org.

Hutchings, F.R. and Unterweiser, P.M. (1981b). Fatigue Failures with Special Reference to Fracture Characteristics. In: *Failure Analysis, The British Engine Technical Report* (ed. G.A. Cottrell). Materials Park, OH: American Society of Metals.

McDaniels, S.J. (1997). Stress corrosion cracking and in An overview of fatigue and other metallurgical failure modes and analysis at the Kennedy Space Center. In: *Flammability and Sensitivity of Materials in Oxygen-enriched Atmospheres:*

Eighth Volume, ASTM STP 1319 (ed. W.T. Royals, T.C. Chou and T.A. Steinberg). ASTM http://www.corrosion.ksc.nasa.org.
Meier, M. (2008). File 231, from Wikimedia Commons (m610); with free use under Creative Commons Attribution – Share-Alike 3.0.
Moyer, S.M. (2012). Lamellar tearing and failure cases, figure 8-17, p. 8-23 of the 15th ed. In: *AISC Steel Construction Manual*. Chicago, IL: American Institute of Steel Construction http://www.failures.wikispaces.com.
Stephens, R.I., Fatemi, A., Stephens, R.R., and Fuchs, H. (2000). *Metal Fatigue in Engineering*, 51. New York, NY: Wiley.
TWI (2000). Defects – lamellar tearing. Job Knowledge 47, modified from the origin article appearing in *Connect*, May/June 2000. www.twi-global.com.

Further Reading

Macdonald, K. (2011). *Fracture and Fatigue of Welded Joints and Structures*, Woodhead Publishing Series on Welding and Other Joining Technologies, 1e. Woodhead Publishing (now part of Elsevier, Ltd).
Maddox, S.J. (1991). *Fatigue Strength of Welded Structures*, Woodhead Publishing Series on Welding and Other Joining Technologies, 2e. Woodhead Publishing (now part of Elsevier, Ltd).
McCafferty, E. (2010). *Introduction to Corrosion Science*, 1e. New York, NY: Springer Publishing Company.

13

Discontinuities in Multi-pass Welds

13.1 Needs for Multi-pass Welding and Welds

Whenever two or more passes are required to produce a weld joint, it is referred to as a *multiple pass weld* or, simply, *multi-pass weld*, and the process is called *multi-pass welding*. Joint configurations can be any of the five shown schematically in Figure 3.1 (i.e. butt, corner, tee, lap, or edge), any of which could require only a single pass or multiple passes. Fusion welding can be used for applying surfacing welds as shown in Figure 13.1d, which always requires multiple passes. Figure 13.2 shows photographs of multi-pass welds in a grooved butt joint (in Figure 13.2a) and as a fillet in T-joint (in Figure 13.2b).

From a metallurgical standpoint, there are a few other instances in which multi-pass welding is used: (i) to deposit a root pass before the fill pass (in what might otherwise be a single-pass weld), (ii) to deposit one or more build-up layers (if required) and one or more cushion or buffering layers beneath a surfacing weld intended to provide improved resistance to wear, i.e. under one or more hard-facing layers, and (iii) to deposit a buttering layer on joint surfaces before filling with either a single pass or multiple passes. The purpose of the *root pass*, regardless of the number of subsequent passes (Figure 13.3), is to ensure that a fusion weld penetrates completely through the backside of a plate, so as to avoid a lack of penetration or a lack of fusion defect and set a sound base for subsequent passes. The purpose of any *build-up layer* is to replace metal usually lost to wear but, possibly, to corrosion, whereas the purpose of *cushion or buffering layers* beneath a hard-facing layer is to render the hard, brittle wear-resisting overlay (of one or more layers) a greater tolerance to impact by strain accommodation (Figure 13.3a). The purpose of a *buttering layer* is to coat the faces of a joint element (or abutting elements) with a metal (e.g. pure nickel) or an alloy (e.g. 50Ni–50Fe and austenitic stainless steel) to prevent pickup of base metal into the fusion zone (FZ) filler metal to cause undesirable dilution (Figure 13.4b).

Such multi-pass welds serve any of several functions, as described in the following sections.

A Practical Guide to Welding Solutions: Overcoming Technical and Material-Specific Issues,
First Edition. Robert W. Messler, Jr.

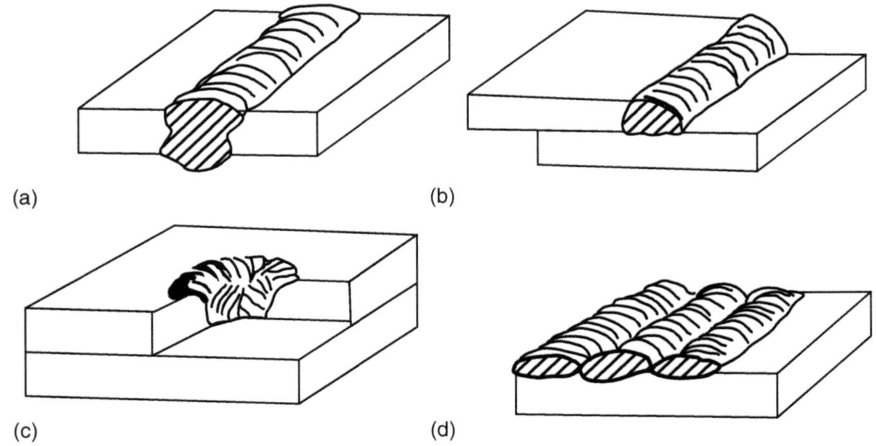

Figure 13.1 Schematic illustrations showing the fundamental types of welds, including (a) butt, (b) fillet, (c) plug, and (d) surfacing. Source: Messler (1993). Used under reciprocal agreement between Wiley and Elsevier for fair use.

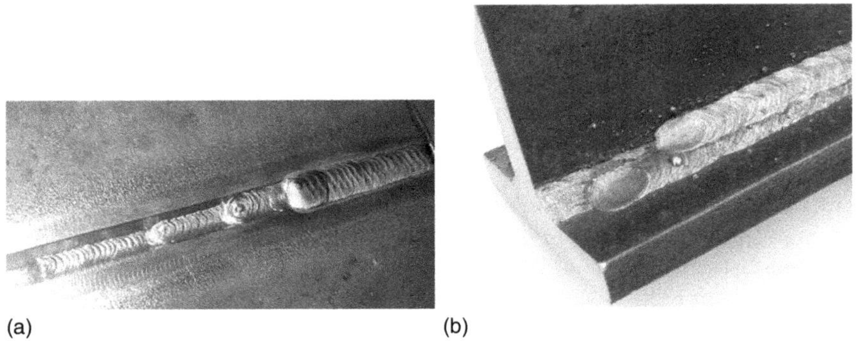

Figure 13.2 Photographs of (a) multi-pass butt weld (using gas tungsten-arc welding [GTAW]). Source: (a) Collier (2015); used with permission of Jody Collier. (b) Fillet weld (using GTMA). Source: Malcolm (2015). Used with permission of Malcolm.

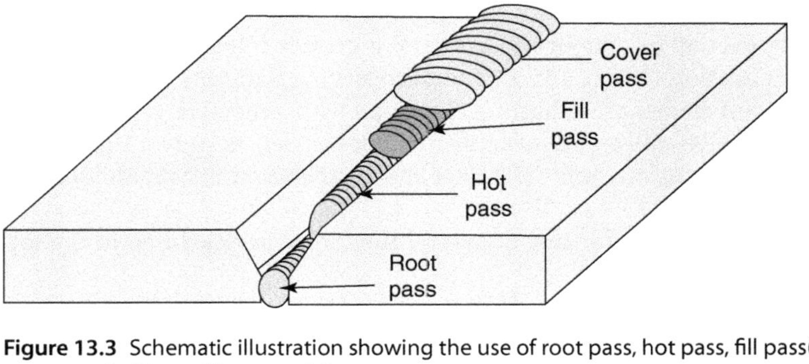

Figure 13.3 Schematic illustration showing the use of root pass, hot pass, fill pass(es), and cover pass(es) in multi-pass welding.

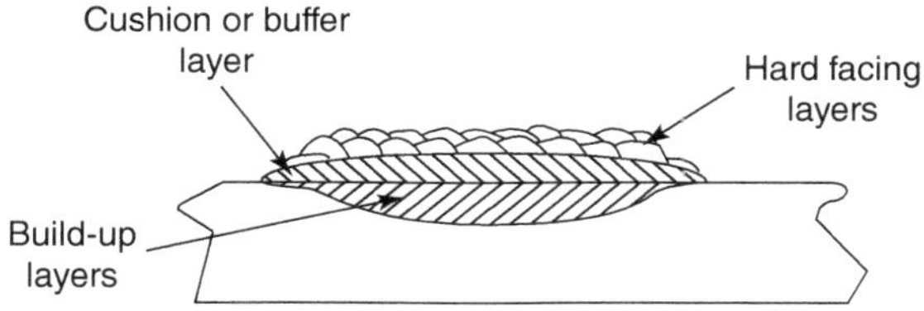

(a)

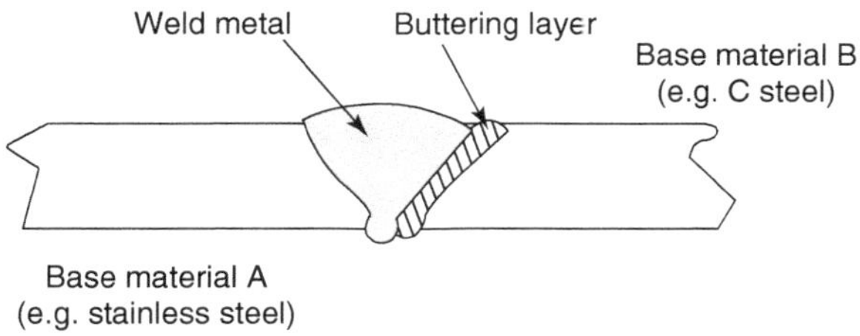

(b)

Figure 13.4 Schematic illustrations showing (a) the use of rebuild layer(s) and buffer layer(s) beneath hard-facing layer(s) and (b) buttering layer(s) on joint element(s) in dissimilar weld joints.

13.2 Various Functions of Multi-pass Welding and Welds

Multi-pass welds fulfill one or more of several functions, as follows:

- First and foremost, to produce a sufficiently large-volume weld to carry service loads within a design allowable of yield or ultimate strength, whether as a butt weld (Figure 13.5) or as a fillet weld (Figure 13.2b), as stress equals load/cross-sectional area

  and/or

- To provide sufficient filler metal/weld metal for thick-section groove welds (Figure 13.5)

  and in addition

- To allow a root pass, if required by code (e.g. ASME Boiler and Pressure Vessel Code) or military or company standard or past practice, followed by a hot pass, fill pass(es), and cap or cover pass(es) (see Figure 13.3)
- To prevent overheating during heavy-section joining by using multiple smaller volume passes deposited with lower net linear heat input; typically referred to as *stringer beads*
- To balance heat input in a thick section and, thus, minimize distortion from volumetric shrinkage and thermal contraction stresses with a one-sided weld using a double-groove joint preparation (Figure 3.3)
- To temper prior passes in a hardenable steel via use of a *temper bead technique*, as explained later in Section 13.4

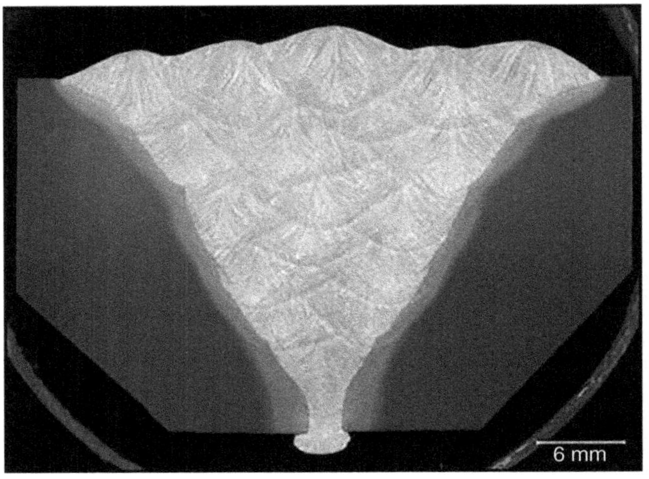

Figure 13.5 Macrograph of a single-sided multi-pass V-groove weld in thick-section steel. Source: Manning (2005). Used with permission of Edison Welding Institute.

- To restore any metal lost to wear (or corrosion) before overlay with a wear-resisting, hard-facing alloy or corrosion-resisting alloy cladding using *build-up layers* (Figure 13.4a)
- To provide a compressive pressure-accommodating *cushion or buffer layer* beneath a hard-facing overlay or surfacing weld (Figure 13.4a) or a buffer layer or layers beneath a corrosion-protective cladding
- To apply wear-resisting *hard-facing* alloys (Figure 13.7a) or corrosion-resisting *cladding* (Figure 13.6b)

(a) (b)

Figure 13.6 Photographs showing multi-pass weld application of (a) hard-facing alloy to wear-prone areas of a grader bucket. Source: From Clean Cut Metal Works, Houston, TX, USA. Used with permission of the owner and the expert welder Scott Raabe. (b) Plasma-transferred arc (PTA) deposits of a corrosion-resisting alloy cladding to protect vessel. Although the hard-facing layer is fancier than usual, it is functional for trapping soil or clay to protect the underlying metal from abrasive wear. Source: From Taskmaster Technologies, Inc. Leduc, AB, Canada. Used with permission of Maren Stevens, President and CEO.

Figure 13.7 Photograph showing an example of the use of shape welding for rapid prototyping or rapid manufacture of near-net shape 3D parts, often from difficult-to-process materials. Source: European Space Agency (2013). Used with permission of Professor Stephen Williams, Cranfield University.

- To overcome dilution of weld metal by melted portions of base metal(s), especially when joining dissimilar base metals/alloys (i.e. by *buttering*, as shown in Figure 13.4b)
- To manufacture complex shapes from difficult-to-fabricate alloys using 3-D rapid-manufacture *shape welding* (Figure 13.7).

Each of these creates different potential problems, as described in the rest of this chapter.

13.3 Defects Found in Multi-pass Welds

The various welding scenarios discussed above produce larger numbers and correspondingly larger total volumes of fusion zones and larger (more extensive) heat-affected zones (HAZs) than single-pass fusion welds. More importantly, they also produce complex metallurgical situations as the HAZ of a subsequent pass produces a HAZ in both a prior fusion zone (versus base material) *and* in the HAZ of a prior weld, that is, a heat effect in a cast (versus usually wrought) material and a heat effect in a previously heat-affected structure. This all confounds interpretation of potentially adverse effects.

As a consequence of multiple deposits of weld metal and/or complex metallurgical interactions, several potential problems can arise with multi-pass welding and welds, in an approximate order of frequency of occurrence, including

- Lack of fusion between abutting deposits or beads
- Entrapped slag between passes for slag-protected fusion welding process
- Cold cracking in the HAZ of some passes, especially, but not only, near the base metal
- Hot cracking in the FZ of individual passes

- o Liquation cracking in the HAZ of prior passes by subsequent abutting and overlapping passes
- o Gas porosity in individual passes
- o Property alteration in overlapping passes.

The origin and efforts to correct the various defects mentioned above are discussed herein. Discussion of property alteration in overlapping passes is discussed in the next section.

Lack of fusion (discussed in Section 3.4) occurs when molten metal in a weld pool fails to wet, melt some portion of the metal, and adhere (i.e. fuse) to abutting solid, whether at the fusion boundary of a single-pass weld or to a previously made, fully solidified pass in multi-pass welds. It can occur for any of several reasons given under lack of fusion in Section 3.4, as well as in Table 3.3 along with potential solutions.

When slag-protected fusion welding processes (e.g. shielded metal arc welding (SMAW), flux-cored arc welding (FCAW), submerged arc welding (SAW), electroslag welding (ESW)) are used to perform multi-pass welding, the opportunity exists for some slag to become trapped between the passes as a *nonmetallic inclusion discontinuity*. With multi-pass welding, this usually occurs because a hardened slag on a previously made pass is not thoroughly removed before the next pass or layer of passes is made (Figure 7.3a). Slag entrapment can occur for any of several reasons given in Section 7.2 and as summarized in Table 7.2, along with potential solutions in both places.

The opportunity for cracking exists with multi-pass fusion welds, just as it does for single-pass fusion welds. Both *hot cracks* in susceptible metals and alloys and

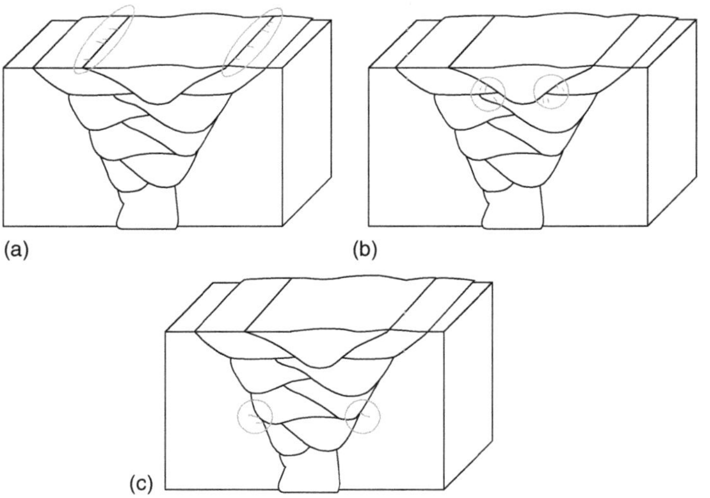

Figure 13.8 Schematic illustration of cracks in multi-pass welds in Alloy 617, a Ni-base Cr–Co–Mo heat-resisting alloy, including (a) transverse hot cracks in the top layer along the bead fusion lines, (b) hot cracks in reheated weld metal in upper layers, and (c) liquation cracks in the HAZ of the base metal. Source: Zinke and Fink (2013). Published by Springer and freely used as Wiley is a STM signatory publisher with SpringerNature.

cold cracks in hardenable steels can occur by the same mechanisms described in Sections 6.3 and 6.4, respectively. *Hot cracking* actually tends to be less problematic, as the preheating created by earlier passes reduces net thermal contraction stresses acting on later passes. *Cold cracking*, on the other hand, can be more severe as final residual stresses can be high and overlapping HAZs can result in greater susceptibility to martensite formation.

Figure 13.8 schematically illustrates the various types and locations of cracking that can occur in multi-pass welds, whereas Figure 13.9 shows a photomacrograph of cold cracks in a multi-pass weld in steel. Potential solutions for each type of cracking are described in Sections 6.3 and 6.4 and are summarized in Table 6.1.

The possibility, as well as the likelihood, of *liquation cracking* in the HAZ of susceptible alloys is often greater with multi-pass welds than with single-pass welds. The reason is the tendency for the partially melted zone (PMZ) and HAZ

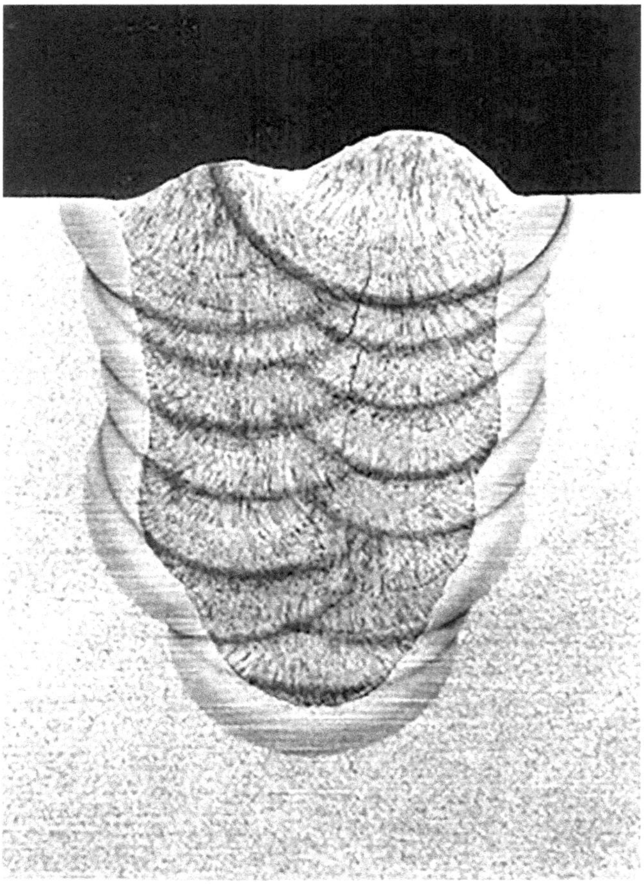

Figure 13.9 Macrograph showing hydrogen-induced cold cracking in a multi-pass weld in steel. Source: Pargeter (2003). Published by and used with permission of the American Welding Society.

at the interface with base metal to be wider (more extensive). The mechanism of liquation cracking, as well as possible causes and solutions, are discussed in Section 10.3 and are summarized in Table 10.2.

Gas porosity can occur in multi-pass fusion welds for the same reasons and from the same potential sources as in single-pass fusion welds. Porosity originating from improper removal of oxide or tarnish layers, as well as from hydrocarbon contaminants, from the surfaces of joint elements, is less problematic than for single-pass welds. However, for joints requiring multi-pass welding, initial cleaning is especially important as both material and labor costs typically high. Obviously, porosity arising from faulty consumables, equipment, technique, or workmanship is still possible. Once again, potential solutions are discussed in Section 5.4 and are summarized in Table 5.3.

13.4 Composition Adjustment with Multi-pass Welding

There are two common applications of multi-pass welding and welds to adjust composition from a base metal element to a final surfacing weld and two to adjust composition from a base metal element of one composition to another base metal element of a significantly different composition. The two former applications involve *buffer layers*, whereas the latter applications involve *buttering layers* and *transition layers*, respectively.

The most common application of surface welding, by far, is to apply a wear-resistant material over a wear-prone base material part or to a wear-prone area or areas of a part or structure (Figure 13.6a). The process is commonly called *weld hard-facing* or simply *hard-facing*, and the wear-resisting material is called a *hard-facing material* or *hard-facing alloy*. As wear occurs by either of two basic mechanisms, metal-to-metal *adhesive wear* or third-particle *abrasive wear*, protection against both generally employs a hard and typically brittle overlay. Such hard, brittle materials are intolerant of impact and often self-crack upon cooling (because of thermally induced tensile stresses) in what is known as "cross checking."

In most cases, the hard-facing alloy is compatible with but radically different in composition than the base material to which it is to be applied (e.g. an Fe-based alloy with large amounts of added Cr and high C to produce large-volume fractions of wear-resisting Cr-carbides in a ferritic or martensitic matrix for overlaying a wear-prone steel). It is often advisable, if not required by specification, that a *buffer layer or layers* be applied beneath the hard-facing layer or layers (Figure 13.4a). The buffer layer serves to (i) provide some strain accommodation, thereby affording the hard-facing layer a greater tolerance for compressive pressure and impact[1] and (ii) prevent dilution of the hard-facing deposit (and compromise of its wear resistance) by mixing in the melted portions of the base metal.[2] A common alloy used beneath hard-facing layers on steel

---

1 Because of this strain accommodation to high compressive pressure and/or impact, the buffer layer(s) beneath a hard-facing overlay is often called *cushion layer(s)*.

2 *Dilution* is treated in detail later in Section 13.4.

is an austenitic stainless steel or Mn-rich austenitic steel. A third reason for a buffer layer could be to provide compatibility between the base material and the hard-facing material. Required compatibility is usually to deal with radically different material compositions but might be to deal with the differences in coefficients of expansion greater than about 20%.

The second-most common application of surface welding is to apply a more corrosion-resistant alloy to a less corrosion-resistant base material in what is known as *weld cladding* or simply *cladding* (Figure 13.6b). For such multi-pass applications, one or more buffer layers may be required to (i) provide compatibility between base material and clad material (as above) or (ii) overcome base material dilution of the cladding, with compromise of the cladding's corrosion resistance. Such weld-applied clad layers are often applied to the linings of easy-to-fabricate and low-cost carbon steel pipes to be used to transport corrosive products or to external areas or surfaces of low-cost mild or low-alloy steel parts to tolerate locally corrosive environments.

The second major way in which multi-pass welding and welds are used to achieve compositional adjustment is in joining details of sufficiently different compositions that serious, unacceptable degradation of one or more key properties of the weld metal would occur *or* the base materials to be joined are sufficiently different in composition, they would be incompatible for direct joining by welding. The former situation requires the use of a *buttering layer* or *layers* to overcome weld metal dilution. The latter situation requires the use of one or more *transition layers*.

It is a fundamental requirement that a fusion welding process melts some portion (small volume) of each joint element's base material in order to obtain proper material continuity and to be considered a *fusion* welding process. For *autogenous welding*, in which joint elements are abutted with straight edges (either in straight or square butt configurations or rarely for welding but commonly for brazing, with angled or scarf joint configurations) without the use of any filler, the composition of the weld fusion zone is essentially the same as the base material.[3] However, whenever a filler metal/alloy is used to join like or dislike base materials, there is some *dilution* of the composition of the filler metal/alloy by the melted base material(s) mixed in by normal weld pool turbulence from convection forces. In changing the composition of the filler by dilution, properties will surely be changed, for better or worse.

The amount (or degree) of *dilution* in the weld must be known whenever the composition of the filler metal is different from either base metal being joined, i.e. whenever a *matched filler* is not used. Figure 13.10 provides a schematic illustration of a single U-groove preparation with an initial, root pass shown during multi-pass welding. Formulas are also given for calculating (i) the percentage of dilution as well as (ii) the percentage of any particular element e in the weld for this most general situation (i.e. a and b, different base materials; c, backing strip; and d, filler). It is important to know both results to know the following:

3 Autogenous welds are virtually always made between base materials of the same composition, e.g. 304 to 304 stainless steel.

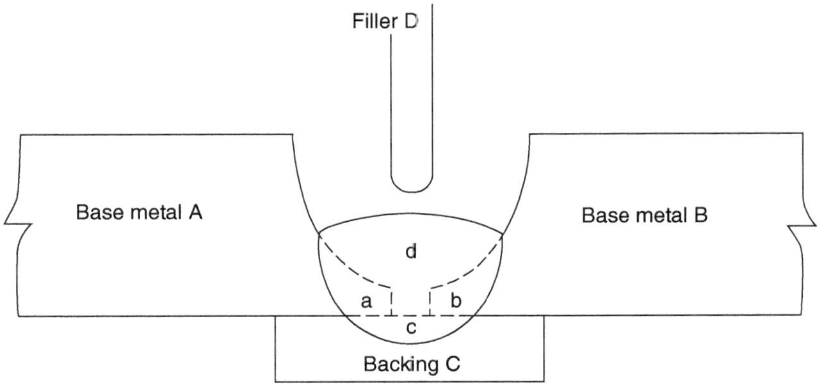

(1) $\% \text{ Dilution} = \dfrac{a+b+c}{a+b+c+d} \times 100$

(2) $\% \text{ Element E in weld bead} = \dfrac{1}{a+b+c+d}\left[a(\% \text{ E in a}) + b(\% \text{ E in b}) + c(\% \text{ E in c}) + d(\% \text{ E in d})\right]$

Figure 13.10 Schematic illustration showing a single U-groove and root pass being made between two different base metals A and B using a third metal backing C and fourth metal filler D, along with formulas for calculating (i) % dilution and (ii) % element E in weld metal. Source: Estes and Turner (1964). Published by and used with permission of the American Welding Society.

- Whether sufficient pickup (by dilution) has occurred where a steel weld metal is rendered unacceptably hardenable (i.e. with too high a value of C_{eq}) that there is risk of as-quenched martensite formation in the FZ and rendering the FZ prone to cold cracking
- Whether a key element in the filler for resisting corrosion in the FZ (e.g. Cr and/or Mo) has been diluted to below what is needed
- Whether a key element needed for strength or ductility or toughness (e.g. Ni in steel) or to avoid brittleness (e.g. C) has been lost in the FZ because of dilution
- Whether pickup of tramp elements from the base metal (e.g. S or P) will render the FZ prone to hot cracking.

Dilution can be minimized by (i) choosing a proper joint preparation, (ii) using a buttering layer (Figure 13.4b), or (iii) properly adjusting the placement of passes in a multi-pass weld overlay (e.g. using more shallow-penetrating stringer beads).

A buttering layer material must always be compatible to both the base material and the filler material between which it is to be used. It is usually also selected for special properties of tolerance to pickup by dilution from the base material to which it is applied (e.g. an austenitic stainless steel, pure Ni, or 50Ni–50Fe used to butter high-C cast iron before joining to structural steel).

Figure 13.11 shows a macrograph of both a multi-pass butter weld and multi-pass cladding [weld] used in the joining of a hardenable alloy steel to a steel pipe with a multi-pass closure weld.

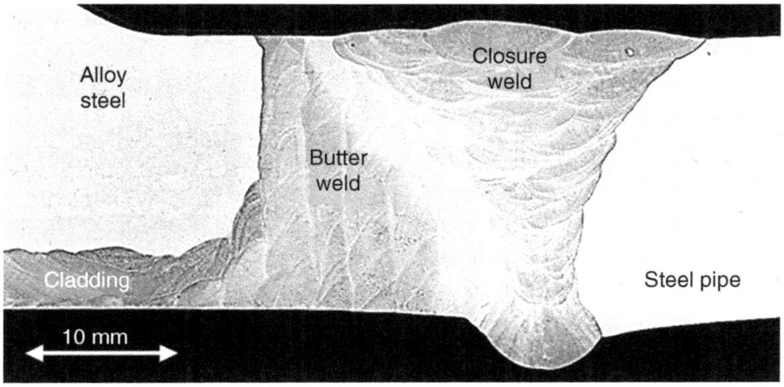

Figure 13.11 Photomacrograph showing use of multi-pass butter weld, multi-pass cladding, and multi-pass closure weld between an alloy steel part and a mild steel pipe. Source: Dodge et al. (2012). Published by and used with permission of the ASME.

The final way in which multi-pass welding and welds are used for compositional adjustment involves the use of *transition layers*. As these are one of the more practical, if not common, methods of joining dissimilar metals and alloys, they will be discussed in Chapter 20.

Obviously, every application of buffering layers is different, so the potential problems that could arise are different as well. The basic types of discontinuities and solutions to them have been well covered earlier in this book, however, so readers should be able to find what they need.

13.5 Property Alteration with Multi-pass Welding

The inextricable interrelationship among structure–property–processing and performance (Figure 2.6) should be a tip-off that multi-pass welding has the potential to alter the properties in complex ways compared to single-pass fusion welding. Besides altering structure and properties in unmelted base material by heat effects (i.e. in the HAZ) associated with single-pass fusion welding, multi-pass welding leads to the following more complex situations:

- Remelting (and solidification) of previously melted regions of earlier passes by subsequent overlapping passes (i.e. a new FZ over an old FZ)
- Creation of a HAZ in the FZ of earlier passes by subsequent overlapping passes (i.e. a HAZ over an old FZ)
- Creation of a new HAZ in the HAZ of earlier passes by subsequent adjacent passes (i.e. a new HAZ in an old HAZ).

Such interactions are not only complex but can be confounding to most metallurgists and materials engineers, no less others. One can only imagine what occurs in the multi-pass welding of heat-treatable alloys, which obviously respond to cycles of heating and cooling, whether intentional and controlled or unintentional and uncontrolled. Multi-pass welds in age-hardening alloys and transformation-hardening steels lead to complex interactions and, not surprisingly, can lead to property alteration. As is usually the case in engineering, the outcomes of controlled events are generally more favorable than the outcomes of uncontrolled events. This is true for multi-pass welding and multi-pass welds as well.

Let us look at an example of each in typical mild steel. The controlled situation involves the temper bead welding technique. The uncontrolled situation involves multi-pass welding and creation of localized brittle zones.

Metallurgists and laypersons alike have long known that welding can have both positive and negative effects on the properties of the base materials being joined as well as on previously deposited weld metal. The historical way in which some of the most deleterious effects were lessened was to post-weld thermally stress relieve the welds and weldments to lower residual stresses from near the yield strength of the alloy to one-third or less of that value. However, aside from the hue and cry of fabricators of welded products and structures as to the "unaffordability" to "impossibility" of post-weld stress relief heat treatment, many modern high-performance steels are designed to be used in the as-welded condition,[4] as many such micro-alloyed steels containing vanadium (V) lose toughness if post-weld heat treated. For such steels, as well as for many other steels prone to hardening during welding, special welding techniques are used, the most popular being *temper bead welding*.

In *temper bead welding*, a weld bead is placed at a specific location in or at the surface of a multi-pass weld being made in steel for the purpose of favorably affecting the properties of the HAZ or fusion zone of previously deposited weld metal. The effect is from the heat beneath the temper bead. Any portion of the HAZ of the previous weld or welds that is heated above the lower critical temperature A_1 on the Fe–Fe_3C phase diagram (Figure 9.2) dividing the lower temperature α ferrite + Fe_3C cementite region from the α ferrite + γ-austenite intercritical region in which austenite first forms during the eutectoid transformation on heating above the eutectoid temperature (723 °C/1333 °F) is subject to changes in microstructure and properties upon cooling. When post-weld cooling exceeds the alloy's critical cooling rate (CRR),[5] brittle as-quenched martensite forms from whatever austenite is present in this so-called inter-critical heat-affected zone (ICHAZ). Any portion of the same HAZ that is heated above the upper critical temperature A_3 that denotes entry to the all γ austenite region[6] will form 100% brittle as-quenched martensite under the same fast cooling typically encountered around welds made or near room

4 Readers are cautioned: When the properties of a medium- or high-carbon or low-alloy steel are controlled by solution-strengthening mechanisms (versus specialized micro-alloying and elaborate thermomechanical processing at the mill, as most are), post-weld heat treatment for stress relief is still advisable!

5 Rapid cooling above CRR following welding is much more likely for (i) thicker versus thinner materials, (ii) welds made with lower versus higher net linear heat inputs (based on current I × voltage E/travel speed TS as amperes × volts/speed in in. or mm/second to yield watt-seconds (w-s) or joules (J) per in. or mm), as appropriate, and weldments that are not preheated.

6 The A_1 and A_3 demarcate the lower and upper critical temperatures for hypoeutectoid steels having less than 0.77 wt% C. For hypereutectoid steels with more than 0.77 wt% C, which are rarely welded, as they are extremely difficult to weld without encountering severe quench cracking and/or cold cracking, the lower and upper critical temperatures are designated as A_{cm} and $A_{1,3}$.

temperature. In the lower temperature region of the γ-austenite region, known as the fine grain heat-affected zone (FGHAZ), martensite formation is slightly less easy, whereas in the coarse grain heat-affected zone (CGHAZ), martensite formation is much easier and more likely.

The favorable effects of a temper bead are twofold: (i) to refine structure in the CGHAZ, to transform it to an FGHAZ, and to make susceptibility to as-quenched martensite less likely; and (ii) to immediately temper any martensite that is formed upon post-weld cooling, taking the edge off its brittleness and rendering it less susceptible to either quench cracking or delayed cold cracking from hydrogen. In short, temper beads reduce problems from as-quenched martensite formation in the CGHAZ of welds made in hardenable steels.

Although deposition of temper beads during multi-pass welding has a favorable effect, there are many instances when multiple passes during the welding of thick-section steels cause a problem with properties. Figure 13.12 schematically illustrates the reason for this.

Except in rare and probably fortuitous rather than planned circumstances, multiple passes leave unrefined coarse grains of austenite in the HAZ of a first layer upon completion of a second layer of weld deposits. Such small regions can add up to create a localized brittle zone (because of high susceptibility to martensite formation during post-weld cooling in CGHAZs), most often found along the fusion line or boundary of the overall multi-pass weld. Such a localized zone can be problematic in and of itself, as it acts as a stress riser during service loading, particularly under cyclic fatigue. Perhaps more insidious, such a region can result in delayed hydrogen-induced cold cracking later on, in service.

The only message that can be given here is that one be aware of the potential for problems and seek guidance from experts (e.g. the American Welding Society, reputable manufacturers of welding consumables, or the technical support services provided by reputable material producers).

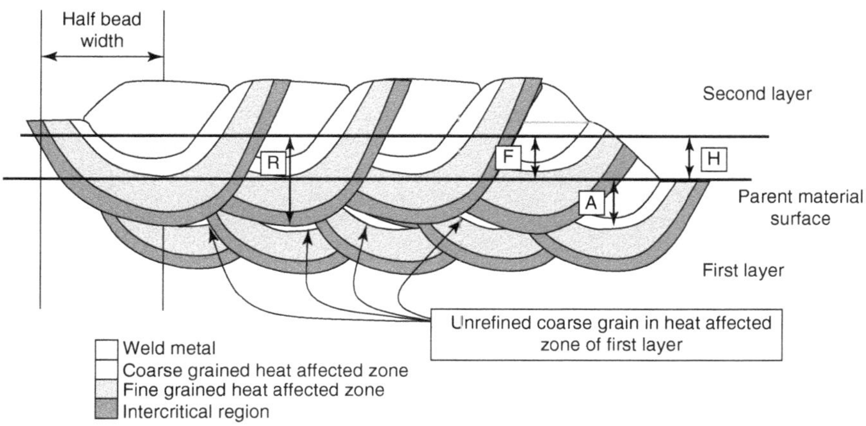

Figure 13.12 Schematic illustration showing the favorable effect of multi-pass temper beads on the structure of complex HAZ to minimize the occurrence of hydrogen-induced cold cracking in unfavorable brittle zones. Source: Lan: et al. (2001). Published by Elsevier, Ltd., and used under reciprocal agreement between Wiley and Elsevier for fair use.

13.6 Troubleshooting Problems in Multi-pass Welding and Welds

Table 13.1 lists the various ways problems can arise in the fusion zone of fusion welds.

Table 13.1 Troubleshooting problems in multi-pass welds.

Distortion of Weldment:	
Imbalanced heat input from the number or total volume of welds	Sequence weld passes to balance heat input Employ double-groove butt joint preparations or employ opposing fillet welds
Defects:	
o Lack of fusion between passes	Employ proper melt-in and wet-in technique
o Entrapped slag with slag-protected SMAW, FCAW, SAW, ESW	Completely remove slag between passes
o Conventional hot cracking	Minimize low-melting residuals in the base metal; use fillers tolerant of residuals; avoid overheating via interpass temperature
o Cold cracking	Employ preheat as recommended; use temper bead technique
o Liquation cracking	Avoid susceptible alloys (per Table 10.1); choose proper filler (per Table 10.2)
o Porosity	Follow guidelines in Table 5.3
Need for Under-Layers:	
o To rebuild lost base metal	Attempt to match filler to base metal
o To provide cushion for impact or compressive pressure	Employ an inherently tough filler alloy beneath hard-facing overlay
o To prevent dilution	Employ buttering layer(s) using a filler tolerant of pickup from base metal (e.g. austenitic SS, pure Ni, and 50Ni–50Fe)
o To facilitate dissimilar metal welding	Employ a filler compatible to both base metals or to facilitate transition by steps
For Intentional Property Alteration:	
To relieve stress in prior passes	Use temper bead welding technique in lieu of PWHT
To Avoid Unwanted Property Alteration:	
Localize regions of brittleness in harden-able steels	Be aware of possibility and seek expert advice; e.g. material provider, trade associations (AISI), and American Welding Society "best practices"

References

Collier, J. (2015). Tig Welding Pictures. www.weldingtipsandtricks.com.

Dodge, M.F., Dong, H.R., Milititsky, M. et al. (2012). Environment-induced cracking in weld joints in subsea oil and gas system – Part 1, figure 2. Proceedings of ASME 2012 31st Conference on Ocean, Offshore, and Artic Engineering, OMAE2012, Rio de Janeiro, Brazil (1–6 July 2012).

Estes, C.L. and Turner, P.W. (1964). Dilution in multi-pass welding of AISI 4130 to Type 304 stainless steel. *Research Supplement Welding Journal* 43 (12): 541s–550s.

European Space Agency (2013). *3D Printed Aeronautics Demonstrator*. Cranfield, UK: Stephen Williams, Cranfield University www.esa.int.

Lant, T., Robinson, D.L., Spafford, B., and Storesund, J. (2001). Review of weld repair procedures for low alloy steels designed to minimize the risk of future cracking. *International Journal of Pressure Vessel and Piping* 78 (11–12): 813–818.

Malcolm (2015/2016). Arc (MMA) Tutorial – Fillet Joints. http://www.mig-welding.co.uk.

Manning, S. (2005). *The Art of Narrow-groove Welding*. Columbus, OH: Edison Welding Institute www.ewi.org.

Messler, R.W. Jr. (1993). *Joining of Advanced Materials*, figure 6.1, p. 149. Stoneham, MA: Butterworth-Heinemann (now part of Elsevier, Ltd.).

Pargeter, R. (2003). Evaluation of necessary delay before inspection for hydrogen cracks, figure 5b. *Research Supplement Welding Journal* 82 (11): 321s–329s.

Zinke, M. and Fink, C. (2013). Welding of nickel-base alloy 617 using modified dip arc process, figure 9. *Soudage Dans Le Monde/Welding in the World* 57 (3).

Further Reading

American Welding Society (AWS) (2001). *Welding Handbook*, Volume 1: Welding Science and Technology, 9e (ed. C.L. Jenney and A. O'Brien). Miami, FL: American Welding Society.

ASM International (1993). *ASM Metals Handbook*, Volume 6: Welding, Brazing, and Soldering, 9e (ed. D.L. Olson, T.A. Stewert, S. Liu and G.R. Edwards). Materials Park, OH: ASM International.

14

Problems with Non-fusion Welding and Non-fusion Welds

14.1 Non-fusion Welding Processes Versus Fusion Welding Processes

In Section 2.1, three distinct mechanisms for obtaining metallic continuity essential for the creation of a weld were listed after Granjon[1]: (i) solid-phase plastic deformation, without or with dynamic recrystallization, (ii) solid-phase diffusion, and (iii) melting and solidification. The latter mechanism is the basis for all fusion welding and the myriad of varied problems that have been exclusively and justifiably discussed to this point, as fusion welding processes dominate welding practice across applications around the world. However, what about the other two mechanisms, which, together, are the basis for *non-fusion welding*?

Non-fusion welding processes accomplish joining by bringing the atoms of metals (or ions of glasses or ceramics, or molecules of thermoplastic polymers) to equilibrium spacing in the solid state. Upon reaching equilibrium spacing, atomic level bonds are formed, and a weld is produced as a direct result of the continuity obtained, with a strength across the former interface equal to the cohesive strength within the base material(s). With non-fusion welding, by definition, occurring below the melting point of a material (i.e. solidus temperature for an alloy), any heat used to facilitate atom movement in the solid state or generated as a consequence of any work done (by plastic deformation or friction) generally has minimal adverse effects on the base material(s). As a result, non-fusion welding offers the possibility, which is impossible with fusion welding, of producing a prefect joint, with the same structure and properties everywhere.

Without requiring melting during weld production,[2] non-fusion welding processes offer a number of advantages compared to fusion welding processes as follows:

1 *Fundamentals of Welding Metallurgy* by H. Granjon, 1991, Abington Publishing, Woodhead Publishing, Cambridge, UK. A metallurgist by profession, the late Henry Granjon was always profoundly interested in education and training throughout his career at the Institut de Soudage (French Welding Institute). This interest was channeled through his course at ESSA (Graduate School of Welding Engineering) where he constantly conveyed his passion for welding metallurgy to his students.

2 There are some embodiments of welding that create the weld without the *need* for melting, but for which some transient melting may occur, with virtually all traces of molten metal being expelled during a pressure cycle that culminates the joining. These include many of the resistance welding processes, e.g. flash welding (FW) and upset welding (UW).

A Practical Guide to Welding Solutions: Overcoming Technical and Material-Specific Issues,
First Edition. Robert W. Messler, Jr.

- General absence of melting and, thus, solidification (so base material structure is retained unaltered)
- Low heat input or generation (minimally disrupts microstructure)
- Wide variety of process embodiments
- Applicability to many materials within a fundamental class (i.e. metals, ceramics, glasses, and polymers) as well as between classes (because there is little or no intermixing), facilitating dissimilar material joining
- High joint efficiency is possible for many situations, whereas the same cannot be said for fusion welding processes.

As for all things in life, not everything is perfect, however. There are shortcomings that include the following:

- Stringent requirements for precleaning of joint faying surfaces[3] for most processes, often more stringent than for fusion welding processes
- Stringent requirements for process control
- Elaborate tooling is required for some process embodiments
- Challenging nondestructive evaluation for assessing joint quality
- Repairing process-induced defects is difficult to impossible, as once joined, there is no way to eliminate a defect by repeating the welding cycle
- Most processes require specialized equipment (some of which must be built rather than bought), are rarely portable, and must almost always be automated.

Later in this chapter, it will be seen that the problems that occur in welds are overwhelmingly the result of improper or inadequate cleaning of joint elements before welding.

14.2 Overview of Non-fusion Processes

There are three principal ways in which non-fusion welding is made to occur: (i) pressure and gross plastic deformation and metal flow, which is generically called *pressure welding*, (ii) friction with either macroscopic or microscopic plastic deformation and metal flow which is generically called *friction welding*, and (iii) solid-state diffusion causing movement of atoms, without any or with some plastic deformation, which is generically called *diffusion welding* (DFW).[4] A fourth way receiving growing attention largely with the advent of 3D printing and other stereolithographic techniques for rapid (one-off) prototyping or, increasingly, rapid manufacturing is solid-state deposition welding using vapor condensation or electrochemical reactions. Table 14.1 lists the eight major non-fusion welding processes along with variations of embodiments within each.

For completeness of this chapter as a reference, a very brief summary of each major types and some their subtypes are discussed in the following sections.

3 A faying surface is one of the surfaces of material elements making up a joint, irrespective of the planned joining method.

4 It is important to note that solid-state diffusion (actually interdiffusion) between base materials or, occasionally, an intermediate to aid with strain-accommodation to force solid, but plastic, metals together, is always involved in non-fusion welding, as it is to some degree in all welding processes, depending on temperature.

Table 14.1 List of the eight major non-fusion welding processes with variations.

1. Cold welding (CW)
   Press welding
   Forge welding (FOW)
   Roll welding (ROW)
   Toggle welding
   Hydrostatic impulse welding
   Shock-wave impulse welding
2. Hot pressure welding (HPW)
   Pressure gas welding (PGW)
   Forge welding (FOW)
3. Roll pressure welding (ROW)
   Cold, warm, or hot welding
4. Explosion welding (EXW)
5. Friction welding (FRW)
   Radial friction welding
   Orbital friction welding
   Rotational friction welding
      Direct-drive welding
      Inertia welding
   Angular(reciprocating)friction welding
   Linear (reciprocating) friction (or vibrational) welding
   Friction stir welding
   Friction surfacing
6. Ultrasonic welding (USW)
   Spot, ring, line, or seam USW
   Microminiature welding
   Microminiature thermosonic welding
7. Diffusion welding (DFW)
   Conventional diffusion welding
   Deformation diffusion welding
   Resistance diffusion welding
   Continuous seam diffusion welding (CSDFW)
   Diffusion brazing (DFB)
   Combined forming welding
      Creep isostatic pressing (CRISP)
      Superplastic forming diffusion bonding (SPF/DB)
8. Solid-state deposition welding
   Electrochemical deposition
   (Physical) Vapor deposition
   Chemical vapor deposition
   Chemical reaction bonding

Details are left to readers to seek out information in other sources (e.g. Messler 1999/2005; Cary and Helzer 2004; AWS 2007).

14.2.1 Pressure Welding Processes

As the name indicates, non-fusion *pressure welding processes* primarily depend on the application of pressure to obtain metallic continuity and produce welds. Pressure welding processes can be performed cold (e.g. cold welding, CW) or hot (e.g. hot pressure welding, HPW). Pressure application can be slow (e.g. pressure gas welding and cold press welding), relatively fast (e.g. cold forge welding, cold roll welding, cold toggle welding, hot forge welding, and warm or hot roll welding), or extremely rapid (e.g. hydrostatic impulse cold welding, shock-wave impulse cold welding, and explosion welding). All of the processes are characterized by the application of significant pressure in order to cause abutting surfaces of joint elements to come into intimate contact over the entire faying surface to obtain metallic continuity. Solid-state interdiffusion is surely involved with those processes that involve heating, with a principal driving force being dynamic recrystallization. Welds typically exhibit high joint efficiency, as a result of the wrought (versus cast, dendritic) microstructure. Cold pressure welding works best with metals or alloys that plastically deform easily without being heated (e.g. those with face-centered cubic crystal structures). Any metal or alloy that plastically softens upon heating can be pressure welded.

14.2.2 Friction Welding Processes

As a group, *friction welding processes* convert mechanical energy into heat at the joint to be welded by using the relative movement between joint elements under moderate compressive pressure. Metallic continuity always involves plastic flow of heat-softened base metals, often on a microscopic scale but, for some variations, involves more gross flow (e.g. friction stir welding, FSW and friction surface welding). Some, but not all, variations of friction welding require an upsetting or forging pressure to be applied to complete the weld once heating, softening, and continuity have been achieved. Even though the faying surfaces are scrubbed free of any light oxide or tarnish layer during the early stage of welding, the surfaces of parts to be joined should be thoroughly cleaned of oxide or tarnish layers before welding is begun. Friction welding works best when at least one of the metals or alloys involved in the joint is easily deformed because it has a face-centered cubic structure.

14.2.3 Diffusion Welding Processes

Because atoms are constantly undergoing 10^{13}–10^{14} vibrations per second that enable them to jump from a lattice site into an adjacent vacant lattice site (by a vacancy diffusion mechanism) or, in those rare instances when they are small enough to occupy interstitial sites between atoms on a crystalline lattice, to jump into an adjacent unoccupied interstitial site (by an interstitial diffusion mechanism), metallic continuity can be achieved to create a weld by solid-state diffusion alone. Although enough pressure is needed to hold mating joint elements in contact while a diffusion weld is being created, pressure need not be high enough to cause plastic deformation except for the deformation DFW subprocess. The driving force for atoms to eliminate the interface between abutting metals is to minimize the surface energy. Metals and alloys that diffusion weld must easily (and produce the best welds) are either free of persistent oxides or tarnish layers or, as in the unusual case of titanium and its alloys, have an oxide layer that dissolves upon heating before the metal or the alloy melts. DFW can be facilitated between some metals and alloys (as well as between some metallic and ceramic materials) using a suitable intermediary.

14.2.4 Solid-State Deposition Welding

Although the underlying solid-state deposition processes have been used for a very long time, the potential for such processes to weld materials together is relatively recent. The basis for solid-state deposition processes is the formation of strong chemical bonds as one material is deposited on another whose surface is atomically clean. Because atomic level cleanliness (i.e. freedom from contaminating surface layers) and intimate contact are the two prerequisites for producing a weld, the possibility of making welds by such processes should have been obvious. A major recent impetus for exploring and exploiting solid-state deposition welding has been stereolithography or 3D printing for use in one-off rapid prototyping or rapid manufacturing of complex parts.

With all these options for making welds that have the potential to be perfect from the standpoint of having properties that are not degraded from the base material(s) by melting and resolidification, what could go wrong.

14.3 Problems with Non-fusion Welds and Non-fusion Welding Processes

Two of the most attractive features of non-fusion welding are as follows: (i) with no need for melting, there is no fusion zone, and (ii) with only modest heat used for or generated during weld formation, there is little of no heat-affected zone (HAZ). With these two attributes, the resulting welds, at least, are relatively free from problems in the form of discontinuities (see Table 3.2). Potential problems are more often associated with the operation of the process than with the quality of the weld, the greatest problems being optimizing the welding parameters and achieving process control in production.

The *single* greatest potential problem that can – and does – occur in non-fusion welds is *lack of fusion* at the faying surface of joint elements. In the case of non-fusion welding, this means achieving proper atomic level metallic continuity between faying surfaces, albeit with melting and wetting, simply by intimate contact. The cause of this defect in non-fusion welding is virtually always lack of adequate cleaning of joint element faying surfaces before welding, leaving a contaminating layer of oxide or tarnish that prevents metallic bond formation.

As listed first under "Shortcomings" in Table 14.1, there are stringent requirements for cleaning joint faying surfaces for most non-fusion welding process embodiments, especially, but not limited to CW, HPW, roll [pressure] welding (ROW), and DFW. For most of the other process embodiments, there is a built-in mechanism for removing light surface oxide or tarnish layers at an early stage of weld production. The two best examples are explosion welding, in which the supersonic jet of air expelled from the closing gap between the explosion-driven joint elements blasts light contaminants out of the gap and some forms of friction welding in which a vibratory action abrades off light oxide or tarnish layers. The only question is whether debris is moved out from between the faying surfaces.

Although always important for the production of quality welds, stringent cleaning before non-fusion welding is vital if sound welds are to be produced. Depending on the metal or alloy, proper cleaning methods could mean simple solvent degreasing of oxidation-resistant metals and alloys, acid pickling or alkali etching (e.g. stainless steels, Cu alloys, Ni alloys, and Ti alloys), or mechanical removal of inherent oxide layers (e.g. Al and Mg alloys and hot-rolled steel). Recommended practices and procedures for cleaning various metals and alloys before welding of any kind are available from material producers, metal trade associations, and the American Welding Society's *Welding Handbook*.

One of the friction welding processes most susceptible to defects in welds is FSW. Invented and experimentally proven in 1991 by The Welding Institute in Cambridge, England, claims for the process compared to conventional methods of welding include (i) lower energy consumption; (ii) no toxic fumes; (iii) no need for shielding gasses, fluxes, or vacuums; and (iv) energy-efficient material joining. The process has been successfully applied in a wide variety of industries, including aerospace, automotive, railway, and maritime. Although notoriously well suited for joining Al and Al alloys, it has also been successfully employed with Cu and Cu alloys, Mg and Mg alloys, and metal–matrix composites of these metals, Pb, some steels, stainless steels, and dissimilar metal combinations.

As shown schematically in the upper portion of Figure 14.1, the process uses a rapidly spinning nonconsumable pin with shoulder tool squeezed between the faying surfaces of pieces to be welded. Frictional heating from rotation softens (but does not melt) the base materials and "stirs" them to produce a weld. The rotating tool moves (translates) along the joint line until the weld has been completed. As the algebraic sum of rotational velocity and translational velocity are different on the advancing and retreating sides of the weld line, the weld tends to be offset toward the retreating side. As for any welding process, it is extremely important to choose the right combination of welding parameters in order to produce welds of suitable quality. As there are many parameters (e.g. pin design, shoulder design, too rotational speed (TR), translational or welding speed (WS), tool tilt angle, and squeezing force) and little shared experiential knowledge, finding the right combination can be a challenge.

The lower portion of Figure 14.1 schematically represents the various microstructural zones found with FSW. The four distinct zones are as follows:

- *Unaffected zone*: This zone is farthest away from the weld proper. Although temperature in this zone may be elevated slightly because of the conduction of

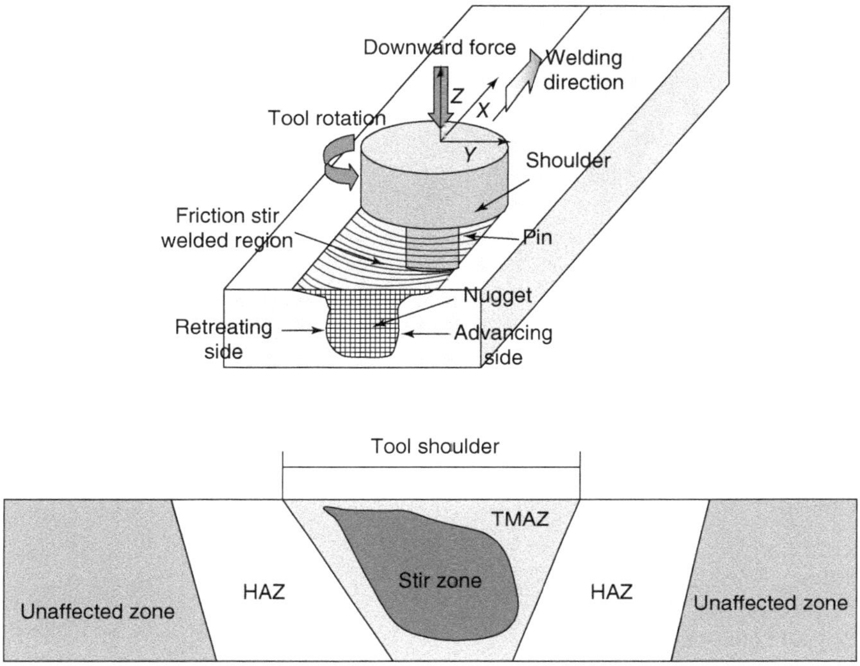

Figure 14.1 Schematic illustration showing details of the friction stir welding (FSW) process (a) and of the generic microstructural zones created during welding (b). Source: Sattari et al. (2012). Published by and used with permission of Scientific & Academic Publishing Co.

frictionally generated heat, the microstructure and mechanical properties are unchanged from before the welding process was performed.

- *Heat-affected zone (HAZ)*: This zone is next to occur in progressing toward the weld center. Herein, the temperature was elevated enough by frictional heat to modify the microstructure and mechanical properties in some way, although no plastic deformation occurred.
- *Thermomechanically affected zone (TMAZ)*: Unlike the HAZ, this zone experienced both heating and plastic deformation during welding. Although not the case with Al or Al alloys, recrystallization occurs in the TMAZ of Ti- and Ti alloys, Cu- and Cu alloys, and austenitic stainless steels.
- *Stir zone (SZ)*: This region within the TMAZ is where, if it occurs, recrystallization takes place.

In order for high-quality friction stir welds, all of the above zones must be developed and the material(s) must be stirred well.

Figure 14.2 shows a photograph of an actual FSW in progress.

Inappropriate heating and mechanical mixing during FSW can lead to the formation of defects, including

- lack of penetration (entirely through the joint, as required)
- lack of fusion
- tunnels

Figure 14.2 Photograph showing a friction stir weld being made in difficult-to-weld oxide-dispersion strengthen (ODS) steel. Source: The Welding Institute (2016). Used with permission of TWI Ltd.

- voids
- surface galling
- nugget collapse
- kissing bonds.

These are commonly divided into categories based on the energy input from combined rotational speed and translational (welding) speed, as shown schematically in Figure 14.3.[5] These categories include, as indicated in the figure, the following:

- a large mass of flash formed because of excess heat input
- cavity or groove-like defects caused by insufficient heat input
- a cavity produced by abnormal stirring.

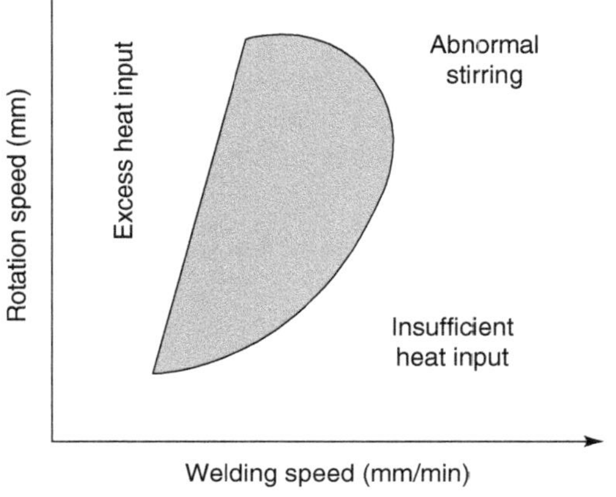

Figure 14.3 Schematic plot showing the combination of rotational speed (rpm) and translational/welding speed (mm/min) that produce welds of acceptable quality (shaded region), with combinations outside this region that produce various defects from excess heat input, abnormal stirring, or insufficient heat input. Source: Kim et al. (2017), figure 8. Published by The Korean Welding and Joining Society and freely used under Creative Commons 3.0.

5 In actuality, plots like this must be developed by the user for each metal/alloy and each tip/shoulder tool combination to be joined by FSW.

Figures 14.4–14.8 are photomacrographs and/or photomicrographs of various defects found in friction stir welds, as described above.

Like any welding process, quality welds require quality design, quality processing, and quality assurance.

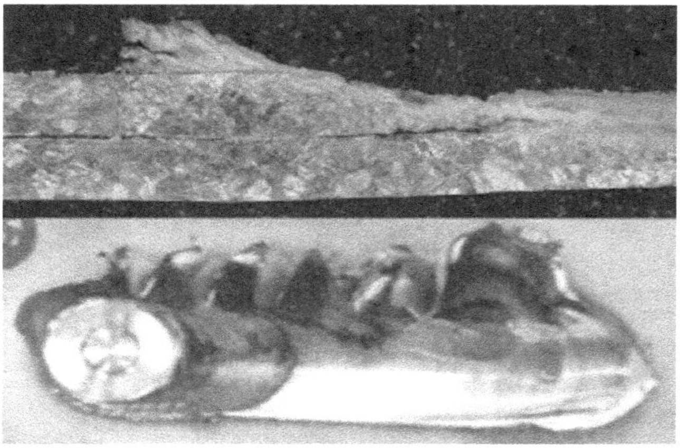

Figure 14.4 Macrograph showing a large mass of flash and of ribbon flash on the edge of the retreating side of a friction stir weld in CP-Ti. Source: Kim et al. (2017), figure 9. Published by The Korean Welding and Joining Society and freely used under Creative Commons 3.0.

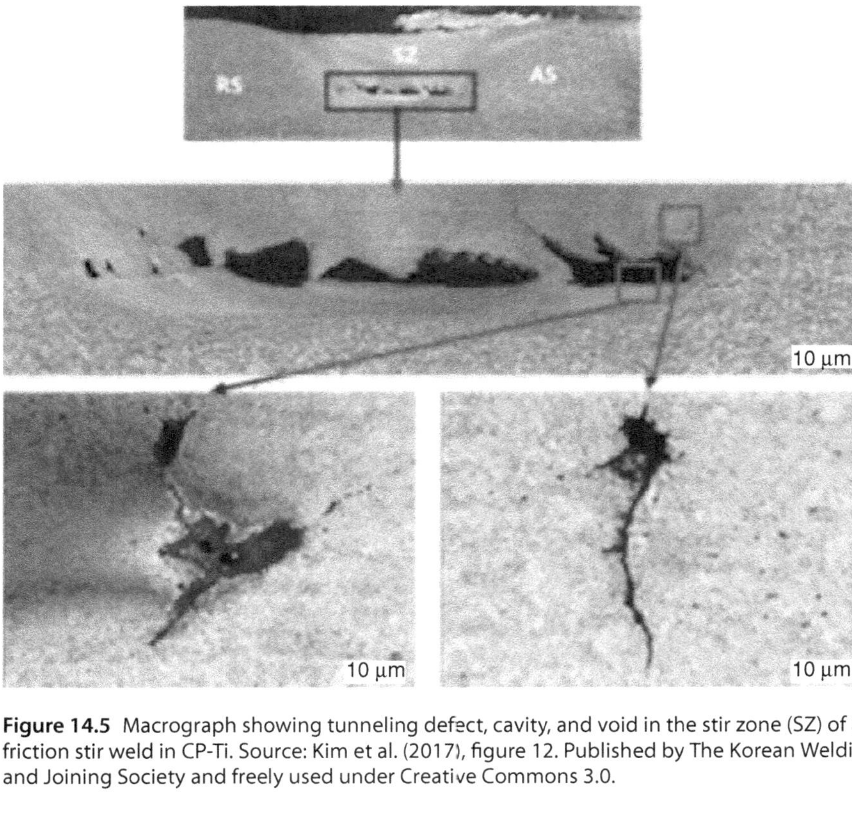

Figure 14.5 Macrograph showing tunneling defect, cavity, and void in the stir zone (SZ) of a friction stir weld in CP-Ti. Source: Kim et al. (2017), figure 12. Published by The Korean Welding and Joining Society and freely used under Creative Commons 3.0.

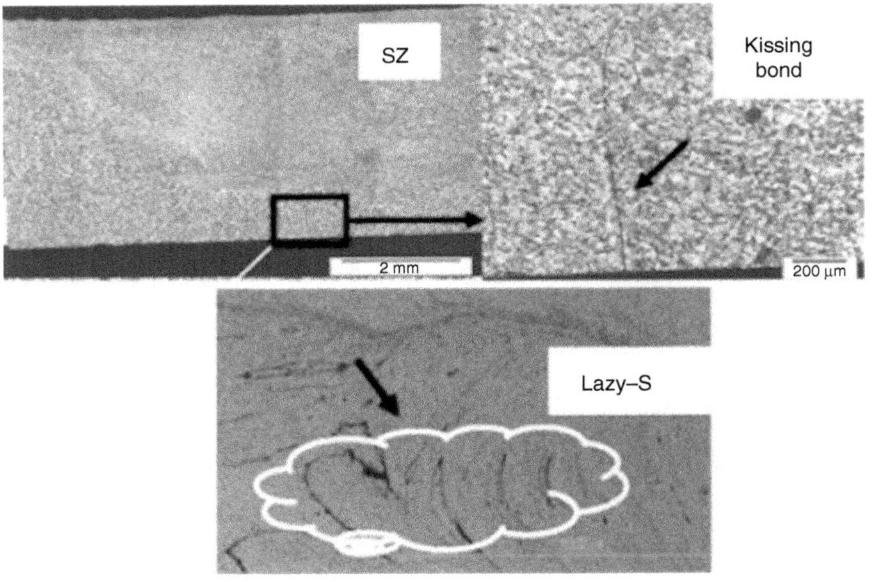

Figure 14.6 Macrographs and micrograph showing a kissing bond and a lazy-S defect in a friction stir weld in CP-Ti. Source: Kim et al. (2017), figure 13. Published by The Korean Welding and Joining Society and freely used under Creative Commons 3.0.

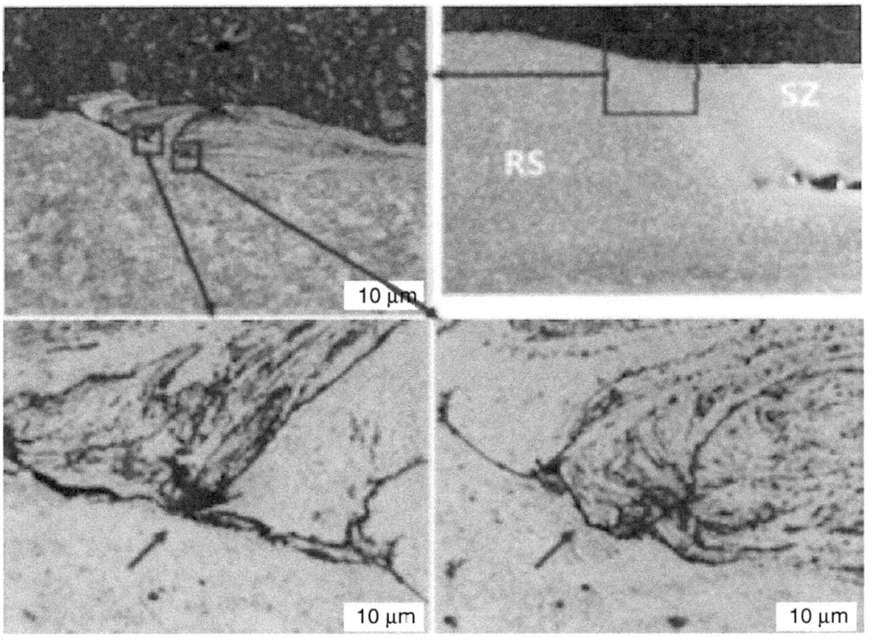

Figure 14.7 Micrographs showing various examples of oxide-induced defects on the edge of the retreating side of the stir zone of a friction stir weld in CP-Ti. Source: Kim et al. (2017), figure 14. Published by The Korean Welding and Joining Society and freely used under Creative Commons 3.0.

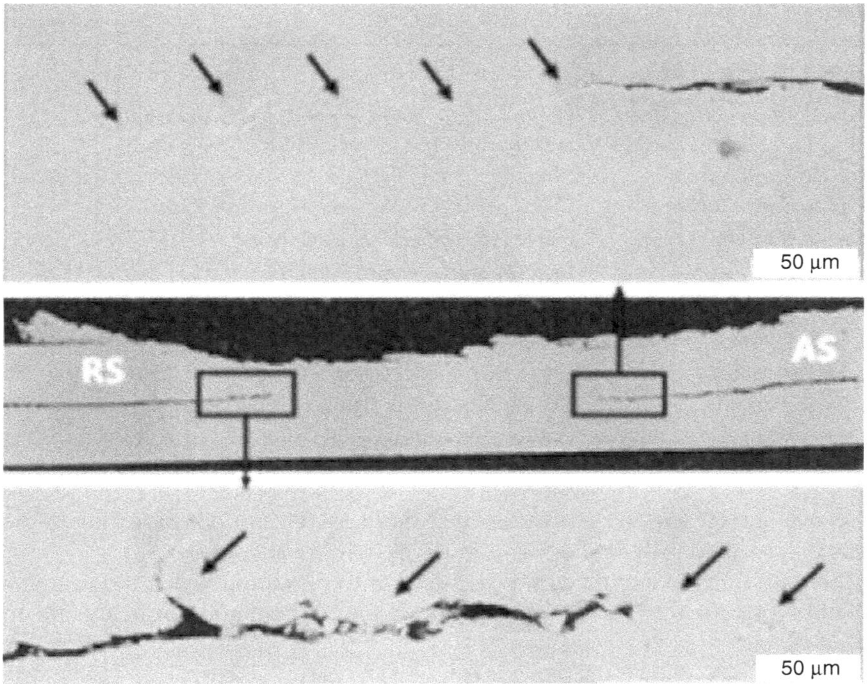

Figure 14.8 Micrographs showing various oxide-induced defects on the edge of the contact area on a lap joint made by friction stir welding in CP-Ti. Source: Kim et al. (2017), figure 15. Published by The Korean Welding and Joining Society and freely used under Creative Commons 3.0.

14.4 Inspection and Repair Challenges with Non-fusion Welds

As a group, non-fusion welding processes pose a special challenge, despite their attractiveness for preserving base material microstructure and properties, as well as for their potential for welding dissimilar materials (e.g. dissimilar metals and/or alloys, metals to ceramics) and advanced materials (e.g. metal–matrix composites and ceramic–matrix composites). The joints produced by non-fusion welding processes are characteristically thin compared to fusion welded joints and, as a consequence, contain thin planar defects and can involve much greater weld area. Together, these two factors make nondestructive evaluation inherently difficult. X-rays tend to have difficulty in detecting defects as these tend to be thinner than the X-ray's resolving power (about 1% of the path length at best, 2% typically). Ultrasonic techniques have difficulty because of the significant proportion of sound that can be transmitted by many very small connecting paths or ligaments.

As if these challenges to inspection are not bad enough, once detected, it is often impossible to repair the defects found. The suitability of many non-fusion welding process embodiments (e.g. explosive welding and DFW) to produce a

Table 14.2 Troubleshooting problems in non-fusion welds.

All Process Embodiments:	
Lack of fusion defect(s)	Employ stringent cleaning of all faying surfaces (per recommendations for metal/alloy)
Friction Stir Welding:	Comply to plot as in Figure 14.3, but specific to material/tool
o Lack of penetration	Check pin choice, plus above (per Figure 14.3)
o Lack of fusion	Employ stringent cleaning methods
o Tunneling	Increase heat input per Figure 14.3
o Voids/cavities	Reduce heat input per Figure 14.3
o Excess flash	Check tool shoulder choice; check rotational speed per above
o Surface galling	Check tool shoulder choice for susceptible materials (e.g. Ti)
o Nugget collapse	Comply to plot as in Figure 14.3, but specific to material/tool
o Kissing bonds	Consider abnormal stirring; comply to Figure 14.3

weld over a large surface area means that there are few options of getting to the defect(s), such as an area of disband in DFW, to make any repair.

The only thing worse than allowing defects to remain in welds because one failed to look for them is having looked and found them and not being able to repair them!

14.5 Troubleshooting Problems with Non-fusion Welds

Table 14.2 provides some guidance for troubleshooting problems encountered with non-fusion welds.

References

AWS (2007). *Welding Handbook*, 9e, vol. 3, Part 2. Miami, FL: AWS.

Cary, H.B. and Helzer, S. (2004). *Modern Welding Technology*, 6e. London, UK: Pearson.

Kim, J.-D., Jin, E.-G., Murugun, S.P., and Park, Y.-D. (2017). Recent advances in friction-stir welding process and microstructural investigation of friction stir welded pure titanium. *Journal of Welding and Joining* 35 (4): 6–15.

Messler, R.W. Jr. (1999/2005). *Principles of Welding: Processes, Physics, Chemistry and Metallurgy*. New York, NY/Berlin, Germany: Wiley/Wiley VCH.

Sattari, S., Bisadi, H., and Sajed, M. (2012). Mechanical properties and temperature distributions of thin friction stir welded sheets of AA5083, figure 1. *International Journal of Mechanics and Applications* 2 (1): 1–6.

The Welding Institute (2016). *Welding the unweldable: TWI friction stir welds ODS steel*. Cambridge, UK: TWI www.twi-global.com.

Further Reading

American Welding Society (AWS) (2004). *Welding Handbook*, Volume 2: Welding Processes, 9e (ed. A. O'Brien). Miami, FL: American Welding Society.
American Welding Society (2014). *AWS Welding Handbook: Resistance and Solid-State Welding and Other Joining Processes*, 2e. New York, NY: Palgrave Macmillan (part of Springer Nature).
Messler, R.W. Jr. (2004/1999). *Principles of Welding: Processes, Physics, Chemistry, and Metallurgy*, Chapter 4. Weinheim, Germany/New York, NY: Wiley-VCH/Wiley-Interscience, Wiley.
Weman, K. (2011). *Welding Processes* (Woodhead Publishing Series on, Welding and Other Joining Technologies, 2e. Woodhead Publishing (now part of Elsevier, Ltd.).

Part III

Material-Specific Weld-Related Problems

The attractiveness of metals and alloys as the usual material of choice in engineering is their special combination of strength, toughness, durability, and ease of fabrication by a wide variety of processes, including welding. However, metals and alloys exhibit different degrees of weldability, and some, despite their attractive engineering properties, are susceptible to their own particular problems. *What* problem(s) particular metals and alloys can exhibit as a result of having been welded is the third clue to their root cause and, thus, solution.

15

Embrittlement of Carbon and Low- and Medium-alloy Steels

15.1 The Importance of Steel

Although a crass play on words, in a very real sense, steel makes the world go "round."[1] More steel is used annually in construction and manufacturing than any other material except cement and concrete.[2] In 2017, global cement production (which includes cement used in concrete) exceeded 2.6 billion metric tonnes or MT (i.e. 2.6×10^{12} kg/5.72×10^{12} lbs[3]). The rate of consumption has been growing by 5.5–5.7% per year for more than a decade, with the growth in China far exceeding that level at around 8.5% annually for each of the past five years. In the same year, global crude steel production[4] was about 1.7 billion MT, with the leading producers in the world (in million MT) being (1) China (803.1), (2) Japan (105.1), (3) India (89.0), (4) United States (82.0), (5) Russia (70.9), (6) South Korea (69.7), (7) Germany (42.7), (8) Brazil (33.3), (9) Turkey (31.5), and (10) Ukraine (23.0).

By several measures, steel is the technological and economic engine of the world. Stated differently: To be a "player" in the modern world, a country must have strong steelmaking capability and capacity. To put the importance of steel into perspective, consider the following:

- Steel accounts for nearly 90% of all primary metals produced annually in the world (see Table 15.1)
- There is profound correlation between strength of a nation's economy and its production of steel (Table 15.2)

1 The less crass and better known saying is that "Love makes the world go round", from the 1958 song "Love Makes the World Go Round" composed by Ollie Jones (1929) and popularized by American singer and TV star Perry Como (1912–2001), known as "Mr. C."
2 Believe it or not, wood (i.e. lumber) and wood products (e.g. plywood, fiberboard, particle board, oriented strand board, glue-lam, and veneers) are the third most widely used materials, although by a significant gap and at a much lower tonnage, as wood has low density compared to both of the other materials (i.e. about 0.9–1.1 for wood versus 8.05 for steel and 2.4–2.6 g/cm^3 or kg/m^3 for cement/concrete) and is not produced or sold by tonnage but by board meters or -feet or volume (as m^3 or ft^3).
3 10^9 is 1 billion, whereas 10^{12} is 1 trillion!
4 "Crude steel" refers to the various iron-rich mineral aggregates from which the starting alloy for all production of all varieties of steel is produced.

A Practical Guide to Welding Solutions: Overcoming Technical and Material-Specific Issues,
First Edition. Robert W. Messler, Jr.

Table 15.1 Global primary metal production in 2017 (in million MT).

Crude steel	1700
Aluminum	64
Copper	26
Titanium	25
Zinc	13.7
Lead	5.2
Nickel	2.5
Magnesium	1.0
Molybdenum	0.23
Tungsten	0.12
Rare earth metals	0.12
Chromium	0.03
All others	~0.03
Total	1840

1 MT = 1000 kg = 2204.67 lbs.

Table 15.2 Ranking as a steel producer versus ranking of economy (by GDP in 2017).

Economy (by GDP in $)	Rank as steel producer
1. United States (19.42T)	4
2. China (11.8T)	1
3. Japan (4.84T)	2
4. Germany (3.42T)	7
5. UK (2.5T)	—
6. India (2.45T)	3
7. France (2.42T)	—
8. Brazil (2.14T)	8
9. Italy (1.81T)	—
10. Canada (1.6T)	—

GDP refers to "gross domestic product"; measure of the value of all produced goods by a nation annually.

- Although the rate of growth in steel production in the very mature economy of the United States has remained around 2.5% annually for the past five years, the annual growth rates in rapidly developing economies such as China, India, and the Association of Southeast Asian Nations have averaged 8–10%, 6–8%, and 5–6%, respectively.

Table 15.3 Comparisons among the United States, China, and the world in terms of population, GDP, and steel production (as of the end of 2017 or estimated for 2018).

	United States	China	World
Population (1st-Q 2018)	326.8M (4.28%)	1.415B (18.53%)	7.64B
GDP ($ end of 2017)	19.739T (23.39%)	12.86T (15.24%)	84.38T
Steel production (2017) (million MT)	78.8 (4.64%)	803.1 (47.24%)	1700

M, million; B, billion; T, trillion.

Steel production in China over the past decade has been – and remains – staggering, testimony to its rapidly changing and growing economic strength and technological presence in the world. Table 15.3 shows this with data comparisons as of the end of 2017 for the population, gross domestic product (GDP), and steel production for the United States, China, and the world. With a population of 4.33 times that of the United States, and a GDP currently 65% that of the United States, China produces 10.2 times as much steel as the United States, and just over 47% of the steel produced in the entire world!

With major uses of steel of all varieties and in all forms[5] in construction (e.g. buildings, bridges, highways, and dams, as rebar), machinery, ship building, automobiles, trucks and buses, power equipment, appliances, railways and railcars and locomotives, and containers, three things should be apparent:

1. No wonder steel is a technological and economic driver for an industrialized county!
2. No wonder China needs so much steel!
3. No wonder any problem encountered welding steel is a PROBLEM!

Although mechanical fastening with nuts and bolts or, less and less, with rivets, is used to join steel details into assemblies and structures, welding, in general, and fusion welding, in particular, dominates.[6] The reasons are simple: High joint efficiency when done properly, leak tightness against fluids (gases and liquids), whether from external or internal pressure, and permanence (i.e. no unintended loosening or disassembly). The key, however, is contained in the caveat "when done properly."

What follows herein is the discussion of the various ways carbon and low- and medium-alloy steels can suffer embrittlement as fusion welding. Various other problems that can and do affect steels were treated in numerous other places, including

5 By *varieties* of steel are meant low- (0.05–0.25 wt%), medium- (0.25–0.8 wt%), and high- (0.8–~1.4 wt%) alloy carbon steels, low-(less than 4–5 wt% alloy content) and medium- (0.5–10 wt% alloy content) alloy steels, high-alloy (greater than 10 wt% alloy content) tool steels, and stainless steels. By *forms* of steel are meant roll product (e.g. plate, sheet, I, H, angle and channel shapes, rod and bar (including rebar), and wire and cables), castings, and extrusions.

6 There is little doubt that the growth of fusion welding, notably arc welding processes, but also resistance welding processes, was driven by the increased use of steel in fabrication, manufacturing, and construction.

- *Gas porosity*, although not a particular problem in steels, is covered in Chapter 5
- *Hot cracking* in steels (from low-melting impurities) is covered in Section 6.3
- *Cold cracking* in steels from hydrogen, tensile stresses, and as-quenched, untampered martensite formed in the fusion zone (FZ) or the heat-affected zone (HAZ) is covered in Section 6.4 and that in the partially melted zone (PMZ) in Section 10.3. It is also covered as a problem in "transformation-hardening steel HAZs" in Section 11.5
- *Nonmetallic slag inclusions* from slag-protected fusion welding processes is covered in general in Section 7.2, for multi-pass welds in Section 13.3
- *Sigma phase embrittlement* of high-Cr, Fe-based stainless steels and some Ni alloys is covered in Chapter 19
- Possible embrittlement from martensite formation because dilution is covered in Section 20.3.

Let us proceed with *embrittlement* in steels.

15.2 Four Causes of Embrittlement in Carbon and Low- and Medium-alloy Steels

Recall that carbon or mild steels are defined as two-component binary alloys of C dissolved interstitially in Fe.[7] They are subdivided into low-, medium-, and high-carbon grades with ranges specified by the American Iron and Steel Institute (AISI) as 0.05–0.30, 0.30–0.8, and 0.8–2.0 (realistically, rarely higher than about 1.5) wt%, respectively.[8] Low- and medium-alloy steels contain additions of other substitutional solutes (such as Mn, Cr, Mo, Ni) principally added to improve through-hardening or hardening through the thickness (i.e. hardenability) and, perhaps, some strong carbide formers (such as V and W) to provide additional second-phase strengthening. Low-alloy steels are typically considered to contain less than about 4 wt% alloy content but may contain up to near 8 wt%, whereas medium-alloy steels are typically considered to contain 8–10 wt% alloy additions.

Above this level of alloying (typically greater than around 12 wt%), steels are considered *high-alloy steels*, with three categories being (i) high-Mn (i.e. 12–14 wt%) austenitic stainless steels or Hadfield steels used for work hardening to form deformation-induced martensite for wear resistance under impact; (ii) *tool steels* intended to provide special combinations of properties (such as

7 In fact, a small amount of manganese (Mn), typically 0.3 to around 0.6 wt%, is always added to tie up any residual (tramp) sulfur (S) as globules of low-melting MnS, as opposed continuous films of low-melting FeS, in grain boundaries, so as to help prevent hot tearing during mill thermomechanical processing and hot cracking during fusion welding. Any Mn above the amount needed for this purpose serves to provide substitutional solid solution strengthening of Fe. A small amount of silicon (Si), typically 0.15–0.3 wt%, is added to remove any oxygen in the melt (i.e. to deoxidize the steel), help refine grains (by forming SiO_2 at grain boundaries), and make the molten metal more fluid for easier pouring. Si is also a good substitutional solid solution strengthener.

8 An ultrahigh carbon grade, with over 2.0 wt%, exists for extreme hardness in some tools. These steels are typically produced by powder metallurgy (PM) techniques, as they are too brittle to cast, forge, or machine.

resistance to wear and shock loading, resistance to hot distortion, retention of high hardness to high temperatures either while glowing red or with repeated heating and cooling cycling, etc.) using 10–15 wt% total additions of Ni (for toughness), W, or Mo for high-temperature service, V for hardness, etc., to a basic 1.0 wt% C steel; and (iii) *stainless steels* for resistance to corrosion by balancing additions of greater than 14 wt% Cr with additions of Ni and/or Mo.

Only problems of embrittlement in carbon or low- or medium-alloy steels are discussed herein; these problems being divided into one of four types by root cause, as follows:

1. *Hydrogen embrittlement*, properly called *hydrogen-induced or hydrogen-assisted cracking in steels*
2. *Secondary hardening of steels* during post-weld heat treatment or during elevated temperature service
3. Ductile-to-brittle transition in steels at lower temperatures
4. *Residual stress embrittlement* under impact or fatigue loading.

Each of these is discussed individually in the following sections.

15.3 Hydrogen Embrittlement: A Misnomer in Steels

Hydrogen embrittlement properly refers to the process by which hydride-forming (i.e. M_xH_y) metals (such as Ti, Zr, V, Ta, and Nb) become brittle and fracture because of the introduction and subsequent extremely rapid interstitial and grain boundary diffusion of atomic hydrogen (H) into the metal. The term itself is actually a misnomer when used to describe the susceptibility of certain steels to cracking from hydrogen. More accurate terms, as they denote the actual process, are *hydrogen-induced cracking*, *hydrogen-assisted cracking*, or *environmentally assisted cracking*. The difference is that steels (and some other alloys) do not form hydrides and become brittle.

When associated with welding, this problem is alternatively called *cold cracking*, as it occurs only once the weld metal or HAZ cools well below the melting point (MP) or solidus temperature, often near room temperature (RT), or *delayed cracking* (as it often takes time to occur; minutes, hours, days, weeks, months, or even years). As details of the causative mechanism are adequately discussed elsewhere (e.g. Section 6.4 for FZ or HAZ, Section 10.4 for PMZ, and Section 11.5 in detail), they are not discussed here. Suffice to say, to be susceptible to such cracking, a combination of three factors, simultaneously, is required: (i) the presence and diffusion of hydrogen, (ii) susceptibility material microstructure (notably as-quenched, untempered martensite in steels), and (iii) tensile strength, whether externally applied or locked-in (i.e. residual). It can occur whenever and wherever the cooling rate following welding is faster than the critical cooling rate (CCR) for the steel, which is usually fast enough in RT or sub-RT steel. Cracking always occurs by a transgranular mode (i.e. does not follow grain boundaries, but crosses grains at random). Figure 6.6a,b show several common locations, i.e. at the toe and root of a fusion weld in the high-temperature HAZ where martensite is most likely to form on cooling.

Elimination of hydrogen from forming, coating, plating, acid cleaning, and finishing operations during fabrication, manufacturing, or construction is essential. Likewise, any potential source of hydrogen from water or hydrocarbons most be eliminated during fusion welding (see Section 6.4). As for susceptible martensite, two general approaches are as follows: (i) prevent formation by preheating a sufficiently hardenable steel or (ii) immediately post-weld temper at steel weldment that might be susceptible and was not preheated during welding. Susceptibility to the formation of martensite (in the HAZ, if not FZ and PMZ) can be assessed using the carbon equivalent (C_{eq}) for steel, given in footnote 11 in Chapter 6 as C_{eq} = wt% C + wt% Mn/4 + wt% Si/4 < 0.45 for carbon steel or C_{eq} = wt% C + wt% Mn/6 + wt% (Cr + Mo + V)/5 + wt% (Cu + Ni)/15 < 0.35 excellent, 0.36–0.40 very good, 0.41–0.45 good, and >0.45 potentially problematic for low- or medium-alloy steels. Recommendations for preheat temperatures can be found in Section 11.5 or on the Internet.

The key is to avoid problems by proper housekeeping before, during, and after welding and adhering to recommended practices for preheating and post-weld heat treatment (PWHT).

Once martensite forms during welding, the race to avoid potential disaster is on!

15.4 Secondary Hardening in Steels

So-called secondary hardening in steels is the result of lost toughness in the alloy because of the precipitation of embrittling carbides, the most common occurring in steels containing alloying additions of Mo (i.e. Mo_2C), V (i.e. VC), or, while rare, Nb (i.e. NbC). These typically form during slow cooling through the range 400–500 °C/750 to ~950 °F, which can obviously occur following solidification of welds made by fusion processes. Figure 15.1 shows a representative plot of hardness (in Rockwell C) for a steel containing various concentrations of Mo.

This type of embrittlement in steels is alternatively called *temper embrittlement* (as it often occurs during tempering) or *blue brittleness*, as it occurs when a blue-colored oxide heat tint appears on the steel from exposure to 400–500 °C/750 to ~950 °F. It is attributed to unacceptably high concentrations of impurity elements (e.g. Sb, P, and As) at grain boundaries leading to brittleness.

Susceptibility to secondary hardening can be assessed using either of two predictive empirical formulas by Watanabe or Bruscato, as follows:

- *Watanabe J factor*: $J = (Mn + Si)(P + Sn) \times 10^6$ wt% ≤ 180
- *Bruscato X factor*: $X = (10P + 5Sb + 4Sn + As)/100$ in ppm < 20.

Obviously, caution must be exercised when susceptible steels are being fusion welded.

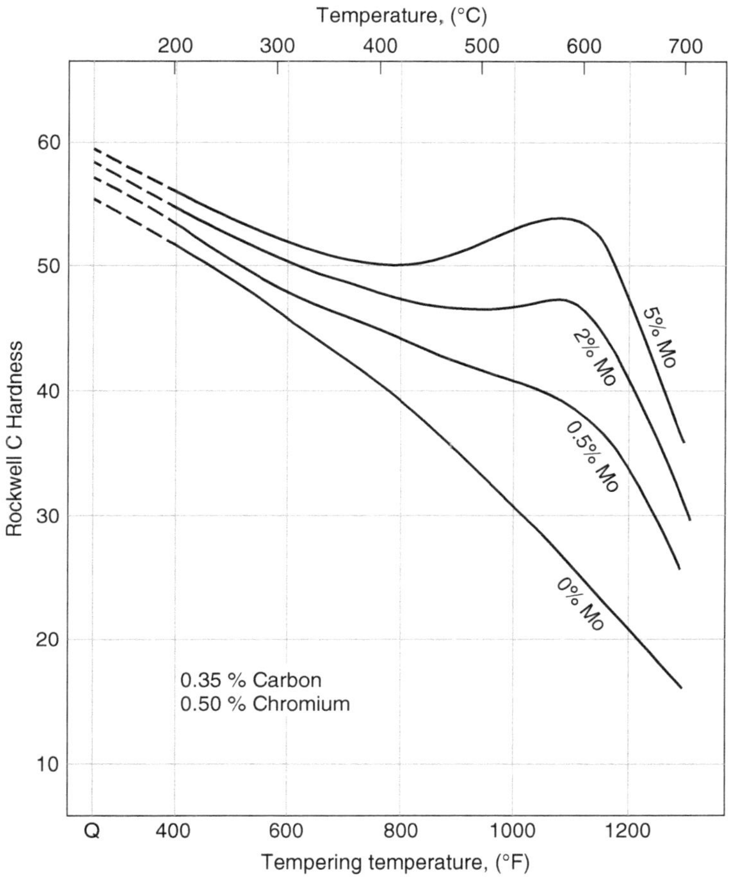

Figure 15.1 Plots showing the phenomenon of secondary hardening during either tempering or in service most notably occurring in steels containing an addition of Mo. Increased susceptibility with increased Mo content is shown, with hardness peaking at around 1100 °F/~590 °C. Source: Bain (1939). Used with permission of the ASM International.

15.5 Ductile-to-Brittle Transition in Steels

Steels, like virtually all metals and alloys having a body-centered cubic (BCC) structure, and some metals or alloys having a hexagonal close-packed (HCP) structure, are prone to *ductile-to-brittle transition* of impact toughness when temperature falls below a certain level, known as the ductile–brittle transition temperature or DBTT (Figure 15.2). Metals and alloys having a face-centered cubic (FCC) structure are immune to such behavior (retaining virtually the same

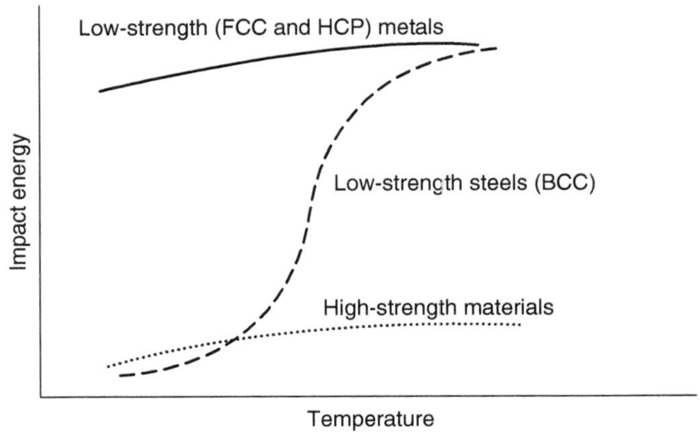

Figure 15.2 Schematic plots showing the three general types of behavior for Charpy V-notch impact energy versus temperature, as influenced by the metal's or alloy's crystal structure. Note the pronounced loss, and often sudden transition, of energy absorption as temperature is lower for BCC metals and alloys, epitomized by steels. Source: Callister (2007a). Used with permission of Wiley.

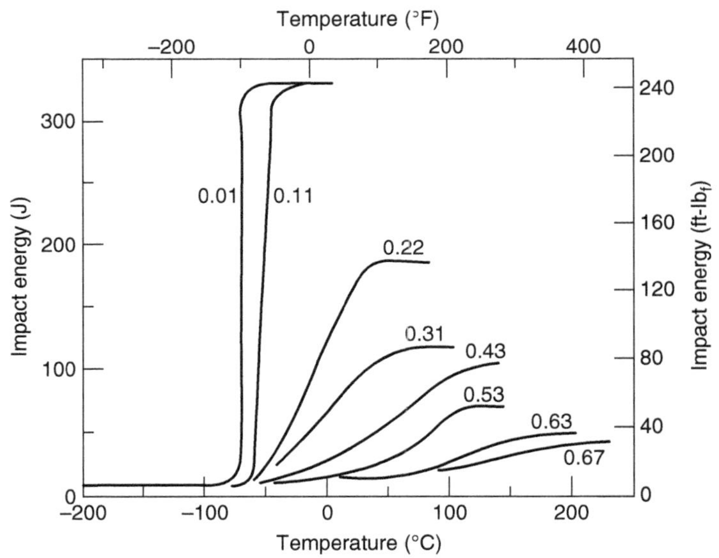

Figure 15.3 Plots showing the effect of carbon content on the Charpy V-notch energy absorption versus temperature in C steels. Source: Callister (2007b). Used with permission of the ASM International.

impact toughness to cryogenic temperatures to absolute zero). This behavior is known to have been the root cause for the catastrophic fracture of the rivets in the hull of the HMS Titanic in the sub-27 °F/−3.3 °C water of the North Atlantic on 14 April 1912.

Loss of impact toughness by carbon steels, as well as low- and medium-alloy steels, is a well-known phenomenon (Figure 15.3). Its particular occurrence in welded structures is often the consequence of welds and their HAZ's cooling faster than the alloy's CCR and/or increased hardenability because of C-enrichment of the weld metal due to base metal (BM) dilution. In either case, fusion welds tend to be more susceptible to DBTT embrittlement than usual.

15.6 Compromise of Fatigue and Impact Behavior by Residual Stresses in Steels

Fusion welds routinely undergo volumetric shrinkage in the FZ and thermal contraction in the HAZ (see Sections 4.3 and 4.4). As a consequence, the development of residual stresses in and around a fusion weld is the rule rather than the exception. Such residual stresses vary with location along (i.e. longitudinally), across (i.e. transverse), and through-the-thickness (of thick section welds). Because applied external stresses add algebraically to locked-in residual stresses, both fatigue and impact behavior are affected, almost always adversely. The problem was briefly addressed in Section 12.4.

If and when a weldment is not post-weld heat treated or otherwise treated to relieve weld-induced residual stresses, both fatigue life (i.e. endurance) and tolerance of impact are compromised (Figure 15.4). Unrelieved residual stresses also compromise the resistance of the weld region (i.e. FZ and HAZ) to corrosion. The particularly insidious aspect of such vulnerability is the inability to predict vulnerability because of the complexity of the analysis.

Figure 15.5 shows the benefit of higher temperatures and longer times for relieving residual stresses to lower and safer levels.

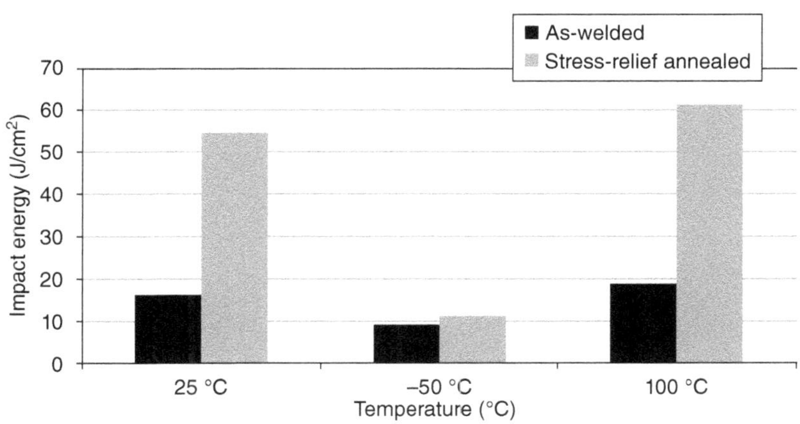

Figure 15.4 Bar graphs showing the effect of post-weld stress relief heat treatment on the impact energy compared to as-welded steels at three different test temperatures. Source: Emre and Kaçar (2015). Used with permission of Materials Research São Carlos.

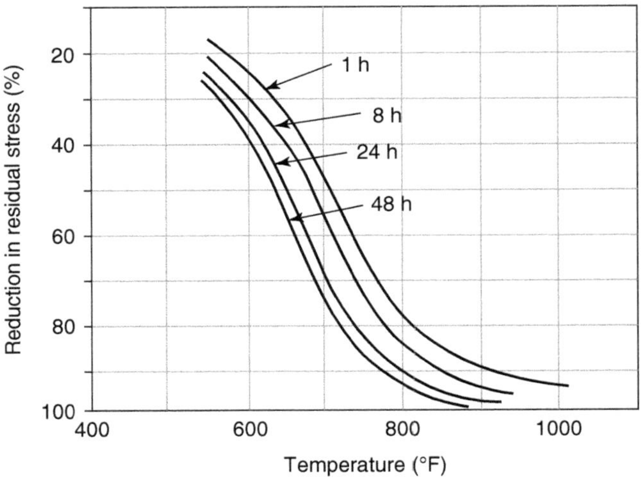

Figure 15.5 Illustrative plots showing the general increased effect of higher temperatures and longer times of PWHT on the reduction of residual stresses.

15.7 Troubleshooting Problems from Embrittlement of Steels by Welding

Table 15.4 lists the various ways problems can arise from embrittlement arising from welding steels.

Table 15.4 Troubleshooting problems from embrittlement of carbon and low- and medium-alloy steels.

Hydrogen-Assisted Cracking:	
Transverse cracking in weld metal (WM) or, more likely, HAZ during cooling after solidification or later	• Eliminate all sources of hydrogen (Section 6.4) • Prevent formation of martensite in C steels by keeping $C_{eq} < 0.45–0.5$; in low- or medium-alloy steels by keeping $C_{eq} < 0.4$ • Preheat as recommended • PWHT to relieve stresses and temper
Secondary Hardening:	
Intergranular cracking in steels containing strong carbide formers (Mo, V, and Nb)	• Check susceptibility using Watanabe *J* factor and/or Bruscato *X* factor • Use PWHT recommendations of manufacturer
Ductile-to-Brittle Transition:	
Loss of impact energy absorption below certain transition temperature (DBTT)	• Avoid using steels below their established DBTT
Compromised Fatigue and Impact Toughness:	
Loss of fatigue or impact toughness w/o stress relief (SR)	• Post-weld stress relieve as recommended by manufacturer

References

Bain, E.C. (1939). *The Function of the Alloying Elements in Steel*, figure 155, p. 248, 1e. Materials Park, OH: American Society for Metals.

Callister, W.D. Jr., (with assistance from David G. Rethwisch) (2007a). *Materials Science and Engineering: An Introduction*, figure 8.15, p. 226, 7. New York, NY: Wiley.

Callister, W.D. Jr., (with assistance from David G. Rethwisch) (2007b). *Materials Science and Engineering: An Introduction*, figure 8.16, p. 227, 7e. New York, NY: Wiley (but originally from J.A. Reinbolt and W.J. Harris, Jr. (1951). Influence of carbon content on the Charpy V-notch energy versus temperature behavior of steel, *Transactions of ASM*, 43.

Emre, H.E. and Kaçar, R. (2015). Effect of post-weld heat treatment process on microstructure and mechanical properties of friction welded dissimilar drill pipe [steels], figure 8. *Materials Research* 18 (3): www.materialsresearch.org.br.

Further Reading

Linnert, G.E. (2009). *Welding Metallurgy: Carbon and Alloy Steels*, Volume 1 – Fundamentals, Volume 2 – Technology, 4e. Miami, FL: American Welding Society.

Ustinovshchikov, Y.I. (2013). Secondary hardening mechanism of alloy steel. *Journal of Metal Science* 18 (7): 337–344.

16

Sensitization or Weld Decay and Knife-line Attack in Stainless Steels

16.1 A Primer on the Metallurgy of Stainless Steels

The so-called *high-alloy steels* (i.e. Fe containing more than 10% by weight of total alloying additions, other than C) are usually divided into three categories: (i) high manganese austenitic steels (once commonly called "Hadfield steels"), (ii) tool steels, and (iii) stainless steels. The first category finds use in wear applications that involve high compressive forces from either friction and/or impact (e.g. railroad switch plates and frogs and rock crusher cones or swing hammers). Additions of 13–14 wt% Mn to Fe containing 0.8–1.25 wt% C render the alloy austenitic, but with the ability of the tough, ductile austenite to transform into strong, wear-resistant martensite by shear deformation, that is to work harden.[1] The second category refers to a variety of carbon and alloy steels (with 0.5–1.5 wt% C) that are particularly well suited to especially demanding applications requiring some combination of hardness and/or retention of hardness at elevated temperatures, resistance to deformation at elevated temperatures, resistance to shock loading, resistance to adhesive or abrasive wear, and/or ability to retain a cutting edge at elevated temperatures. Such hard-to-achieve combinations of properties are commonly required in tools (e.g. high-speed cutters, dies and molds, and chisels). The third category refers to alloys especially designed for resistance to corrosion, particularly, but not only, involving wet environments.[2]

Resistance to corrosion is the result of adding enough Cr to Fe as a substitutional solute, with greater resistance generally afforded as the concentration of Cr increases above around 14 wt%. Chromium preferentially reacts with oxygen from the atmosphere compared to iron,[3] forming a translucent layer of Cr_2O_3

1 Deformation-induced martensite transformation can occur in austenite that is metastable, as occurs in high-Mn Hadfield steels and in austenite that is retained in steels with an M_f temperature below room-temperature and is subjected to strain. Such a transformation is logical in that austenite transformation to martensite always involves massive shears based on Kurdjumov–Sachs or Nishiyama–Wassermann models of the crystallographic orientation relationship between FCC austenite and BCC or BCT martensite.

2 Corrosion is typically divided into dry and wet corrosion. Dry corrosion typically occurs in the absence of water, in various gaseous reactants, and involves surface oxidation. Wet corrosion typically involves an aqueous environment and involves electrochemical reactions.

3 The more negative the heat of formation ΔH_f° for a chemical reaction, the more readily that reaction occurs. The values of ΔH_f° for various oxidation reactions in stainless steel

A Practical Guide to Welding Solutions: Overcoming Technical and Material-Specific Issues,
First Edition. Robert W. Messler, Jr.

just a few molecules thick, but with tremendous adhesion to the underlying iron (i.e. it is tenacious) and imperviousness to oxygen atoms from the atmosphere diffusing in and to iron atoms from the substrate diffusing out. As such, it provides a passivating layer on the surface of the stainless steel alloy that renders the underlying iron unreactive, thereby protecting it from corrosive attack. As it forms naturally, this passivating layer is self-healing if scratched.

All stainless steels contain a relatively low concentration of C (typically, less than 0.08 wt%)[4] to provide interstitial strengthening without extending the so-called gamma-loop/γ-loop on the Cr–Fe equilibrium phase diagram (Figure 16.1), within which face-centered cubic (FCC) γ-Fe is stable over body-centered cubic (BCC) α-Fe. Stainless steels are classified into four major categories or types or grades based on their primary microstructure: (i) ferritic, (ii) martensitic, (iii) austenitic, and (iv) duplex (e.g. ferrite + austenite).[5] Each type

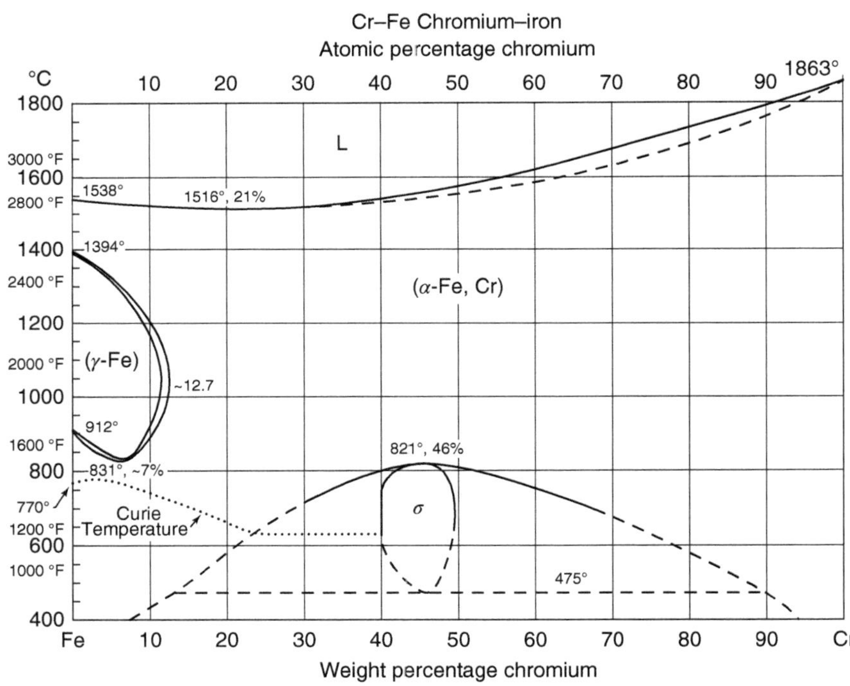

Figure 16.1 Schematic illustration of the Cr–Fe equilibrium phase diagram. Source: ASM International (2016). Published by and used with permission of the ASM International.

are −1128 kJ/mol for Cr_2O_3, −822.2 kJ/mol for Fe_2O_3, −226.5 kJ/mol for FeO, and −244.3 kJ/mol for NiO. Hence, Cr_2O_3 forms preferentially to any oxide of iron.

4 The exceptions are martensitic stainless steels, which have higher concentrations of C (typically, 0.15–0.25 wt%) to (i) extend the γ-loop of the Cr–Fe equilibrium phase diagram to higher Cr contents in order to allow heat treatment into the austenite region to then allow martensite formation upon quenching and (ii) for increased strength and, particularly, hardness, as a principal use of martensitic stainless steels in surgical instruments and cutlery.

5 There are also precipitation-hardenable (PH) stainless steels, which can be artificially aged to increase the strength of certain ferritic, austenitic, or martensite types.

has its preferred areas of application based on the relative degree of resistance to corrosion versus strength and/or hardness. The various types are created by manipulating and balancing the relative proportions of γ-austenite-stabilizing solutes (e.g. principally achieved with Ni) versus α-ferrite-stabilizing solutes (principally Cr, secondarily Mo). Austenitic stainless steels are the most corrosion resistance as a result of their higher Cr contents, with some compromise of strength as a result of low C content and the FCC structure of austenite.

16.2 Sensitization of Austenitic Stainless Steels by Welding

Austenitic stainless steels containing more than about 0.06–0.065 wt% C are often rendered susceptible to intergranular corrosion in the HAZ of any welds. The general phenomenon that can be the result of any prolonged exposure[6] in the range 500–800 °C/~925–1475 °F is known as *sensitization* for the way the material is rendered sensitive to corrosive attack and as *weld decay* for the way the sensitized region heated above about 800 °C/1472 °F in the HAZ of a weld is eaten away by corrosion. In fact, sensitization of the microstructure leads to weld decay around the welds.

Sensitization is caused when mixed $M_{23}C_6/M_7C_3$ Cr-rich Cr,Fe carbides preferentially precipitate along inherently high strain–energy grain boundaries in austenite with prolonged exposure in the sensitization range. Although formation can begin at as low as 426 °C/799 °F, it occurs exponentially faster[7] as the temperature is raised above this level. In welding, risk becomes especially high when the temperature in the HAZ is between 600 and 850 °C (about 1100 and 1550 °F), the so-called *sensitization range*. When Cr-rich carbides form as Cr and C in a solid solution in the austenite diffuse to grain boundaries, Cr content is depleted or "denuded" in regions surrounding the newly formed grain boundary carbides (Figure 16.2). Once the Cr content in the solution drops below around 12 wt%, the corrosion resistance of the material is drastically reduced. In fact, the Cr-depleted region forms a galvanic couple with the nondepleted, higher Cr region, becoming relatively anodic and susceptible to corrosive attack (Figure 16.3).

Sensitization in austenitic stainless steels is more severe under the following conditions:

- the higher the carbon content above about 0.06 wt%
- the greater the net linear heat input to a weld
- with any residual cold work done before welding
- in the absence of any alloying additions (i.e. substitutional solutes) with greater affinity for C than Cr.

When it occurs, sensitization occurs only in a narrow band (or temperature range) in the HAZ (curve b in Figure 16.4) because for cooling rates above

6 Exposure can be from heat treatment or elevated temperature service, not only from the heat of welding.

7 As a rule of thumb, the time required for a formation reaction or transformation that occurs at an exponential rate is halved for each 30 °C/50 °F increase in temperature.

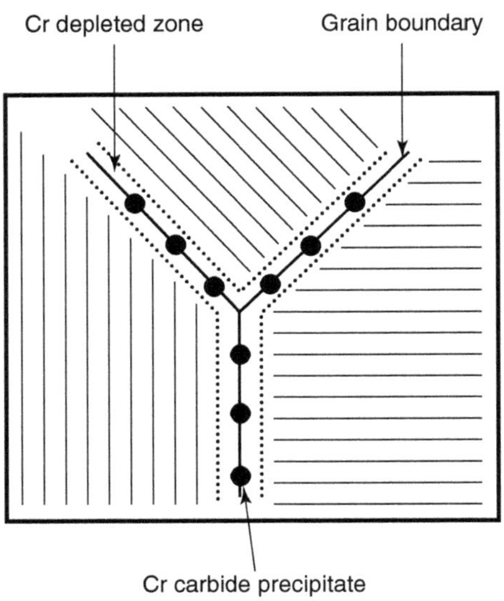

Figure 16.2 Schematic representation of the depletion of Cr in the region of grain boundaries in austenitic stainless steel caused by the preferential precipitation of $M_{23}C_6$- and M_7C_3-type Cr,Fe carbides in these regions. This depletion renders the grain boundary regions anodic relative to the nondepleted grain interiors, thereby sensitizing the area to galvanic corrosive attack in a phenomenon known as "sensitization." Source: Kou (1987). Published by and used with permission of John Wiley & Sons.

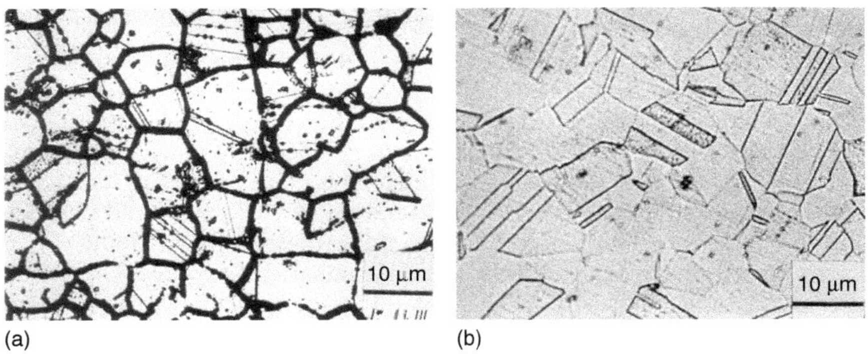

Figure 16.3 Photomicrograph showing intergranular corrosive attack from sensitization and, hence, weld decay, in the HAZ of a weld in a susceptible austenitic stainless steel, here 304 (a). Normal, unsensitized 304 is shown for comparison (b). Source: From Wikipedia under "Intergranular Corrosion," with the two images being free for use under Creative Commons Attribution ShareAlike CC BY-SA 3.0.

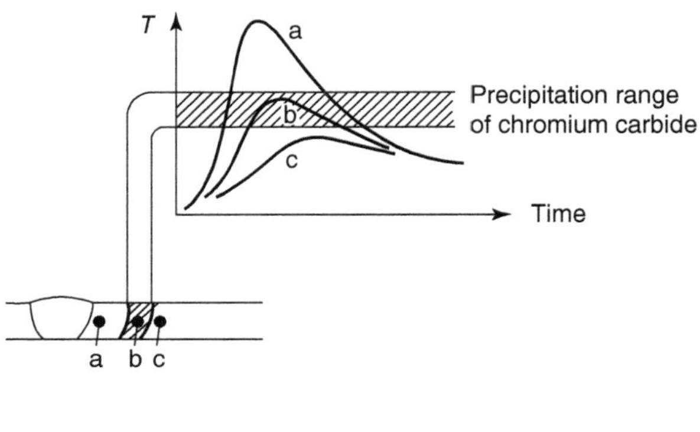

Figure 16.4 Schematic plots showing weld thermal cycles at various locations in the HAZ of a weld and the range of temperatures in which sensitization can occur in austenitic stainless steels. Source: Kou (1987). Published by and used with permission of John Wiley & Sons.

a certain level (curve a in Figure 16.4), the time available for diffusion in the sensitization range is too short to allow enough carbides to form to denude grain boundary regions. Also, if the peak temperature of a welding thermal cycle is too low (curve c in Figure 16.4), precipitation cannot occur because nucleation does not occur.

Knowing the mechanism by which sensitization occurs in the HAZ of welds made in solid solution in austenite stainless steels, there are several ways the problem can be avoided and one by which it can be resolved once it has occurred. The methods for avoiding the problem, in order of preference, are as follows.

In the first method, the carbon content of the selected stainless steel can be lowered to reduce the total amount of Cr,Fe carbides that can be formed, thereby reducing the total amount of Cr that can be removed from the solid solution matrix phase. When problems from sensitization were first reported, material producers responded by producing the so-called low-carbon or L grades (e.g. 304L and 316L), with carbon content limited to 0.035 wt% max. versus 0.08 wt% max. specified for standard grades, which routinely contained more than 0.06–0.07 wt% C. This lowered the strength of the alloy from less interstitial solid solution strengthening by C but overcame the immediate problem of general sensitization and weld decay. To offset the loss strength caused by reducing the carbon content, small amounts (0.01–0.03 wt%) of nitrogen (N) were added, as N is also a potent interstitial solid solution strengthener, as well as an austenite stabilizer, like C. Grades with intentionally added N are designated by an N post-script, such as 304LN.

With modern steelmaking practice, austenitic stainless steels typically contain 0.035–0.040 wt% in standard grades and, because of other controls on composition and thermomechanical processing, do not suffer any loss of strength. In reality, it is hard to find modern standard grades of austenitic stainless steels that sensitize to the degree resistance to corrosion in most traditional environments for the alloy is degraded.

In a second and often preferable method, a small amount of a metallic element having greater affinity for C (and, thus, carbide formation) than Cr can be added to a austenitic standard-grade stainless steel. The two commercially available options involve addition of Ti to type 304 to produce type 321 and Nb to type 304 to produce type 347.[8] Formation of either TiC in type 321or NbC in type 347/347H prevents the depletion of Cr and, thus, sensitization, with the alloys being said to have been *stabilized.*

An always advisable third method to attempt to prevent weld decay in susceptible grades of austenitic stainless steel is to keep the net linear heat input as low as possible for producing welds of the required depth and width or volume. To the degree it is feasible, efforts should be made to keep the peak temperature in the HAZ to under 700 °C/1300 °F and to cool through the sensitization range 600 and 850 °C/1100 and 1550 °F as quickly as possible, perhaps using forced cooling, copper chill blocks, heat damming materials, or, for alloys not susceptible to hydrogen-induced cracking, moistened towels.

---

8 The specification for type 321 calls for Ti additions of $5 \times (C + N)$ min. to 0.70 max. wt%. The specification for type 347 calls for Nb additions of $10 \times (C + N)$ min. to 1.00 max. wt%, whereas for the higher C, more creep-resistant type 347H calls for $8\times(C + N)$ min. to 1.00 max. wt%.

Once it is suspected that sensitization has occurred in the HAZ of a weld made in austenitic stainless steel, it is theoretically possible, but generally not practical, to post-weld heat treat the weldment to redissolve the precipitated Cr,Fe carbides. This *solutionizing treatment* or *solution anne*al requires heating to 1000–1100 °C/1800–2000 °F for a minimum of 1 h/in. (25 mm) of thickness (in the thickest region near the weld), followed by rapid quenching (usually into water) to retain the redissolved Cr in the solid solution matrix phase of the alloy. The obvious limited practicality of this approach relates to the likelihood of distortion to large and/or geometrically complex weldments while at such high temperatures (where they plastically soften) and/or during rapid cooling by quenching. If possible, using restraining fixtures, however, the method is very effective.

16.3 Sensitization of Other Grades of Stainless Steel

Some grades of ferritic (e.g. 430 and 446) as well as duplex ferritic–austenitic stainless steels (e.g. 2205) can exhibit sensitization, in general, and HAZ weld decay following welding. In both types, the temperature range for sensitization is much higher than for the austenitic grades, typically, above 925 °C/1750 °F. Hence, when weld decay occurs, it occurs very near the fusion zone (FZ) boundary. Also, unlike for the austenitic grades, lowering the carbon content even below 0.010 wt% fails to prevent sensitization, although addition of elements such as Ti or Nb is helpful for stabilizing Cr in solution.

Immunity to preferential intergranular corrosive attack can be restored following welding by solution annealing at 650–815 °C/1200–1500 °F for 30–60 minutes.

Prevention by keeping net linear heat input as low as practical is, as usual, advisable.

16.4 Knife-line Attack in Stabilized Austenitic Stainless Steels

Resorting to stabilized grades of austenitic stainless steels (i.e. 321, 347, or 347H) to avoid problems with weld decay is not a sure-fire cure-all. One may be swapping weld decay for *knife-line attack*. *Knife-line attack*, like weld decay, is the result of preferential precipitation of Cr,Fe carbides along the grain boundaries, with an attendant depletion of Cr in these regions and subsequent corrosive attack upon exposure to certain corrosive environments. Unlike weld decay, knife-line attack occurs only in stabilized grades of austenitic stainless steels, and nowhere else. The term "knife-line attack" comes from the fact that corrosive attack occurs in a very narrow, knife-line-like band in the area immediately outside the fusion boundary.

Figure 16.5 schematically illustrates key aspects of the mechanism responsible for knife-line attack. In the schematic plot of thermal cycles at two locations outside the weld FZ thermal cycle "a" is for a local very close to the fusion boundary, where the peak temperature is very high and the cooing rate after the

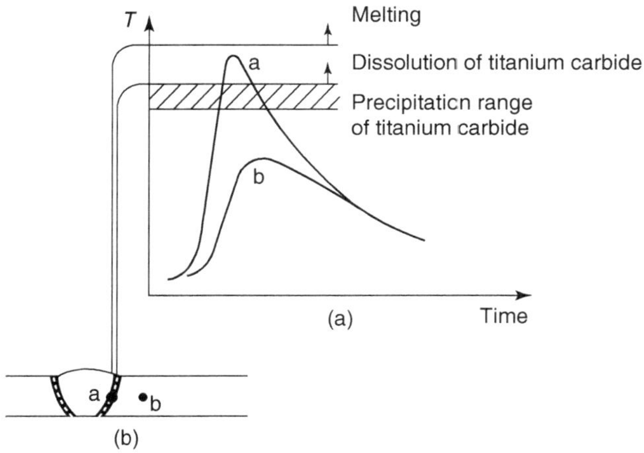

Figure 16.5 Schematic plots showing weld thermal cycle (a) responsible for the occurrence of knife-line attack in the region of an HAZ where carbides with the stabilizing solute addition are dissolved during welding, allowing for the formation of Cr,Fe carbides at lower temperatures during post-weld heat treatment or during elevated-temperature service. Farther from the weld, weld thermal cycles (b) do not reach peak temperatures high enough to cause precipitation of the more refractory carbides. Source: Kou (1987). Published by and used with permission of John Wiley & Sons.

peak (i.e. following welding) is very rapid. As the peak temperature is above the solvus temperature of either Ti or Nb carbide (i.e. TiC or NbC), these carbides are dissolved[9] in the solid solution matrix phase. Because of rapid cooling through its precipitation range (shown cross-hatched in the figure), new Ti or Nb carbide does not reprecipitate as cooling occurs, thus leaving abundant C in solution. Upon reheating into the lower temperature range in which Cr,Fe carbide formation occurs during either post-weld stress relief heat treatment or during elevated temperature service, Ti or Nb carbide does not form because the temperature is not high enough. Cr,Fe carbide, in the other hand, does form at grain boundaries, rendering them susceptible to corrosion because of Cr depletion. Wherever the peak temperature in the HAZ is below the solvus temperature for either Ti or Nb carbide (for example, thermal cycle c in the figure), these carbides do not dissolve.

Multi-pass welding can also lead to knife-line attack, when one pass dissolves Ti or Nb carbide and a subsequent pass causes precipitation of Cr, Fe carbides along the grain boundaries. For this reason, care must be exercised in deciding on welding sequence in locations on weldments where exposure to corrosive environment is expected.

The likelihood of knife-line attack can be reduced by (i) additions of certain rare-earth elements (e.g. La and/or Ce) to stabilized grades of austenitic stainless steel, as accelerated precipitation of carbides of these metals leaves less carbon

9 *Dissolved* is different than *melted*. Dissolved means a carbide, for example, here, TiC or NbC, dissociates into its component atoms, as they diffuse into a host solvent (here, the solid solution matrix austenite of the stabilized-grade of stainless steel). Both dissociation and subsequent diffusion occur entirely in the solid state. Melting means a solid transforming into a liquid without changing its chemical form (e.g. pure solid Fe transforming into pure liquid Fe or pure solid sugar transforming into molten liquid sugar of the same type, e.g. sucrose, $C_{12}H_{22}O_{11}$).

in the solution to form unwanted Cr,Fe carbides during later heating, (ii) using low-C grades versus stabilized grades, and (iii) very high temperature post-weld solution annealing and quenching, when practical.

If final applications justify the time, effort, and cost, susceptibility to both weld decay and knife-line attack can be evaluated by exposing sensitized materials to corrosive agents using either the Huey test (with boiling 65% nitric acid) or the Strauss test (using much safer copper sulfate solution). Attack can be detected and assessed by metallographic examination or by testing the exposed piece by listening for a ring when struck by a metal object or when dropped. If there is no ring with a tap or a thud or shattering with a drop, the test sample has been severely attacked.

16.5 Troubleshooting Problems from Sensitization or Knife-line Attack

Table 16.1 provides some guidance for troubleshooting problems encountered with sensitization or weld decay and knife-line attack in stainless steels.

Table 16.1 Troubleshooting problems with sensitization or weld decay and knife-line attack in stainless steels.

Sensitization or Weld Decay:	
▪ General sensitization occurs after prolonged exposure at 500–800 °C/925–1425 °F in aust.-SS with >~0.06 wt% C	▪ Limit exposure above 500 °C/925 °F ▪ Switch to aust. SS with <0.045 wt% C ▪ Switch to stabilized 321 or 347 ▪ Solution anneal weldment per text
▪ Weld decay occurs in the HAZ of above where peak temperature exceeds ~800 °C/1425 °F	▪ Switch to aust. SS with <0.045 wt% C ▪ Switch to stabilized 321 or 347/347H ▪ Lower net linear heat input ▪ Use short-cycle welding process ▪ Solution anneal weldment per text
▪ General sensitization or weld decay occurs in ferritic 430/446 and duplex 2205 with exposure >925 °C/1250 °F	▪ Minimize net linear heat input ▪ Use short-cycle welding process ▪ Solution anneal weldment per text
Knife-line Attack in type 321 or type 347:	
▪ Occurs in very narrow band of HAZ near FZ ▪ Can occur with multi-pass welds	▪ Employ 321 or 347/347H with additions of La or Ce ▪ Switch to low-C grade; <0.045 wt%

aust. SS, austenitic stainless steel.
Test susceptibility for any of the above using Huey test or Strauss test.

References

ASM International (2016). *Metals Handbook*, Volume 3: Alloy Phase Diagrams, 9e. Materials Park, OH: ASM International.
Kou, S. (1987). *Welding Metallurgy*. New York, NY: Wiley.

Further Reading

Lippold, J.C. and Kotecki, D.J. (2005). *Welding and Weldability of Stainless Steel*, 1e. New York, NY: Wiley-Interscience/Wiley.
Sedriks, A.J. (1996). *Corrosion of Stainless Steel*, 2e. New York, NY: Wiley.

17

Stress Relief Cracking of Precipitation-Hardening Alloys

17.1 Different Names, Same Phenomenon

As out of place as it may seem in a book on resolving welding problems, when the subject of stress relief cracking is addressed, an apropos well-known quote from Shakespeare's tragic play *Romeo and Juliet* comes to mind[1]:

> 'Tis but thy name that is my enemy;
> Thou art thyself, though not a Montague.
> What's Montague? It is nor hand, nor foot,
> Nor arm, nor face, nor any other part
> Belonging to a man. O, be some other name!
> What's in a name? That which we call **a rose**
> **By any other name would smell as sweet;**
> So Romeo would, were he not Romeo call'd,
> Retain that dear perfection which he owes
> Without that title. Romeo, doff thy name,
> And for that name which is no part of thee
> Take all myself.

The meaning of this wonderful line is that what matters is what something is, not what it is called.

When it comes to *stress relief cracking*, alternatively referred to as *stress relaxation cracking*, *reheat cracking*, and *strain-age cracking*, what matters is not what it is called, but that it is bad! Different names, however, tend to lead the uninitiated to believe that the four are different. They are not!

A major cause of failures in creep-resistant precipitation-hardening (PH) alloys such as ferritic alloy steels, stainless steels, and Ni-based superalloys is cracking during post-weld stress relief. Because the purpose of such post-weld heat treatment (PWHT) is twofold, i.e. (i) to relieve weld-induced residual stresses and (ii) to develop maximum strength by aging, two logical names are used to describe the phenomenon. The first is *stress relief cracking* and the second is *strain-age cracking*. Not illogically, because thermal stress relief results in stress

1 William Shakespeare (1564–1616) wrote the play *Romeo and Juliet* in 1600, telling of forbidden and doomed love between members of rival families in Verona, i.e. Juliet Capulet and Romeo Montague.

A Practical Guide to Welding Solutions: Overcoming Technical and Material-Specific Issues,
First Edition. Robert W. Messler, Jr.

relaxation, any cracking that results during the PWHT is sometimes referred to as *stress relaxation cracking*. Because all cracking that occurs during PWHT occurs upon reheating the weldment following welding, the final common name is *reheat cracking*. Four names, one phenomenon. A rose is rose is a rose is a rose – by any other name!

High creep-resistant[2] alloys are used in many applications in fossil fuel, nuclear, and even solar concentrator[3] power generation plants, chemical- and petroleum-processing plants, and elsewhere where superheated steam, for example, is generated and used. As but one example, steam pipes and headers in power plants carry superheated steam, under pressure, up to 600 °C/~1100 °F for the lifetime of the plant (typically, designed for 40 years, but, in practice, often up to 50 for fossil fuel plants and expected to be 50–70 for nuclear plants). During such long service lives, sections of pipes typically need repair to fix creep thinning in high-flow-rate regions, with damaged sections being cut out and replaced by welding in a new section. PWHT is required to relieve stresses in the weld heat-affected zones (HAZs) *and* develop strength in the alloy used in the new section by thermal aging, so it is not an option to not post-weld heat treat these alloys. Stress relief cracking usually (but not always) initiates in the HAZ of the new welds and can, propagating by an intergranular mode, extend into the unaffected base metal.

As but a brief primer on the phenomenon of creep in metals and alloys, the following should suffice, leaving more details to interested readers to seek information elsewhere:

- Creep is a progressive form of permanent plastic deformation under constant stress
- It occurs at stresses below the yield stress/yield strength for a material once sustained (or constant) temperature exceeds about $0.4T_{MP}/0.4T_{solidus}$ on an absolute temperature scale (K or R)
- It occurs in three stages as depicted in Figure 17.1, with the onset beginning at the material's elastic limit at temperature, and the transitions between each stage occurring at shorter times, the slope of the steady-state region increasing (i.e. becoming steeper), and final fracture by rupture occurring at lower stresses and shorter times as temperature increases
- During first-stage primary creep, the creep rate decreases with time because of strain hardening (see Section 18.1)
- During second-stage secondary or steady-state creep, the creep rate is constant (i.e. creep rate = $\Delta\varepsilon/\Delta t$) and occurs at a minimum rate because of simultaneous strain hardening and thermal recovery processes (see Section 18.1)

---

2 *Creep* in a material refers to continued, usually slow strain under a sustained (usually constant) stress at temperatures above about $0.4T_{MP}/0.4T_{solidus}$ on an absolute temperature scale (K or R). Creep occurs faster as temperature increases above this minimum, always ending in ductile rupture. Although creep can occur at or near RT for low-MP metals (e.g. Pb) and alloys (e.g. solders), resistance to creep at higher temperatures favors alloys based on solvent metals with a high MP, such as, in ascending order, Ni (1453 °C/2647 °F), Co (1498 °C/2728 °F), Fe (1536 °C/2797 °F), Ti 1668 °C/3020 °F), Nb (2415 °C/4379 °F), Mo (2610 °C/4730 °F), and W (3410 °C/6170 °F).

3 In a solar concentrator plant, water is heated to produce superheated steam using sunlight concentrated to a high energy density. Arrays of mirrors are typically used to focus the Sun's rays onto a steam generator.

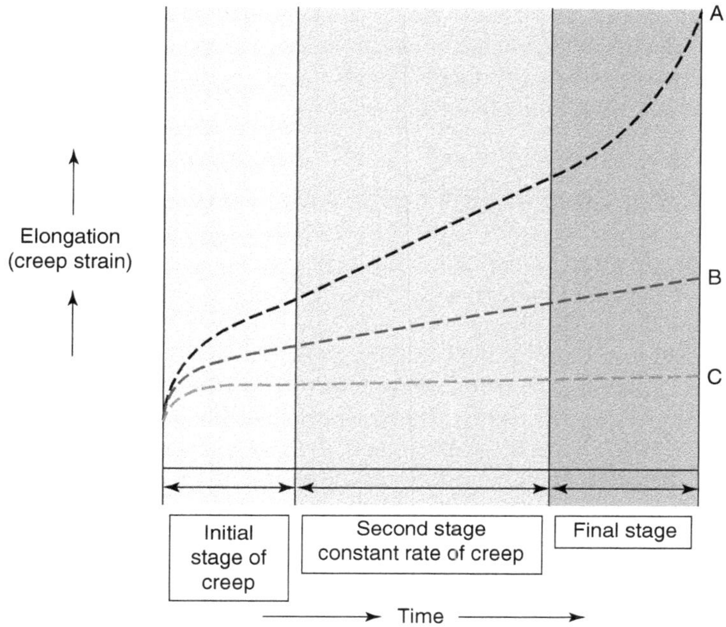

Figure 17.1 Schematic plot of typical creep behavior in metals or alloys, with the three characteristic stages and features labeled, and shown for a high (Curve A), intermediate (Curve B), and low (Curve C) temperature. Source: Messler (2011). Used with permission of Jones & Bartlett Learning.

- During third-stage tertiary creep, the creep rate increases with time leading to localized necking (reduction in cross-sectional area), true stress (load over actual or instantaneous area) increases, and fracture by overload with the following applying:

  $$\text{Creep rate} = C\sigma^{n} \exp\left(\frac{-Q_C}{\text{RT}}\right)$$

  and

  $$\text{Rupture time } t_r = K\sigma^{m} \exp\left(\frac{Q_m}{\text{RT}}\right)$$

  where R is the gas constant; T is the temperature in Kelvin; C, K, n, and m are constants for the material; Q_C is the activation energy for creep; and Q_r is the activation energy for rupture.
- Creep stress rupture data are typically represented by the Larsen–Miller parameter (P) given by

  $$P = T[\log\, t_r + C]$$

  where T is the temperature in Kelvin, t_r is the stress rupture time in hours, and C is a constant of the order of 20.
- At a given stress level, $[\log t_r + C]$ remains constant for a given material.

Although the mechanism of all four-named cracking phenomenon is fundamentally the same, details differ enough between the three most susceptible alloy classes, i.e. (i) ferritic alloy steels, (ii) stainless steel, and (iii) Ni-based superalloys, to merit separate discussion, in the following sections.

17.2 Stress Relief Cracking in Ferritic Alloy Steels

As described in Section 17.1, creep is a long-term mechanism of failure that generally occurs at elevated temperatures.[4] Among Fe-based alloys, a particularly important family of these alloys possess a ferritic structure; i.e. the primary matrix phase is a solid solution of high-melting-point substitutional solutes in the body-centered cubic (BCC) ferrite form of iron (MP = 1536 °C/2797 °F). As such, they are called *ferritic alloy steels*. The two principal alloying elements are Cr (MP = 1875 °C/3407 °F) and Mo (MP = 2610 °C/4730 °F), but V and Nb are sometimes added. Rather than going into solid solution in the ferrite, as strong carbide formers, these latter two form second-phase carbides that strengthen as they precipitate. Within this family of alloys is an important subset known as high-strength low-alloy (HSLA) steels.

As already mentioned by reference to the common addition of V and/or Nb to ferritic alloy steels, creep-resistant steels all contain strong carbide- and/or nitride-forming solutes. Not only are these additions intended to provide a fine dispersion of refractory carbides that increase strength but also to impede the formation of microvoids at grain boundaries, which lead to decohesion[5] or rupture. A key function of the major addition of 2–13 wt% Cr in ferritic alloy steels is to reduce oxidation or scaling of the steel at high temperatures. Being ferritic, as opposed to martensitic or austenitic, ferritic alloy steels also have resistance to hydrogen attack and corrosion by sulfur-bearing hydrocarbons. For all of these reasons, they are widely used in fossil fuel-fired power generation and in oil and gas industries.

Each grade of ferritic alloy steels has a creep limit (i.e. a stress and temperature above which it should not be used) but also a similar limit on oxidation resistance. The limits of both increase with the alloy content, enabling the more highly alloyed grades to be used up to 650 °C/1200 °F.

Table 17.1 lists some of the more popular creep-resistant ferritic alloy steels, along with their compositions and key mechanical properties.

As the alloy content of ferritic alloy steels increases, so does the ease with which they form martensite (i.e. hardenability) when cooled rapidly from their austenitic condition at higher temperatures. Not surprisingly, this suggests that

4 Creep tends to occur in metals and alloys above around $0.4T_{MP}/0.4T_{solidus}$ on an absolute temperature scale (*K* or *R*), and above around half the absolute melting point of ceramics (i.e. crystalline compounds of metals and nonmetals), which often exhibit so-called congruent melting. An analogous phenomenon of stress relaxation known as "cold creep" or "cold flow" occurs in polymeric materials above their glass transition temperature, which is about 0.4 times the absolute temperature at which thermosetting types either char or decompose and thermoplastic types essentially melt.

5 *Decohesion*, in this context, refers to separation, especially along grain boundaries.

Table 17.1 Nominal composition, mechanical properties, and recommended maximum service temperature of common creep-resistant ferritic alloy steels.

	Composition (nominal wt%)					Mechanical properties (typical)				
Steel grade	C_{max}	**Cr**	**Mo**	**V**	**Nb**	**UTS (N/mm²)**	**0.2% Proof (N/mm²)**	**Elongation (%)**	**Charpy-V (J/°C min)**	**Maximum temperature (°C/°F)**
C ½Mo	0.3	—	0.5	—	—	520	320	35	27@20	470/875
½½¼CrMoV	0.14	0.5	0.6	0.25	—	585	330	25	N/A	540/1000
1¼Cr½Mo	0.13	1.2	0.5	—	—	530	350	35	27@20	593/1100
2¼Cr1Mo	0.13	2.25	1.0	—	—	555	350	35	27@20	593/1100
5Cr½Mo	0.13	5.0	0.5	—	—	690	475	28	27@20	649/1200
9Cr1Mo	0.13	9.0	1.0	—	—	675	475	30	N/A	649/1200
9CrMoVNb (9Cr mod or P91)	0.13	8.75	1.0	0.23	0.08	650	480	30	40@20	649/1200

1 N/mm² = 1 MPa = 0.145 ksi.

the problem of hydrogen-induced cold cracking can arise. As a precaution, the use of low hydrogen welding processes is essential, especially ensuring the use of low-hydrogen coated or covered electrodes for shielded metal arc welding (SMAW) or shielding gases of high purity and moisture content low enough to yield a dew point of less than 50 °C/°F for inert-gas-shielded processes (e.g. gas-shielded flux-cored arc welding [FCAW], gas tungsten-arc welding [GTAW], and gas-metal arc welding [GMAW]). Use of preheat is also essential for most ferritic alloy steels. A complication is that the normal (e.g. International Institute of Welding [IIW]) carbon equivalent method for assessing hardenability and the need for preheating is not valid for these grades of steel, and few welding specifications give much guidance pertaining to recommended preheat temperatures. Fortunately, the American specification ASME B31.3 and the British specification EN 1011 both contain recommendations. Table 17.2 contains data adapted from EN 1011, but it is valid only for processes where hydrogen limit is between 5 and 10 ml in 100 g of weld metal.

Of course, of particular relevance herein, creep-resistant ferritic alloy steels can experience problems with cracking during reheating (i.e. generic reheat cracking), which might occur when the weld/weldment is post-weld heat treated or placed into service at high temperatures without having had PWHT. Because PWHT is intended to relieve stresses almost inevitably induced by welding (i.e. for stress relief), the phenomenon in ferritic alloy steels is most commonly known as *stress relief cracking*. With most applications of these alloys being in pressure vessels, components, or piping, post-weld stress relief is almost always mandated by codes (e.g. ASME boiler and pressure vessel code) or specifications.

Table 17.3 lists typical PWHT temperatures and times for the most common ferritic alloy steels.

Table 17.2 Recommended preheat and interpass temperatures for common creep-resistant ferritic alloy steels.

Steel grade	Thickness (mm/in.)	Minimum preheat (°C/°F)	Maximum interpass (°C/°F)
	≤15/0.6	20/68	
C ½Mo	>5/0.6 ≤ 30/1.2	75/167	250/482
	>30/1.2	100/212	
½½¼CrMoV	All	150/302	300/572
1¼Cr½Mo	≤15/0.6	100/212	300/572
	>15/0.6	150/302	
2¼Cr1Mo	≤15/0.6	150/302	350/662
	>15/0.6	200/392	
5Cr½Mo	All	200/392	350/662
9Cr1Mo	All	200/392	350/662
9Cr1MoVNb	All	200/392	350/662

Interpass temperature refers to the maximum temperature the area immediately surrounding an in-progress weld should reach, as measured using a thermocouple or temperature-indicating wax sticks (e.g. TempilStix™).

Table 17.3 Typical PWHT temperatures and times for common ferritic alloy steels.

Steel grade	Temperature range (°C/°F)	Soak time (hours)
C½Mo	630–670/1160–1238	1 per 25 mm/1 in.
½½¼CrMoV	650–680/1202–1256	1 per 25 mm/1 in.
1¼Cr½Mo	650–700/1202–1292	1 per 25 mm/1 in.
2¼Cr1Mo	680–720/1256–1328	2, minimum
5Cr½Mo	710–750/1310–1382	2, minimum
9Cr1Mo	730–760/1346–1400	2, minimum
9Cr1MoVNb	730–760/1346–1400	2, minimum

In brief, the phenomenon occurs as a result of the conflict between the rate at which weld-induced residual stresses are relieved (or relaxed) and the rate at which various carbides precipitate to strength and harden grain boundaries relative to grain interiors. General details are as follows:

- Ferritic alloy steels containing Cr, Mo, V, and/or Nb are susceptible, with Mo–V and Mo–B steels being the most susceptible
- Cracking occurs primarily in the coarse-grained heat-affected zone or CGHAZ (see Section 11.5), but may, and usually does, propagate toward and into the unaffected base metal by an intergranular fracture mode
- Cracking susceptibility is greater in thick-section weldments

- Cracking is exacerbated by anything that concentrates residual (or service) stress, especially notch effects at the toes of welds
- Cracking initiates at prior austenite grain boundaries by cavity or void nucleation around incoherent Fe-rich M_3C carbide precipitates that are preferentially formed there and grew large enough to lose any coherency (i.e. become incoherent)
- As a result of nucleation of cracking at cavities or voids formed around carbides in grain boundaries, the fracture surfaces exhibit topological features consistent with ductile rupture (e.g. dimples)
- Removal of both carbon and substitutional solid-solution solutes (e.g. Cr, Mo, V, and Nb) weaken the region in and around these prior austenite grain boundaries, rending it prone to plastic flow and ductile fracture
- Although earlier research suggested segregation of tramp elements (e.g. S, P, As, Bi, Pb, and others) to grain boundaries exacerbated cracking, the most thorough studies[6] found no detectable evidence of such elemental segregation playing any role
- The phenomenon can occur anywhere in the range of most post-weld stress relief heat treatments (i.e. 500–775 °C/~900–1450 °F) or in service at similar temperatures when there has been no PWHT.

Figures 17.2 and 17.3 present photomicrographs that reveal the intergranular path of stress relief cracks in ferritic alloy steels, as well as the formation of voids and cavities (around grain boundary carbides) that initiate cracks.

Avoiding or preventing stress relief cracking (by any name, in any material) is the only viable action. As it only appears once a weldment has been completed and PWHT needed to reduce weld-induced residual stresses *and* develop alloy properties is conducted, the cost of cracking is always high. The added value to the weldment is near its peak, and repair, if possible, is very expensive.

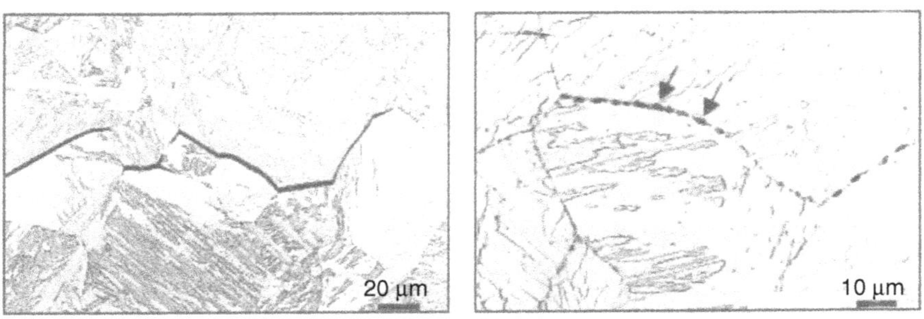

(a) (b)

Figure 17.2 Light optical micrographs showing failed samples of HCM2S creep-resistant ferritic alloy steel, clearly showing the intergranular mode of crack propagation (fracture) along prior austenite grain boundaries (a) and (as indicated by arrows) the void formation at grain boundary carbides (b). Source: Nawrocki et al. (2003), figure 2. Used with permission of The American Welding Society.

6 See paper entitled "The mechanism of stress-relief cracking in a ferritic alloy steel" by J.G. Nawrocki, J.N. DuPont, C.V. Robino, J.D. Puskar, and A.R. Marder, *Research Supplement of the Welding Journal*, Vol. 82, No. 2, 25s–35s, 2003.

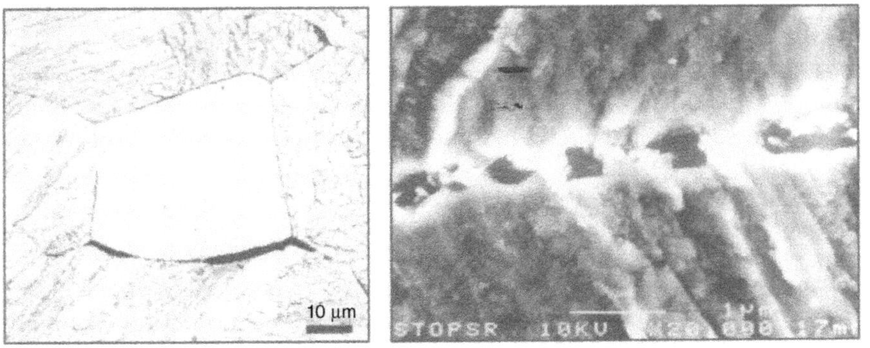

(a) (b)

Figure 17.3 An example of crack formation (a) and void formation (b) at prior austenite grain boundaries in the creep-resistant ferritic alloy steel HCM2S. Cavities or voids were found to have nucleated around incoherent M_3C carbides. Panel (a) is from an optical micrograph and (b) is from an SEM micrograph. Source: Nawrocki et al. (2003), figure 12. Used with permission of The American Welding Society.

The first option and avenue to avoid stress relief cracking in ferritic alloy steels is to check susceptibility for a material to be welded (especially if never welded before!) using some popular predictive empirical formulas presented earlier in Section 11.6 but repeated here for convenience, as follows:

- $\Delta G1 = 10C + Cr + 3.3Mo + 8.1\,V - 2$ (using wt%) < 2
- $P_{SR} = Cr + Cu + 2Mo + 10\,V + 7Nb + 5Ti - 2$ (using wt%) < 0
- $R_S = 0.12Cu + 0.19S + 0.10As + P + 1.18S + 1.49Sb$ (using wt%) < 0.03.

The first two equations have been shown to be useful, whereas the third, while of debatable utility, is probably worth consideration, as the impurities included are never desirable.

Once the risk is known from predictive equations, the following are recommendations to avoid stress relief cracking:

- Use the PWHT for stress relief recommended by the alloy producer
- Avoid large weld beads, as these promote coarse-grained HAZ, which is less likely to be refined by subsequent passes during multi-pass welding
- Keep net linear heat input as low as possible while still providing the needed weld profile
- Design to avoid restraint, to keep residual stresses to a minimum
- Dress weld toes to remove stress concentrations
- Specify a maximum level of impurities as low as practical, as a prudent precaution that may or may not be of significance.

For ferritic alloy steels, weld filler (from bare filler wires for GTAW, wire for GMAW or submerged arc welding (SAW), or from covered electrodes for SMAW or cored wires for FCAW) that matches the base metal is usually readily available, except for the ½½¼ CrMoV steel, which is usually welded using a 2¼Cr1Mo filler. When these alloys are welded to a dissimilar alloy within the group or with a carbon–manganese steel, the filler should be matched to the less highly alloyed steel. Selecting a PWHT temperature for dissimilar metal joints requires a compromise between overtempering the lower alloyed steel and undertempering the more highly alloyed steel.

Because of their widespread use, the phenomenon of stress relief cracking in ferritic alloy steels has been the subject of considerable research for more than 60 years. The general cause(s) of stress relief cracking is well known and not a subject of debate. Underlying details, on the other hand, are debated. What has been presented herein is a summary of what is believed (by the author) to be most useful.

17.3 Stress Relaxation Cracking in Stainless Steels

Increasing use of high-temperature, heavy-section stainless steel is increasingly commonplace as engineering designers push the envelope of stress and temperature, as they are inclined to do. The precipitation hardening of stainless steels, especially (but not only) in PH grades, enable attainment of tensile strengths of 850–1700 MPa (123–247 ksi) and yield strengths of 520–1500 MPa (75–218 ksi), which is three to four times that of austenitic stainless steels like 304 and 316. Special H grades of several austenitic stainless steels, such as 304H, 316H, 321H, and 347H, contain higher carbon contents (up to a maximum of 0.10 wt% in most, but up to 0.14 wt% in 321H, versus a maximum of 0.08 wt% in standard grades, with a typical content of around 0.05 wt%) to afford better performance at elevated temperatures (i.e. higher strength at elevated temperatures and improved resistance to creep).

As a consequence, there have been a growing number of cases worldwide where fusion zone (FZ) and, especially, HAZ cracking has occurred after very short periods of service and, overwhelmingly, only on weldments with residual stresses due to welding of cold-worked alloys. So-called *stress relaxation cracking* related to precipitation hardening has been observed in those austenitic stainless steels that typically undergo precipitation hardening to optimize properties for service. Table 17.4 lists susceptible stainless steels.

These grades of austenitic stainless steels, including PH grades, contain alloying elements that form carbides, such as Cr, Mo, Ti, and Nb (usually with Ta, as separation is very difficult both technologically and economically). Carbide precipitation is more pronounced, and strength is greater, particularly at elevated temperatures where creep can occur, in higher carbon H grades.

Stress relaxation cracking occurs at elevated temperatures in susceptible alloys when creep ductility is insufficient to accommodate the strains required for the relief of residual or applied stresses by localized internal yielding. As usual, cracks are intergranular, show little evidence of deformation, and are usually confined to the coarse-grained region of the HAZ (i.e. CGHAZ). As for previously discussed ferritic alloy steels (and for soon-to-be-discussed Ni-based superalloys), post-weld stress relief is code-required for applications involving pressure (e.g. steam pipes and headers, vessels). In almost all instances, cracks are initiated at some source of stress concentration, and propagation is along grain boundaries

Table 17.4 Austenitic and PH stainless steels used for their creep resistance (compositions in wt%; minimum/maximum)

Alloy	C	Mn	Si	P	S	Cr	Ni	Mo	N	Other	Soln./SR temperature
304	—	—	—	—	—	18.0	8.0	—	—		ST 1010–1120 °C/
	0.08	2.0	0.75	0.045	0.030	20.0	10.5		0.10		1850–2050 °F; AC/Q
304H	0.04	—	—	—	—	18.0	8.0				SR avoid 480–900 °C
	0.10	2.0	0.75	0.045	0.030	20.0	10.5				900–1650 °F
316	—	—	—	—	—	16.0	10.0	2.00			ST 1040–1120 °C/
	0.08	2.0	0.75	0.045	0.030	18.0	14.0	3.00			1900–2150 °F; AC/Q
316L(N)	—	—	—	—	16.0	10.0	2.00	—			SR, as above
	0.10	2.0	0.75	0.045	0.030	18.0	14.0	3.00	0.10		
321	—	—	—	—	—	17.0	9.0		—	5(C + N)Ti	ST 950–1100 °C/
	0.08	2.0	0.75	0.045	0.030	19.0	12.0			0.70Ti	1750–2050 °F; AC/Q
321H	0.04	—	—	—	—	17.0	9.0			4(C + N)Ti	SR 500–700 °C/
	0.10	2.0	0.75	0.045	0.030	19.0	12.0			0.70Ti	950–1300 °F; AC
347	—	—	—	—	—	17.0	9.0			10(C + N)Nb ST 1010–1093 °C/	
	0.09	2.0	0.75	0.040	0.030	20.0	13.0			1.0Nb	1850–2000 °F; AC
347H	0.04	—	—	—	—	17.0	9.0			10(C + N)Nb SR 427–816 °C/	
	0.10	2.0	1.0	0.040	0.030	20.0	13.0			1.0Nb	800–1500 °F; AC
17-4 PH	—	—	—	—	—	15.0	3.0			3.0 Cu; 0.15 (Nb + Ta)	
	0.07	1.0	1.0	0.040	0.030	17.5	5.0			5.0 Cu; 0.45 (Nb + Ta)	
17-10P	<0.07	<0.75	<0.50	0.28	<0.03	17.0	10.5				

For solutionizing and aging heat treatments of 17-4 PH and 17-10P PH, check data sheets.

that have been weakened relative to grain interiors. Because cracking requires the presence of high stresses, cracking is more likely to occur in thick-section weldments.

The usual PWHT for stress-relieving austenitic stainless steels (i.e. to cause stress relaxation) is 550–750 °C/1022–1382 °F for a minimum of one hour per in./25 mm of thickness in weld areas, with air cooling being normal. Temperatures up to 875 °C/1600 °F followed by rapid cooling (by quenching) are sometimes used to circumvent cracking problems.

The mechanism for crack formation is identical to stress relief cracking of previously discussed ferritic alloy steels and soon-to-be-discussed Ni-based superalloys.

Minimizing problems from stress relaxation cracking involves the following:

- Considering alloy chemical composition to preclude or minimize the formation of carbide precipitation during PWHT
- Keeping net linear heat input to a minimum that allows a suitable weld profile to be produced
- Keeping grain size as small as possible in the HAZ (i.e. avoid CGHAZ), typically keeping pre-weld ASTM grain size under 3–4
- Using double heat treatments recommended by alloy producers (typically 980 °C/1800 °F soak before welding, PWHT at 875 °C/1600 °F for three hours, with rapid cooling).

17.4 Strain-age Cracking in Ni-Based Superalloys

As the generic name "superalloys" implies, alloys in this category offer high performance in properties, usually in some combination, including excellent mechanical strength, retention of strength to high temperatures and resistance to thermal creep, good dimensional stability, resistance to corrosion and oxidation resistance, and, for one alloy class, excellent resistance to wear. Three alloy systems predominate, i.e. Fe-based alloys, Ni-based alloys, and Co-based alloys.[7] Fe-based superalloys offer the lowest service temperature capability and lowest resistance to oxidation and corrosion, whereas Ni-based superalloys offer the highest service temperature capability. Co-based superalloys offer the best resistance to wear at both low and high temperatures, with slightly lower temperature capability than Ni-based alloys.

All the three major categories of superalloys derive their strength from multiple mechanisms, building on solid solution strengthening using solutes with high melting points (e.g. Cr, Mo, W, V, Nb, and Ta) using a variety of second-phase strengthening mechanisms (e.g. precipitation-hardening and dispersion-strengthening). The secondary phases tend to have stability to high temperatures and often form preferentially at grain boundaries to pin grains against grain boundary sliding during creep. These phases include a variety of

7 Some of the all-β phase Ti-based alloys have utility to elevated temperature and could be considered superalloys. Nb-based superalloys offer the highest temperature capability but are largely under development with only limited application to date.

carbides (e.g. M_3C, M_7C_3, and $M_{23}C_6$ types), which tend to form in the order listed: g′ Ni_3Al, and g″ Ni_3Nb in Ni-based alloys and topologically close-packed (TCP) phases including sigma phase (see Chapter 19), chi (χ) phase, and Laves phases. It is the formation of some of these phases (especially carbides) on a nanoscale within grains that cause the bodies of grains to be stronger than grain boundaries, wherein precipitated carbides coarsen and become less effective at strengthening, and depletion of Cr (and other strong carbide formers) weakens the denuded network.

To optimally age a weldment of Ni-based superalloys that derive their optimum properties from precipitation hardening, the weldment must first be solutionized[8] (typically 1000–1100 °C/~1800–2000 °F), quenched, and then aged. Depending on the specific alloy, aging can be anywhere in the range 450–750 °C/~850–1400 °F. During the solutionizing step, the residual stresses in the weldment are relieved or relaxed (hence the apropos terms for any cracking that occurs, i.e. stress relief cracking or stress relaxation cracking). It is quite possible – and likely – that some aging takes place in the regions of HAZ that were previously solutionized by the heat of welding, especially because aging occurs at lower temperatures than stress relief. When this occurs, cracking can result as the material in this region age hardens, becoming stronger but less ductile, while still being acted upon by residual stresses. As a result, cracking can occur, with obvious designation as *strain-age cracking* or, more generically, *stress relief cracking*.

The mechanism of strain-age cracking is shown schematically in Figure 17.4.

Several mechanisms have been proposed for the low ductility in the HAZ of Ni-based superalloys that lead to strain-age cracking. These include

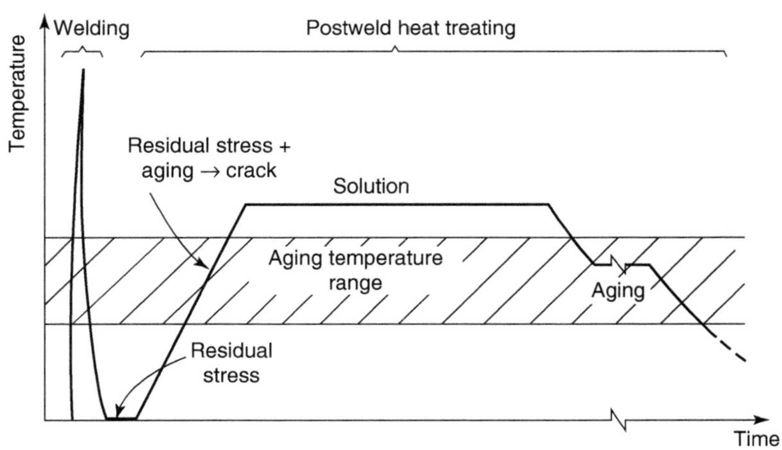

Figure 17.4 Schematic illustration depicting how post-weld heat treatment reheat or strain-age cracking occurs in heat-treatable Ni-based superalloys. Source: Kou (2002). Used with permission of John Wiley & Sons.

8 *Solutionized* is the past tense action of (i.e. the past tense verb for) solution treatment within heat treatment.

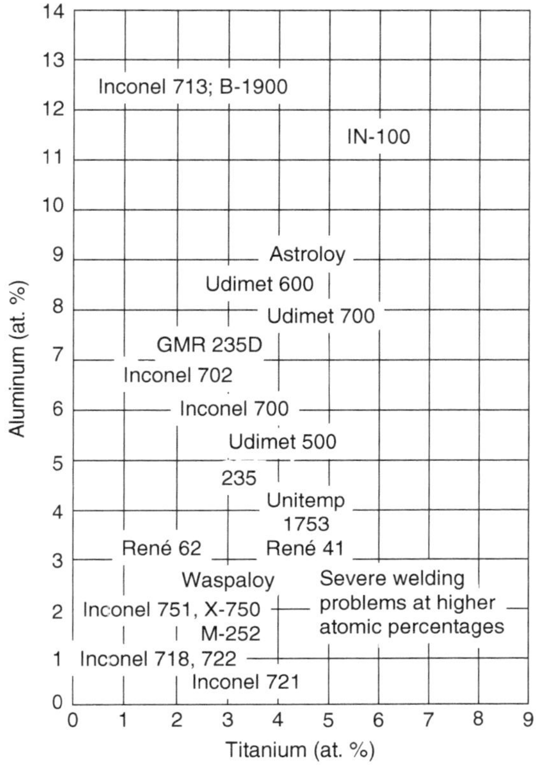

Figure 17.5 Plot of Al versus Ti content indicating susceptibility to strain-age cracking in Ni-based superalloys when (Al + Ti) exceeds about 6 at.%. Source: Olson et al. (1993). Used with permission of ASM International.

(i) embrittlement of grain boundaries because of either liquation or solid-state reactions during welding, (ii) embrittlement of grain boundaries by oxygen during heat treatment, and (iii) a change in the deformation mode from transgranular slip to grain boundary sliding. In all cases, the source of high stress/strain in the HAZ are (i) the ever-present thermal expansion and contraction associated with fusion welding and (ii) contraction accompanying the precipitation process in N-based alloys.

Susceptibility to strain-age cracking in Ni-based superalloys has been found to increase with increased Al + Ti content used to develop the γ' precipitate responsible for imparting elevated temperature strength, which is becoming problematic above around 6 wt% (Figure 17.5).

Strain-age cracking can be avoided in Ni-based superalloys by any of the several methods, including (i) proper selection of base material to keep (Al + Ti) below 6 wt%, (ii) pre-weld overaging via multistep aging processes (recommended by the alloy producer) or by slow cooling from a pre-weld solutionizing treatment, (iii) use of specialized solutionizing and aging treatment cycles for particular alloys (available from alloy producers), (iv) use of vacuum or inert gas atmospheres during heat treatment to preclude grain boundary embrittlement by oxygen, (v) use of low-restraint joint designs, and (vi) welding procedures (e.g. sequencing).

17.5 Troubleshooting Problems from Stress Relief or Strain-age Cracking

Table 17.5 lists the various problems that can arise in alloys prone to stress relief cracking, by any name.

Table 17.5 Troubleshooting the various manifestations of stress relief cracking in creep-resistant alloys.

Stress Relief Cracking in Ferritic Alloy Steels:	
Initiates in the CGHAZ from voids around M_3C carbides of Cr, Mo, V, or Nb in prior austenite grain boundaries, with intergranular propagation	Comply to restraints of predictive equations
	Use manufacturer-recommended PWHT for stress relief
	Keep net linear heat input to minimum
	Minimize restraint
	Dress weld toes to eliminate stress concentration
Stress Relaxation Cracking in Stainless Steels:	
Initiates at δ-ferrite/γ-austenite interfaces	Consider compositions that minimize or prevent
	with less than ~5 vol.% δ-ferrite
	Keep net linear heat input to minimum
	Keep GS to under ASTM 3–4 in HAZ
	Consider manufacturer-recommended double heat treatments
Strain-Age Cracking in Ni-Based Superalloys:	
Worse when (Al + Ti) exceeds about 6 at.%	Keep (Al + Ti) below 6 at.%
	Consider overaging treatments
	Use specialized ST + age cycles from the manufacturer
	Use vacuum or inert atmosphere during PWHT to prevent oxygen pick-up at GBs
	Minimize restraint

Stress relief cracking, by any name, occurs in alloys that undergo concurrent relief of weld-induced stresses and increased resistance to creep by precipitation during aging, whether during PWHT or in service at elevate temperatures. It always initiates in the HAZ and propagates by an intergranular mode.

References

Kou, S. (2002). *Welding Metallurgy*. New York, NY: Wiley.
Messler, R.W. Jr. (2011). *The Essence of Materials for Engineers*, figure 7.22, 196. Sudbury, MA: Jones & Bartlett Learning.
Nawrocki, J.G., DuPont, J.N., Robino, C.V. et al. (2003). The mechanism of stress-relief cracking in a ferritic alloy steel. *Welding Journal (Research Supplement)* 82 (2): 25s–35s.
Olson, D.L., Siewert, T.A., Liu, S., and Edwards, G.R. (1993). *ASM Handbook*, Vol. 6: Welding, Brazing, and Soldering, 8e. ASM International.

Further Reading

Leuk, J.K.L., Shek, C.H., and Lo, K.H. (eds.) (2018). *Stainless Steels: An Introduction and their Recent Developments*. Sharjah: Bentham Science Publishers.
Lippold, J.C. and Kiser, S.D. (2009). *Welding and Weldability of Nickel-Base Alloys*, 1e. New York, NY: Wiley.

18

Loss of Properties in Cold-Worked Metals and Alloys

18.1 Cold Work, Recovery, Recrystallization, and Grain Growth

The combination of high strength *and* good ductility typically found in metals and alloys aids engineers in both design and fabrication in several ways, including (i) general ability[1] to sustain relatively high stresses without undergoing unwanted permanent dimensional or shape change (i.e. plastic yielding), (ii) general ability to resist sudden failure by fracture (i.e. catastrophic overload), (iii) general ability to tolerate shock, impulse, or impact loads without fracture[2], and, by no means incidentally, (iv) ready processing into needed shapes by one or the other or both of cold or hot deformation processes (e.g. rolling, drawing, forging, or forming). Pure metals exhibit high strength by the strength of atom-to-atom primary chemical (metallic) bonds.[3] These same pure crystalline metals exhibit good ductility by the ease with which one close-packed layer of metal atoms can slip over another close-packed layer without disrupting the overall long-range order of the crystalline structure, aided by imperfections in the crystal known as dislocations. *Dislocations* allow slip to take place sequentially – essentially atom by atom – rather than simultaneously, thereby requiring much less stress and making deformation processing easier.[4] By adding foreign

1 The modifier "general" is used as metals and alloys exhibit a wide range of mechanical properties, which only adds to the material's value in engineering. Although most are strong and ductile, some are relatively weak (e.g. lead and solders) and some are quite brittle (e.g. chromium and some cast irons).
2 The area under a stress–strain curve to the point of fracture represents the energy a material can absorb from impact without fracturing, i.e. its *toughness*.
3 Metallic bonds differ widely in strength, thereby imparting a wide range of mechanical strengths, in the form of tensile strength, and thermal strengths, in the form of melting points, to pure metals. Two examples are pure tungsten (*W*) with a tensile strength of 689–3447 MPa/100 000–500 000 psi and an MP of 3422 °C/6192 °F and pure lead (Pb) with a tensile strength of 12–17 MPa/1740–2470 psi and an MP of 327.5 °C/621.4 °F. Metallic bonds are generally weaker than ionic bonds and covalent bonds.
4 A nice animation of how an edge dislocation (extra half-plane of atoms as an imperfection in a crystal lattice) moves to allow one layer of atoms to slip over another adjacent layer one-by-one, rather than all-at-once, is shown in a YouTube video found as the first item with a Google search of "dislocation animation," lasting 0.06 seconds. It is followed by a more detailed description of dislocations, if one wishes.

A Practical Guide to Welding Solutions: Overcoming Technical and Material-Specific Issues,
First Edition. Robert W. Messler, Jr.

atoms (known as solute atoms) to a pure metal (known as the solvent) to create solid solution alloys, the metal is made stronger as the strain field surrounding each foreign atom interferes with the movement of dislocations.

Besides the extremely attractive combination of strength and ductility offered by metals and alloys made possible by the combination of the strength of the bonds in the solid and the presence of dislocation imperfections in their crystal structure, metals and their alloys have another interesting and valuable behavior. When a metal or metallic alloy is made to undergo plastic strain by being stressed beyond its yield strength (YS) at temperatures less than about 0.35–0.5 of its absolute melting point (in degrees Kelvin or Rankine), that is, *cold-worked*, new imperfections are generated in its crystalline lattice and the existing imperfections interact to alter the properties. Point imperfections[5] in the form of vacancies (i.e. vacant lattice sites) and crowd ions (i.e. atoms forced off lattice sites into interstices) are generated, which lower electrical conductivity by causing conduction electrons to scatter, while having little effect on any mechanical properties. Existing line imperfections (i.e. dislocations) in the crystal structure are caused to move and produce plastic deformation by slip.

As available dislocations become unable to accommodate all of the strain required by imposed stresses, new dislocations are generated from various sources, such as free surfaces, grain boundaries, Frank–Reed sources, and others. Pre-existing dislocations and newly generated dislocations interact with obstacles that have associated strain fields (e.g. grain and phase boundaries and interfaces between a matrix-phase and second-phase non-coherent dispersoid and, especially, coherent precipitate particles), piling up to create a back-pressure. Dislocations moving on different paths that intersect form immobile jogs or tangles. Once dislocations are pinned by obstacles, jogs, or tangles, further movement is hindered, and the resistance of the metal or alloy to further strain is increased; that is, the material is strengthened. Pure metals or metallic alloys strengthened by such cold work are said to have undergone *work hardening* or *strain hardening*.

Figure 18.1 schematically illustrates strain hardening or work hardening in both generic interstitial strengthened ferrous alloys (e.g. steels) and substitutional strengthened non-ferrous alloys (e.g. Al, Cu, Mg, Ni, Ti alloys, and others).

Strain hardening during cold-working of a metal or an alloy is a two-edged sword. In certain cold-working operations, such as cold rolling sheet, cold drawing of wire, and cold forming of parts, strain hardening is a hindrance because it makes continued plastic deformation progressively more difficult as the amount of cold work increases. Too much strain hardening may require unacceptably high forces to secure desired deformation and can ultimately lead to cracks or fracture. On the other hand, strain hardening can sometimes be advantageous because it increases the elastic limit[6] of the material. Two examples where this is

---

5 Point, line, and planar or area imperfections are alternatively referred to as zero-dimensional (0-D), one-dimensional (1-D), and two-dimensional (2-D) imperfections. 3-D imperfections can also exist as voids or regions where crystallinity has been lost (i.e. amorphous regions) in a crystalline material.

6 The *elastic limit* of a metal or an alloy is the point on a stress–strain curve where strain under loading is not fully recoverable, i.e. where strain becomes permanent. It demarcates the end of

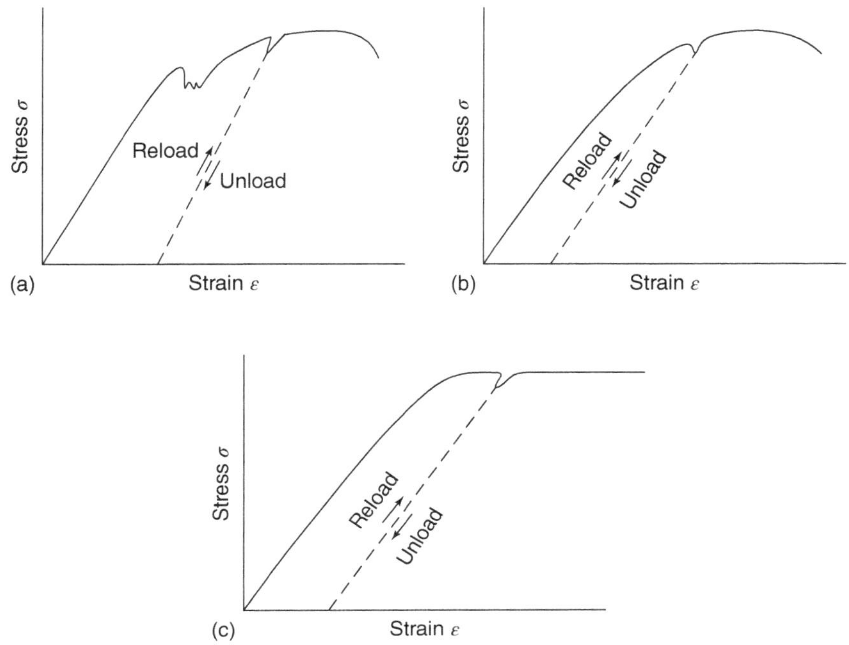

Figure 18.1 Schematic plots of engineering stress (σ) versus engineering strain (ε) showing strain or work hardening in a generic (a) ferrous alloy (e.g. steel) and (b) non-ferrous metal or alloy. Note how the stress to cause continued strain increases, and how removing the load at any point causes the stress needed to cause further plastic deformation (i.e. yielding) upon reloading to be increased. For completeness, the behavior of a hypothetical material that does not exhibit strain hardening is shown in (c). Source: Messler (1999/2005a). Used with permission of John Wiley & Sons.

advantageous are (i) in the manufacture of springs, where direct proportionality between applied stress and resulting strain[7] makes division of the stress scale simple (i.e. linear) and (ii) to eliminate the presence of an upper and lower yield point in ferrous alloys[8] (as seen in plot a in Figure 18.1). Elimination of the upper and lower yield points via strain hardening (typically by prestretching past yield) allows cold forming of sheet-steel parts (e.g. automobile body panels) that are free of "orange peel" stretch marks (called "Lüder's bands") that show flaws in lustrous body finishes.

---

elastic behavior and the beginning of plastic behavior. In practice, a 0.2% offset yield stress is usually used. The elastic limit of *all* metals or alloys that exhibit any ductility lies before the proportional limit, where the stress–strain curve is already nonlinear.

7 The end of the straight-line portion of a stress–strain curve for a metal or an alloy, within which Hooke's law, $\sigma = E\varepsilon$, is obeyed is defined as the material's *proportional limit*.

8 An *upper and lower yield point* in steels is the consequence of interstitial atoms of C accumulating as "atmospheres" in the regions of tensile strain of edge dislocations in order to lower the net strain energy of the paired imperfections (i.e. interstitials and dislocations). It takes extra stress to pull dislocations free of such an atmospheres (hence, an upper yield point), with less stress needed once separation occurs (hence, a lower yield point).

Cold working raises the internal energy of the material due to the 3–8% of the work done being stored as strain energy, rather than being lost as heat, as the other 90+% is lost. This stored energy creates an energetically unfavorable condition that is metastable at best. Hence, unless it is desirable – or tolerable – to keep a metal or an alloy in its strain-hardened condition for the added strength (and, perhaps, hardness) provided, it is usually preferred to employ heat treatment to restore the grain structure to a smaller, equiaxed condition with restored pre-cold-worked properties. Such a heat treatment is properly known as *recovery and recrystallization* or, simply, *recrystallization.*[9]

The upper portion of Figure 18.2 schematically illustrates how the normally equiaxed grain structure of a metal or alloy is elongated by cold work in the direction of working, such as rolling, and then begins to change with the formation of new, small equiaxed grains in the regions of highest stored strain energy (e.g. at triple points of grains and along grain boundaries). Once the strain-elongated grains have been completely replaced by small, equiaxed, strain-free grains, grain growth occurs with a prolonged exposure or higher temperatures.

When a cold work-hardened metal or alloy is exposed to temperatures greater than approximately $0.4T_{\text{MPabsolute}}$ (or $0.4T_{\text{solidus}}$ in absolute degrees K or R), deformed grains begin to nucleate new strain-free (lower energy) grains (as shown in the illustrations at the bottom of Figure 18.2). The driving force for such recrystallization[10] is the reduction of stored strain energy.[11] With the onset of recrystallization, mechanical properties change significantly. Yield and ultimate strength, as well as hardness, decrease, whereas ductility and toughness increase (as shown in the plots in the upper portion of Figure 18.2). Changes in electrical (here, resistivity) and thermal properties (which also originate at the electron level of materials) begin during the lower temperature recovery stage of heat treatment, known as *recovery,*[12] with both an increase in electrical and thermal conductivity and a decrease in electrical and thermal resistivity.

Because recrystallization (like preceding recovery) occurs by nucleation and growth, the extent of both is controlled by (i) temperature and (ii) time at temperature, increasing as one or the other or both increases.

Once recrystallization is complete, the newly formed small, strain-free equiaxed grains grow in order to reduce the surface energy associated with grain boundaries, which are 2-D imperfections. Like recovery and recrystallization, grain growth also increases with increased temperature and time at temperature. Hence, like the earlier stages of recovery and recrystallization, grain growth, in particular, is greatest in the heat-affected zone (HAZ) of a weld closer to the

---

9 Although the general term for any – and all – heat treatments that soften a metal or an alloy is *annealing,* as with specifying what one mean's by an order for "ice cream" by saying "cup" or "cone," "vanilla" or "chocolate," etc., one needs to be more specific. In the case of heat treatment for the purpose of recovery and recrystallization following cold working, the proper term is *recrystallization anne*al or, simply, *recrystallization.*

10 The term "recrystallization" should not be interpreted as meaning re-formed as crystals from a melt! Rather, it refers to new grains formed from previous deformed grains by a solid-state process of nucleation and growth.

11 Strain energy is principally stored in piled-up and tangled 1-D dislocations.

12 The driving force for recovery is reduction stored energy associated with 0-D point imperfections, such as vacancies and crown ions.

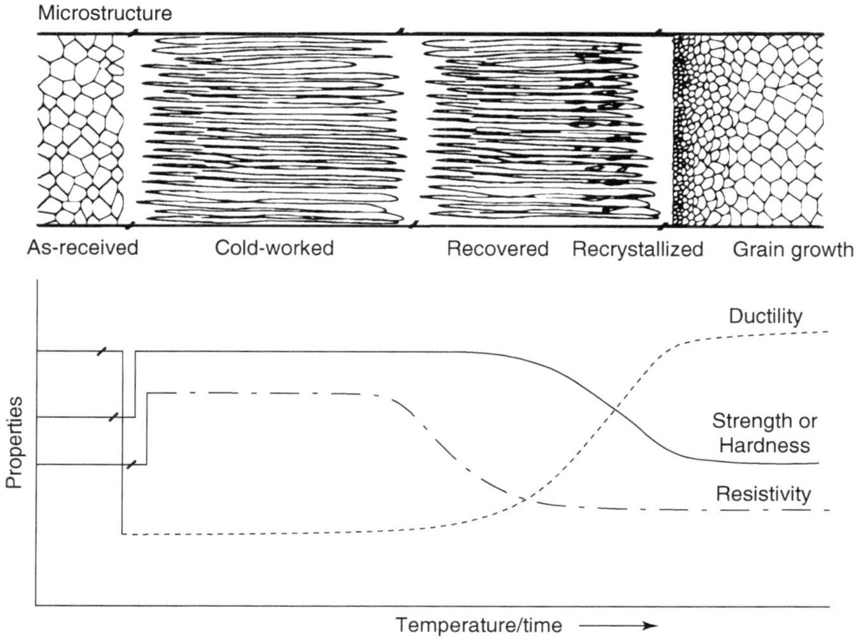

Figure 18.2 Schematic illustration showing plots of various properties as a function of temperature during heat treatment (lower portion of the figure), along with schematic representations of the corresponding grain structure of the hypothetical metallic material as affected by cold work and upon subsequent heat treatment to cause recovery, recrystallization, and grain growth (upper portion of the figure). Only electrical and thermal conductivity are affected during the recovery stage, with each increasing to near-original levels by the end of the recovery process. Source: Messler Jr. (1999/2005b). Used with permission of John Wiley & Sons.

fusion boundary, and all occur to a greater extent as the net linear heat input to a weld is increased. As opposed to recrystallization, which has pronounced effects on mechanical properties, grain growth has less pronounced effects. This said, yield strength is reduced as grain growth increases, impact toughness is decreased, and fracture toughness tends to increase slightly in many alloys.

When a pure metal or an alloy is deformation-processed (e.g. rolled, drawn, extruded, or formed) at temperatures above about $0.4T_{\text{absoluteMP}}/0.4T_{\text{solidus}}$ in absolute degrees K or R, which is known as *hot working*, the processes of recovery and recrystallization and, to a lesser extent, grain growth occur almost simultaneously with deformation. The phenomenon is properly known as *dynamic recrystallization*. Hot working does not lead to strain hardening, which is one of the two reasons, along with lowering the stress (and, hence, power) needed to deform a metal or an alloy, that it is often chosen over cold working, especially for large amounts of deformation. Another distinct difference between hot working and cold working is the former results in equiaxed microstructure, as opposed to elongated microstructure. Equiaxed microstructure results in mechanical properties that are less different in orthogonal directions (i.e. more

Table 18.1 Recrystallization temperatures for some important metals.

Metal	Approximate minimum recrystallization temperature
Lead (Pb)	Below room temperature (20 °C/68 °F)
Tin (Sn)	Below room temperature (20 °C/68 °F)
Cadmium (Cd)	Room temperature (20 °C/68 °F)
Zinc (Zn)	Room temperature (20 °C/68 °F)
Magnesium (Mg)	150 °C/300 °F
Aluminum (Al)	150 °C/300 °F
Copper (Cu)	200 °C/390 °F
Gold (Au)	200 °C/390 °F
Silver (Ag)	200 °C/390 °F
Iron (Fe)	450 °C/840 °F
Platinum (Pt)	450 °C/390 °F
Nickel (Ni)	600 °C/1100 °F
Titanium (Ti)	650 °C/1200 °F
Beryllium (Be)	700 °C/1290 °F
Molybdenum (Mo)	900 °C/1650 °F
Tantalum (Ta)	1000 °C/1830 °F
Tungsten (W)	1200 °C/2200 °F

Source: Jeffers and Archer (1924).

isotropic), whereas elongated microstructure results in mechanical properties that are much different in the worked direction (i.e. anisotropic). In fact, one of the reasons cold working is used is to impart higher yield and ultimate strength and slightly higher modulus of elasticity in the working (e.g. rolling) direction.

Table 18.1 lists the minimum temperatures for recrystallizing the most widely used metals.

The temperature at which recrystallization begins to occur is affected by several factors as follows:

- Alloying raises the minimum recrystallization temperature over pure metals, more so as solute content increases
- Greater degrees of cold work (as measured by percent reduction in cross-sectional area) lowers the minimum recrystallization temperature for pure metals and their alloys
- Longer times-at-temperature allow the recrystallization temperature to be lower following an exponential relationship that, as a rule-of-thumb, a 30 °C/50 °F reduction of temperature doubles the needed time at temperature.

18.2 Cold-Worked Metals and Alloys in Engineering

Metals and alloys tend to be hot worked, or wrought, into the final product form (e.g. plate, sheet, foil, bars, rods, angles, channels, I- or H-beams, wires, and

Table 18.2 Examples of pure metals and alloys commonly used in the cold-worked condition (along with yield strength and ultimate tensile strength in MPa/ksi).

	Cold-worked		Annealed	
	YS	UTS	YS	UTS
Al and Al alloys				
EC-H19	165/24	186/27	28/4	83/12
1xxx-H12, H14, H16, H18	124–152/18–22	131–165/19–24	26–35/4–5	69–90/10–13
3xxx-H18	186–248/27–36	200–283/29–41	41–69/6–10	110–179/16–26
5xxx-H18, H38, H113	228–255/33–37	290–317/42–46	90–145/13–21	193–276/28–40
Cu and Cu alloys				
Pure Cu	414–441/60–64	552–565/80–82	55–62/8–9	427–441/62–64
½-hard	255–317/37–46			207–262/30–38
Hard	292–359/43–52			
Spring hard	345–400/50–58			
Extra spring-hard	359 + 52+			
Mn–bronze	310–414/45–60	517–552/75–80	207/30	448/65
70Cu–30Zn (yellow/ cartridge brass)				
Hard	434/63	524/76	11717	338/49
Spring hard	448/65	621/90		
85Cu–15Zn (red brass)				
Hard	393/57	483/70	97/14	283/41
Spring hard	434/63	579/84		
Cupronickel (70Cu–30Ni)	483/70	517/75	172/25	414/60
Be–Cu (approximately 2% Be)[a]	483/70	738/107	172/25	496/72

(Continued)

Table 18.2 (Continued)

	Cold-worked		Annealed	
	YS	UTS	YS	UTS
Fe and Fe-alloys				
Low-C steels				
1/4-hard/hard	379/55	552/80		310/45
Austenitic 304 stainless steel			210/30	584/85
1/4-hard	517/75	862/125		
1/2-hard	758/110	1034/150		
Full hard	965/140	1276/185		
Mg and Mg alloys				
AZ31B (and others)	165–221/24–32	262–290/38–42	165/24	262/38
Ni and Ni alloys				
Pure Ni/Nickel 200 (cold drawn)	345–896/50–130	689–1034/100–150	207–414/30–60	621–827/90–120
Monel (70Ni–30Cu)	379–827/55–120	586–862/85–125	172–276/25–40	483–586/70–85
Precious metals and alloys				
Gold/Au alloys	Nil	131/19	207/30	221/32
Platinum/Pt alloys	Nil	117–179/17–26	N/A	267–359/38–52

a) Be–Cu alloys can also be artificially aged to even higher strengths, e.g. 1378 MPa/200 ksi.

forgings) by metal producers (e.g. mills or forgers). The primary reason is that elevating a metal or an alloy's temperature lowers its yield or flow strength, thereby reducing the amount of force or pressure (and power) needed to achieve the desired deformation, speeding throughput, with time equating to money. A secondary reason is the creation of more nearly isotropic properties. The two most significant compromises are that hot working results in (i) less precise finished dimensions (e.g. greater thickness variation in plate or sheet) and (ii) more heavily oxidized and less smooth and clean surface finishes. Should more precise dimensional tolerance and/or smoother, cleaner surfaces be required, mills will often perform the final rolling or drawing passes cold (i.e. they will cold finish). Cold working, most often in the form of forming, is usually the preferred choice of fabricators or manufacturers. In both cases, the end goal is the shape and dimensions of the finished item.

There are situations, however, where metals or alloys are purposely cold-worked and left in the strain-hardened condition, either by the metal producer (i.e. mill) or product-form producer (e.g. forger and wire manufacturer). Three examples are in descending order of use: (i) production of the so-called "spring tempers" (e.g. Cu–Zn brasses, plain low- and high-carbon steels, and austenitic stainless steels) or, in the case of Al alloys "strain tempers," with an H suffix in the alloy designation; (ii) strain hardening of pure metals (e.g. Cu, Al, Ni, Ti, Mo, and W) to increase the strength without degrading electrical conductivity to the degree alloying does; and (iii) to develop directional properties (or anisotropy) with textures (e.g. in silicon steels for transformers to reduce eddy current losses, aeroelastic tailoring of Ti alloys in aerospace to increase directional elastic modulus to resist flutter).

Finally, a proper combination of cold work before precipitation hardening (i.e. aging) can result in strength and hardness unattainable by either process alone. This is most commonly done in beryllium–copper sheet materials, but can also be done with age-hardenable Al alloys to obtain more uniform precipitates.

Table 18.2 lists examples of pure metals and alloys commonly used in the cold-worked condition, along with yield strength and ultimate tensile strength (UTS) in cold-worked and annealed conditions.

18.3 Avoiding or Recovering Properties Loss from Fusion Welding

Any process that heats a cold-worked metal or alloy above the minimum temperature needed to cause recrystallization can cause the material to recrystallize and lose strength and hardness (i.e. soften). Of course, as for any solid-state process involving nucleation and growth, the time at temperature, not just temperature, is an important factor. Thus, although the heat of welding surrounding a fusion or even a non-fusion weld can lead to recrystallization in the HAZ, the effect may not be complete recrystallization, except in the highest temperature regions of the HAZ.

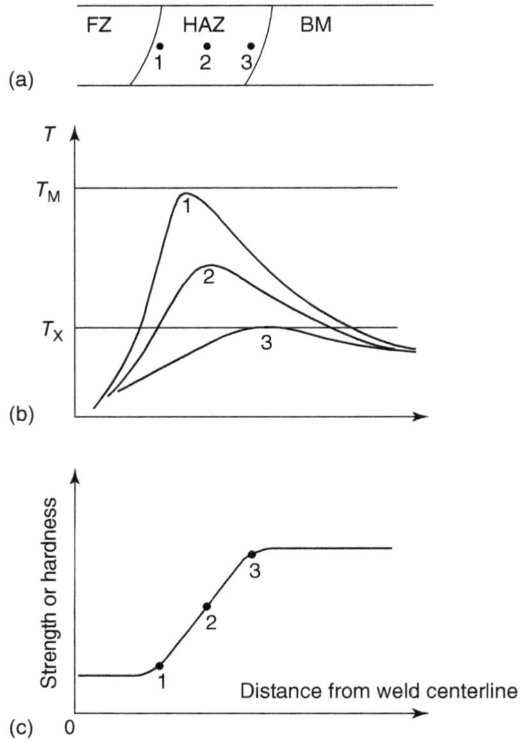

Figure 18.3 Schematic plot showing the softening at three locations in the HAZ of a work-hardened metal or an alloy as a result of welding. Locations in the HAZ are shown in (a), thermal cycles at these locations in (b), and plot of strength or hardness in (c). Source: Kou (1987). Used with permission of John Wiley & Sons, Inc.

Figure 18.3 is a schematic plot showing the softening of a cold work-hardened metallic material due to welding at three locations in the HAZ for three corresponding weld thermal cycles. Note that closer to the fusion boundary, where the peak temperature of the thermal cycle is higher and where the time above the minimum temperature to cause recrystallization (shown as T_x in the figure) is longer, hardness is reduced to a greater degree.

The most obvious factor in HAZ softening of a cold-worked metal or an alloy because of recrystallization is the width of the HAZ, with the width of the HAZ being most influenced by the net linear heat input to the weld. The greater the net linear heat input, the wider the HAZ, all other factors being the same. The wider the HAZ, the more extensive any softening due to recrystallization will be. However, there are other factors that must be considered to fully account for the needed temperature and time-at-temperature effects. In other words, *the higher* the peak temperature at various distances from the weld centerline or fusion boundary, *the faster* the recrystallization will occur. Alternatively, *the longer* any point in the HAZ remains above the minimum temperature required for recrystallization to occur, *the greater degree* to which recrystallization will take place. Other important factors include the following:

- The higher the net linear heat input to a weld, the greater the degree to which recrystallization will occur, all other factors being equal. (This generally means keeping current as low as possible to produce a weld of adequate depth and width and keeping travel speed as fast as possible while meeting the same requirement.)

- The more intensive the welding heat source,[13] the lower the degree to which recrystallization will occur, all other factors being equal
- The shorter the heating cycle inherent to the welding process, the lower the degree to which recrystallization will occur, all other factors being equal. (Stored energy resistance welding processes such as capacitor–discharge welding keeps the time at temperature very short.)
- The lower the melting point of the metal or the alloy, the greater the degree to which recrystallization will occur, all other factors being equal
- The higher the thermal conductivity of the metal or alloy, the greater the degree to which recrystallization will occur, all other factors being equal
- The smaller the thermal mass of the weldment, the greater the degree to which recrystallization will occur, all other factors being equal
- The higher the temperature of the base material at the start of welding (i.e. low ambient temperature versus preheat), the greater the degree to which recrystallization will occur, all other factors being equal.

Figure 18.4 is a schematic plot showing the effect of net linear heat input on the final hardness (after welding) in the HAZ of a cold-worked Al-5Mg alloy (AA5356).

As always in welding, as in most things, there are two fundamental approaches to minimizing adverse effects from the heat of welding on base material microstructure (in the HAZ): (i) preventing or avoiding the problem and (ii) remediating the problem once it has occurred. Prevention of problems in

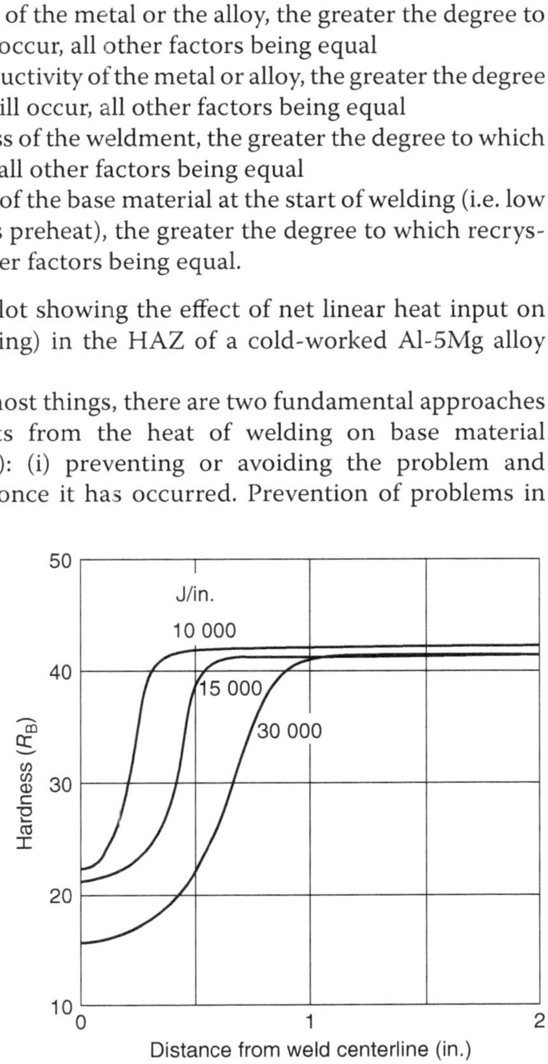

Figure 18.4 Schematic plots showing the effect of net linear heat input (in J/in.) on the hardness in the HAZ in cold-worked AA5356 (nominally, Al–5Mg) Al alloy. Source: White et al. (1960). Used with permission of The American Welding Society.

13 The most significant factor here is the energy density of the heat source in W/m^2. The typical energy densities are oxy-fuel sources 10^2–10^4, arc sources 5×10^6–5×10^8, plasma sources 5×10^7–5×10^9, laser beam sources 10^8–10^{11}, and electron beam sources 10^{10}–10^{12}.

engineering, in general, and welding, in particular, as in medicine and law, is fundamentally preferable to remediation. This may be no truer than for welding work-hardened metals or alloys. For once a cold-worked structure has undergone any recrystallization and lost strength and hardness as a result, the only way to attempt to restore these properties is to thoroughly rework the affected area by mechanical means. Although not always impossible, it is always very difficult and less effective than the original working.

The options for restoring cold work in a weld include (i) rolling or roll-planishing the raised weld crown bead and root bead, as well as the surrounding HAZ, reducing the thickness of regions by cold processing and (ii) locally cold working the weld region (fusion zone, FZ + HAZ) by hammer peening (often using a pneumatic riveting gun) or aggressively shot peening. Such rework is generally limited to thinner section welds and, in most cases, will alter the shape and dimensions of the weldment.

The more practical approach is to prevent, or at least minimize, the loss of mechanical properties by minimizing the degree of recrystallization as described above. Of course, it is possible to prevent recrystallization of work-hardened metals and alloys by not allowing temperature to rise above approximately 0.4 of the material's absolute melting point or solidus temperature. The options include (i) cold non-fusion processes or (ii) low-temperature alternatives to welding, such as brazing or soldering.

18.4 The Worked Zone in Pressure-Welded Metals and Alloys

Although not usually a problem, when pressure is used to bring the atoms of metals or alloys together to achieve metallic continuity to form a weld without fusion, there is a localized effect on the microstructure because of the mechanical work done. If the non-fusion process involves heat to facilitate weld formation, or generates sufficient heat in the process of making the weld, dynamic recrystallization can occur in the worked zone. The mechanism and result are exactly the same as what occurs in any hot working process. In short, as soon as dislocations are caused to move during the pressure cycle of non-fusion welding, pile up or entangle, and multiply by generation from various sources, the strain energy of the material's crystal lattice is increased, the driving force for recrystallization is present, and the process of recrystallization occurs almost all at once, but, in actuality, in a rapid sequence.

The consequence of dynamic recrystallization accompanying hot pressure non-fusion welding processes (as well as some resistance welding processes involving mechanical upset) is often evidenced by a fine-grained structure in the vicinity of the original interface between the materials being joined. This is shown schematically in Figure 18.5.

The differing structure observed in a traverse across a hot pressure non-fusion weld would be reflected in a microhardness traverse.

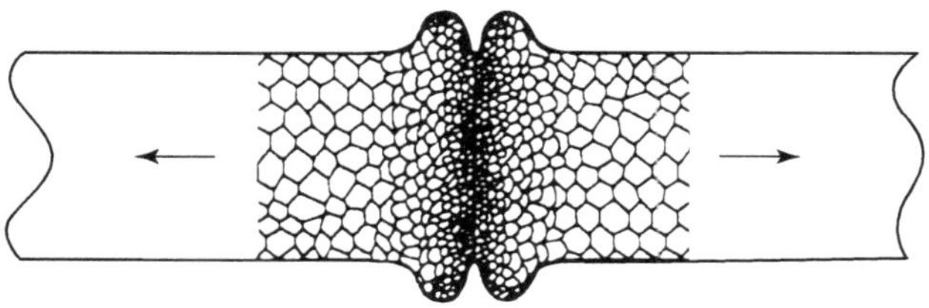

Figure 18.5 Schematic illustration showing the occurrence and effect of dynamic recrystallization in the vicinity of the original interface between two metallic pieces welded using pressure at temperatures above approximately $0.4T_{MP}$ or $0.4T_{solidus}$ in absolute degrees *K* or *R*. Note the region of grain refinement in the worked zone. Source: Messler (1999). Used with permission of John Wiley & Sons, Inc.

Table 18.3 Troubleshooting problems in cold-worked metals and alloys.

Loss of cold-worked structure and properties in HAZ	
Prevention:	Employ cold-pressure non-fusion welding as the best measure
	Employ low-temperature brazing or soldering, if feasible
Minimization:	Use lowest net linear heat input acceptable for weld
	Use high energy density welding process (e.g. LBW, EBW)
	Use process with very short heating cycle (e.g. capacitor–discharge resistance welding)
Recovery:	Heavily roll planish crown and root beads and HAZ
	Intensively shot peen weld zone (FZ + HAZ)

18.5 Troubleshooting Welding Problems in Cold-Worked Metals and Alloys

Table 18.3 lists the various problems that can arise in welds made in cold-worked metals and alloys.

References

Jeffers, Z. and Archer, R.S. (1924). *The Science of Metals*, 86. McGraw-Hill Book Company.

Kou, S. (1987). *Welding Metallurgy*. New York, NY: Wiley.

Messler, R.W. Jr. (1999). *Principles of Welding: Processes, Physics, Chemistry, and Metallurgy*. New York, NY: Wiley.

Messler, R.W. Jr. (1999/2005a). *Principles of Welding: Processes, Physics, Chemistry, and Metallurgy*, 516. New York, NY/Berlin, figure 16.1: Wiley/Wiley VCH.

Messler, R.W. Jr. (1999/2005b). *Principles of Welding: Processes, Physics, Chemistry, and Metallurgy*, 519. New York, NY/Berlin, figure 16.3: Wiley/Wiley VCH.

White, S.S., Manchester, R.E., Moffatt, W.G., and Adams, C.M. (1960). Plastic properties of aluminum–magnesium weldments. *Research Supplement Welding Journal* 39 (1): 10s–19s.

Further Reading

Messler, R.W. Jr. (2004/1999). *Principles of Welding: Processes, Physics, Chemistry, and Metallurgy* (Chapter 16). Weinheim/New York, NY: Wiley-VCH Verlag/Wiley.

19

Embrittlement with High-chromium Contents

19.1 Phase Formation and Structure

The formation of embrittling phases in metals or alloys anytime is problematic but can be especially so as a result of welding. As discussed in Section 6.4 and in Chapter 15, brittle martenite that can be formed in weld metal but is more often formed in the heat-affected zone (HAZ) of welds in hardenable steels can lead to strain-induced quench cracking even before there is even a chance to post-weld temper to prevent hydrogen-induced cold cracking. Obviously, if hydrogen from any source is present during or after welding, there is the possibility of delayed hydrogen-assisted cracking if the as-quenched martensite is not recognized to exist or if a suspect weldment is left untempered. Chapter 15 also discusses other ways carbon- or low-alloy steels can suffer from the formation of embrittling phases. Other alloy systems suffer from other embrittling phases, including hydride formation in Ti and Ti alloys, hydrogen embrittlement in Zr and Zr alloys, interstitial (i.e. H, C, O, and N) embrittlement in refractory metals and alloys (e.g. Group VA V, Nb and Ta and their alloys, and Group VIA Mo and W and their alloys), and others.

One of the most complex and exasperating forms of embrittlement plagues some of the most corrosion-resistant stainless steels during or after casting, rolling, forging, welding, or aging. The culprit is σ-phase formation.

As shown in the Fe–Cr (often improperly listed as Cr–Fe[1]) equilibrium phase diagram in Figure 16.1 and reproduced in Figure 19.1 for convenience, σ-phase forms alone or together with α-ferrite (α) as Cr content exceeds about 22 wt% and, as a practical matter, up to about 30 wt%.[2] If fact, this upper limit is almost certainly fixed to avoid intolerable problems from σ-phase either during the initial casting of billets during the production of the alloy or during subsequent production of roll product or forgings.

1 Convention dictates that a phase diagram is designated by the symbol for the solvent followed by the symbol for the solute element or elements, e.g. for carbon in iron, Fe–C. Hence, following convention, Cr in Fe is given by the Fe–Cr phase diagram.

2 Specified chromium (Cr) contents of wrought as well as cast grades of stainless steels range from a low 10.5 wt% to a high 30 wt%, with one exception at 32 wt% as follows: ferritic 10.5 wt% in 409 to 30.0 wt% in 29-4/29-4-2; martensitic 11.5 wt% in 409 to 18.0 wt% in 440A, B, or C; austenitic 16.0 wt% in 301 and 316 to 30.0 wt% in 329; PH 11.75 wt% in Custom 455 to 18.50 wt% in HNM; duplex 18.0 wt% in UNS S31500/3RE60 to 30.0 wt% in UNS S32906.

A Practical Guide to Welding Solutions: Overcoming Technical and Material-Specific Issues,
First Edition. Robert W. Messler, Jr.

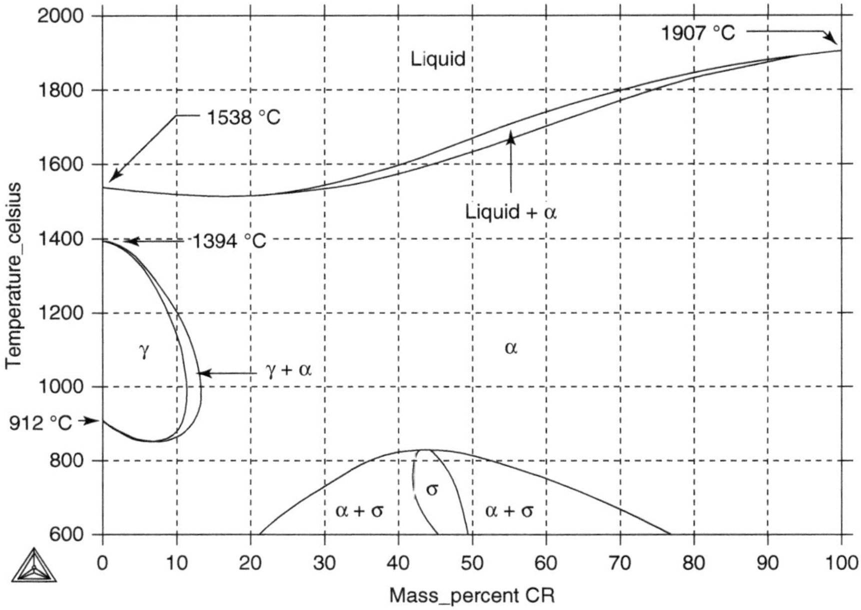

Figure 19.1 The Fe–Cr equilibrium phase diagram (shown earlier in Figure 16.1 as Cr–Fe) showing the existence of σ-phase as an intermetallic Fe–Cr compound alone or with α-ferrite (α). This particular version of the phase diagram was calculated with Therm-Calc®, coupled with PBIN thermodynamic database. Source: Lucas et al. (2007). As part of Elsevier, Ltd. Used under reciprocal agreement between Wiley and Elsevier for fair use.

σ-phase is an ordered intermetallic Cr–Fe compound[3] with a tetragonal crystal structure. It is alternatively designated as $D8_b$/tP30 by SPS/PS and as **$P4_2$/mnm** (D^{14}_{4h}) by SG#136. For Cr–Fe (σ-phase), the values of the lattice parameter a (i.e. orthogonal edges of the front and rear faces of the unit cell shown in Figure 19.2) and the lattice parameter c (i.e. orthogonal depth of the unit cell shown in Figure 19.2) are $a = 0.87995$ nm and $c = 0.45442$ nm. Pearson's tables list 84 intermetallic compounds with this structure type.

σ-phase is formed by solid-state transformation of body-centered cubic (BCC) δ-ferrite, the elevated temperature version of lower temperature BCC α-ferrite, to σ + face-centered cubic (FCC) austenite, i.e. $\delta \Leftrightarrow \sigma + \gamma_2$, which is essentially a eutectoid reaction. In fact, the σ-phase precipitates in Cr-rich regions of δ-ferrite, doing so preferentially at phase boundaries with austenite, including around dendrite arms, growing into the δ-ferrite once nucleated (Figure 19.3).

3 In an ordered phase, such as an intermetallic compound (i.e. a compound consisting only of metallic elements), the atoms of each element involved occupy particular sites in the crystal structure. For example, in the gamma prime (γ′), Ni_3Al ordered the intermetallic phase is one of the primary strengtheners of Ni-based superalloys; all atoms of Al are located at the corners of the face-centered cubic unit cell, whereas all atoms of Ni are located at the center of the faces. With each of the eight corner atoms being shared by eight unit cells that touch at a common corner, $1/8 \times 8 = 1$ equivalent of Al atom for each unit cell. With each of the six face-centered atoms being shared by two abutting unit cells, $½ \times 6 = 3$ equivalent of Ni atoms for each unit cell.

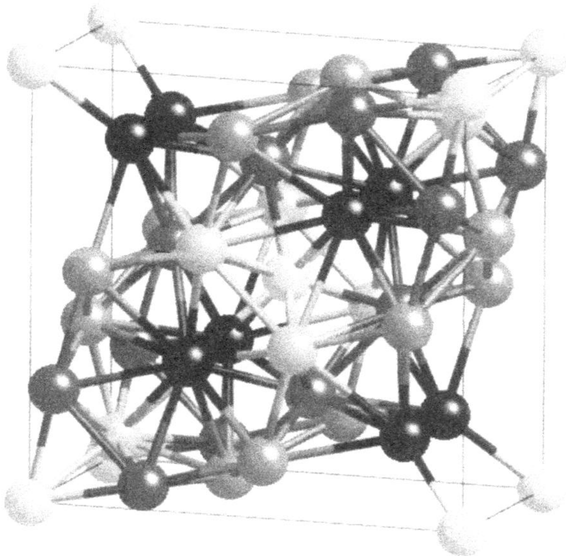

Figure 19.2 Schematic illustration of the complex unit cell for Cr–Fe (σ) phase found in the Fe–Cr system above about 22 wt% Cr. Source: Pearson et al. (1985). Used with permission of ASM International.

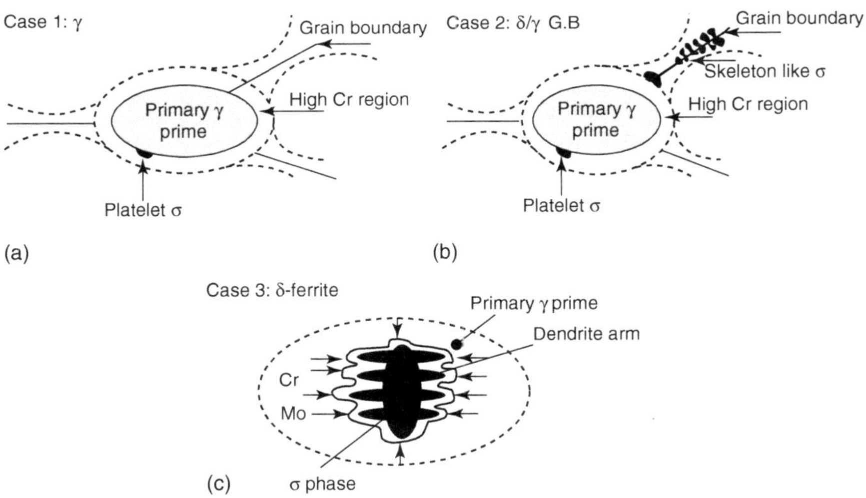

Figure 19.3 Schematic diagrams depicting three different morphologies for σ-phase nucleation and growth observed by Na et al.: (a) Case 1, at high Cr regions of δ-ferrite phase boundaries with primary γ into the δ-ferrite; (b) Case 2, additionally with a skeleton-like form in the δ/γ grain boundaries; and (c) Case 3, at the Cr- and Mo-rich interface of δ-ferrite dendrites within γ-austenite. Source: Na et al. (2000). Used under reciprocal agreement between Wiley and Elsevier for fair use.

Because δ-ferrite is a BCC solid solution phase, as is usually the case in metallurgy, BCC solutes, such as Cr, tend to segregate to it from FCC austenite. The transformation tends to occur in the range of 600–1000 °C/~1100–1800 °F, with the rate of transformation peaking around 900 °C/1650 °F. Both susceptibility to σ-phase formation and the rate at which transformation takes place increase as Cr content increases.

Figure 19.4 shows micrographs of σ-phase in a couple of susceptible alloys, notably at phase boundaries with δ-ferrite.

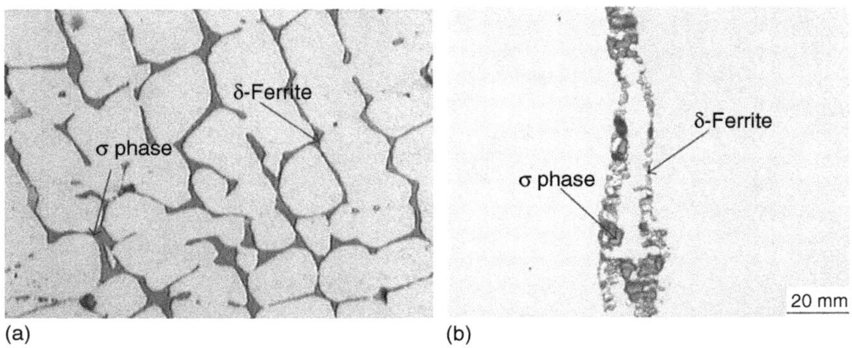

Figure 19.4 Micrographs showing σ-phase in representative susceptible alloys, including (a) in as-cast 309 LSi and (b) in hot-rolled 316 austenitic stainless steel. Note that the brittle Cr,Fe intermetallic compound tends to form at the interfaces with delta ferrite. Source: (a,b) Hsieh and Wu (2012). Used with permission of Hindawi Publishing Corporation and Hindawi Ltd.

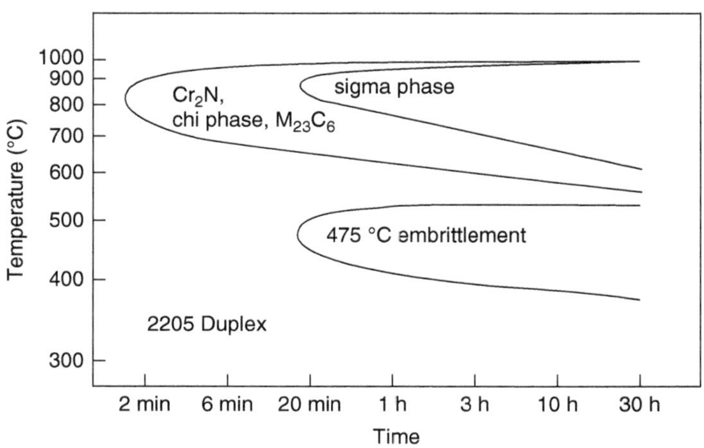

Figure 19.5 Schematic plot of C-curves showing the kinetics of the formation of various embrittling phases, including σ-phase, in 2205 ferrite–austenite duplex stainless steel. Source: Zucato et al. (2002). Used with permission of *Materials Research* at University, Dep de Engenharie de Materials.

Figure 19.5 is a schematic plot of showing the kinetics of various precipitating phases that can form in high-Cr 2205 duplex (ferrite + austenite) stainless steel with 22 Cr-3.2 Mo-5-6 Mo, including the σ-phase. Note how σ-phase formation is more sluggish than other phases.

Susceptibility to σ-phase formation begins once the content of ferrite-stabilizing Cr exceeds about 20 wt% and increases dramatically as the concentration reaches 25–30 wt%. Susceptibility has also been found to increase with additions of other ferrite-stabilizing elements, notably Mo and Si. The effect is logical as all of these solutes promote the formation of δ-ferrite known to trigger the transformation at phase boundaries with austenite. Although the kinetics for σ-phase formation can be slow (e.g. thousands of hours for some alloys exposed to elevated temperatures), of pertinence here, nucleation and formation can occur very quickly, in a matter of seconds, during welding of some alloys.

19.2 Adverse Effects of σ-Phase

Ordered intermetallic compounds (like ordered ceramic compounds) are inherently brittle. Not surprisingly, the σ-phase degrades the mechanical properties of susceptible alloys, increasing hardness, decreasing percent elongation, drastically decreasing impact toughness, providing initiation sites for fatigue, and degrading resistance to corrosion. As bad as these other adverse effects are, the degradation of resistance to corrosion is arguably the most serious for alloys chosen for their corrosion resistance. Resistance to intergranular corrosion, pitting corrosion, and crevice corrosion have all been observed to be compromised. For each of these forms of corrosion, maintaining Cr content above a minimum level is essential. As with sensitization due to the precipitation of Cr,Fe carbides along grain boundaries (see Section 16.2), the precipitation of Cr-rich σ-phase depletes Cr in the region of phase boundaries between δ-ferrite and austenite. The σ-phase and the γ-austenite phase induce a galvanic effect, with the normally highly corrosion-resistant γ-austenite corroding preferentially. The finer the σ-phase precipitates, the worse the degradation effect, as the finer particles tend to form a nearly continuous (albeit lacy) network structure at the interfaces.

Of particular relevance here, the susceptibility of σ-phase formation adversely affects the weldability of susceptible alloys, of which, as will be discussed in the following section, there are many.

19.3 Susceptible Alloys

A comprehensive review (by others) identified 50 transition alloys showing susceptibility to σ-phase formation. Most significant within this list, for their widespread use, are the following alloys containing more than 20 wt% Cr: Cr–Ni, Fe–Cr–Ni, Fe–Cr–Mo, Fe–Cr–Mn, Fe–Cr–Ni–Mo, and Fe–Cr–Si. Of particular note are the following susceptible alloys:

- Austenitic stainless steels containing 4–8 vol.% δ-ferrite to promote weldability by precluding hot cracking during fusion welding (including 308/308CRE, 309/309L, 310, 312, and 316/316L). (Recall that δ-ferrite acts as a scavenger of S, P, and other tramp elements that promote hot cracking in the austenite phase and, so, is often made a constituent of the weld deposit from filler wires through higher contents of ferrite-stabilizing solutes.)
- Cast grades of austenitic stainless steel containing more than 22 wt% Cr are especially susceptible
- Ferritic and super-ferritic stainless steels, which are prone to the formation of many secondary phases (e.g. χ-phase, Lave phases, etc.), notably, Fe–29Cr–8W
- Duplex stainless steels, including 2205 (22Cr–5Ni–3.2Mo) and A905 (26Cr–5.8Mn–4.3Ni–2.2Mo).

Figure 19.6 shows the region of the Schaeffler constitution diagram for stainless steels in which alloys are susceptible to σ-phase formation because of the presence of significant δ-ferrite with γ-austenite. The frustrating irony, not really so rare in engineering, is that in attempting to solve one problem

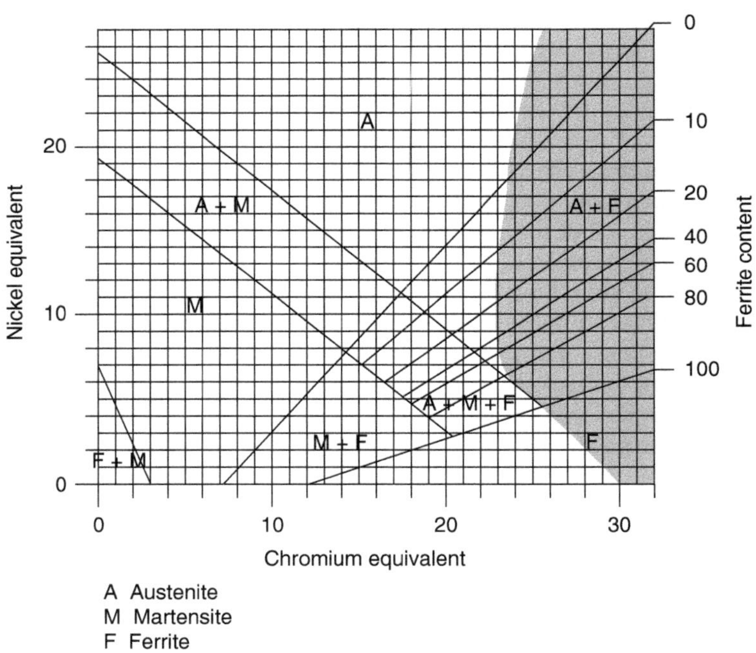

Figure 19.6 The Schaeffler diagram of microconstituents found in stainless steels as a function of nickel equivalent (Ni_{eq}) and chromium equivalent (Cr_{eq}.), showing the region at high Cr levels susceptible to σ-phase formation by shading. Source: The Schaeffler diagram of 1948 is in the public domain and used freely.

(i.e. susceptibility to hot cracking), another problem is created (i.e. susceptibility to σ-phase formation).

Finally, σ-phase formation is commonly encountered in fusion-welded Inconel 718 superalloy (50.00–55.00 Ni, 17.00–21.00 Cr, 4.75–5.5(Nb + Ta), 2.80–3.30 Mo, 0.20–0.80 Al, 0.065–1.15 Ti, 1.00 max. Co, 0.08 max. C, 0.35 max. Si, 0.015 max. S, 0.015 max. P, 0.006 max. B, 0.30 max. Cu, and bal. Fe) and Udimet 720/720 Li (55.16–59.705 Ni, 15.5–16.5 Cr, 14.0–15.0 Co, 4.75–5.25 Ti, 2.25–2.75 Al, 2.75–3.25 Mo, 1.00–1.50 W, 0.025–0.050 Zr, 0.0100–0.0200 B, and 0.010–0.020 C).

The rate of σ-phase formation in some of these alloys is slow, only being encountered during post-weld heat treatment (e.g. aging and stress relief) or in service involving prolonged exposure to elevated temperatures. In others, on the other hand, that σ-phase can form quickly enough to occur during welding or post-weld cooling.

Identification of σ-phase is nontrivial, typically requiring microprobe elemental analysis and, perhaps, X-ray or electron diffraction. Making matters even more complicated is the fact that the σ -phase can form with any of several morphologies (i.e. locations and shapes within a microstructure). Four morphologies have been observed and classified as (i) grain boundary or phase boundary precipitation, (ii) triple-point precipitation (where three grains meet at a point), (iii) corner precipitation (where δ-ferrite tends to be most rich in Cr), and (iv) cellular precipitation (in which the σ-phase and secondary austenite (γ_2) precipitate in a laminar form, suggesting formation by a eutectoid reaction).

Therefore this σ-phase is complicated stuff. How does one predict it, avoid it, or resolve it should it occur?

19.4 Guidelines for Avoiding or Resolving Problems from σ-Phase

As with most problems in welding, one needs to know two things about σ-phase formation to be successful welding susceptible alloys. First and most prudent: How to avoid the problem in the first place. Second, should the problem occur, how to resolve it.

Chemical composition can be used to predict the precipitation of σ-phase. One popular method uses an empirical formula (derived from data, not from first principles) proposed by Gow and Harder in 1942[4]:

$$\text{Ratio factor} = \%\text{Cr} - \frac{16\%\text{C}}{\%\text{Ni}}$$

If the ratio factor is >1.7, the σ-phase can precipitate in stainless steels. When the C content is much higher, the ratio factor is much lower, and the tendency for σ-phase to precipitate is decreased.

4 "Balancing the composition of cast 25% chromium – 12% nickel-type alloy" by J.T. Gow and O.E. Harder, *Transactions of the American Society for Metals*, Vol. 30, 855–935, 1942.

Another predictive formula for σ-phase formation in stainless steels by Woodyatt in 1984[5] uses the alloy's electron vacancy number N_v from composition in at.% as given by:

$$N_v = 0.66\text{Ni} + 1.71\text{Co} + 2.66\text{Fe} + 4.66(\text{Cr} + \text{Mo} + \text{W}) + 5.66\text{V} + 6.66\text{Zr} + 10.66\text{Nb}$$

When the electron vacancy number exceeds 2.52, σ-phase can be formed in the stainless steel. A concern, however, is the absence of any accounting for any interstitial C and/or N.

The most widely used empirical formula by Hull in 1980[6] assesses chromium equivalent Cr_{eq} and includes many alloying elements (using wt%) as follows:

$$\text{Cr}_{eq} = \text{Cr} + 0.31\text{Mn} + 1.76\text{Mo} + 0.97\,\text{W} + 2.02\text{V} + 1.58\text{Si} + 2.44\text{Ti} + 1.70\text{Nb} - 0.266\text{Ni} - 0.177\text{Co}$$

The higher the value of Cr_{eq} above about 24, the more likely the σ-phase will be formed.

A useful guide on susceptibility to σ-phase formation can be seen in the full color version of the Schaefller diagram found in Job knowledge 103 Marder: the yellow region located in the middle right, in the region labeled "A + F" between 5% vol.% and about 90 vol.% ferrite (F) above a Cr_{eq} of 24.

Although prediction of susceptibility is useful and prudent, more important is knowing what can be done to attempt to retard the formation of the σ-phase. Fortunately, the formation of σ-phase is normally quite sluggish in the range 500–900 °C/932–1652 °F, at least until Cr content exceeds around 25 wt%. However, if experience has shown susceptibility to σ-phase formation in weldments or as Cr content rises above about 22 wt%, consider the following:

- Limit the content of δ-ferrite in the alloy and any filler to 2–5 vol.% (as higher amounts promote σ-phase formation)
- Limit post-weld stress relief heat treatments, leaning toward lower temperatures and shorter times, always consulting alloy producers for recommendations.

Another recommendation to retard the precipitation of the σ-phase during welding and remove any formed after welding is to employ a solution heat treatment above 1050 °C/1922 °F for at least one hour per inch (25 mm) of thickness. Although this is a severe heat treatment, it is effective in that any σ-phase diffuses into the γ-austenite matrix.

In the case of σ-phase embrittlement, forewarned is forearmed.

19.5 Troubleshooting Problems with σ-Phase Associated with Welding

Table 19.1 lists the various problems that can arise with σ-phase formation during welding of stainless steels and some Ni-based superalloys.

---

5 *Austenitic Stainless Steels: Microstructures and Mechanical Properties* by L.R. Woodyatt, 1st ed., 1984, Springer, London.

6 "The effect of δ-ferrite on the hot cracking of stainless steel" by F.C. Hull, *Research Supplement Welding Journal*, Vol. 46, 399s–409s, 1980.

Table 19.1 Troubleshooting problems with embrittling σ-phase formation in stainless steels and Ni-based superalloys.

Austenitic and Duplex Ferrite–Austenite Stainless Steels:	
Forms along the $\delta - \gamma$ phase interfaces especially when δ-ferrite >~5–8 vol.% and for Cr 25–30 wt%	Keep ratio factor (using wt%) = %Cr-16×%C/%Ni < 1.7 Keep N_v (using at.%) = 0.66Ni + 1.71Co + 2.66Fe + 4.66[Cr + Mo + W] + 5.66V + 5.66V + 0.66Zr + 10.66Nb < 2.52
especially	Keep Cr_{eq} (using wt%) = Cr + 0.31Mn + 1.76Mo + 0.97W + 2.0V + 1.58Si + 2.44Ti + 1.70Nb − 0.266Ni − 0.517Co < ~24
	Limit δ-ferrite to 2–5 vol.%
	Limit post-weld SR to short time at low temperatures
	Solution treat at 1050 °C/1922 °F for 1 hr./in.
Ni-base Superalloys:	
Forms as Ti + Al content gets high	Check susceptibility using SIGMA-SAFE phase diagram approach
	Check with alloy producer for guidance

References

Hsieh, C.-C. and Wu, W. (2012). Overview of intermetallic sigma (σ) phase precipitation in stainless steel. *ISRN Metallurgy* 2012: 732471, 16 pages, figure 5.

Lukas, H., Fries, S.G., and Sundman, B. (2007). *Computational Thermodynamics: The Calphad Method*. New York, NY: Cambridge University Press www.calphad.com.

Marder, G. Welding of austenitic stainless steels. http://www.twi-global.com.

Na, Y.S., Park, N.K., and Reed, R.C. (2000). Sigma morphology and precipitation mechanism in Udimet 720 Li. *Scripta Materialia* 43 (7): 585–590.

Pearson, W.B., Villars, P., and Calvert, L.D. (1985). *Pearson's Handbook of Crystallographic Data for Intermetallic Phases*. Metals Park (now Materials Park), OH: American Society of Metals (now ASM International).

Zucato, I., Machado, I.F., Moreira, M.C., and Lebrão, S.G. (2002). Microstructural characterization and the effect of phase transformations on toughness of the UNS S31803 duplex stainless steel aged treated at 850 °C. *Materials Research* 5 (3): 385–389, figure 1.

Further Reading

Hsieh, C.-C. and Wu, W. (2012). Review article. Overview of intermetallic sigma phase precipitation in stainless steel. *International Scholarly Research Notices (ISRN Metallurgy)* 2012: 1–16.

Lippold, J.C. and Kotecki, D.J. (2005). *Welding and Weldability of Stainless Steel*, 1e. New York, NY: Wiley-Interscience/Wiley.

Vitek, J.M. and David, S.A. (1986). The sigma (σ) phase transformation in austenitic stainless steel. *Welding Journal (Research Supplement)* 64 (4): 106s–111s.

20

Weld Dilution and Chemical Inhomogeneity

20.1 The Designer's Druthers[1]

Engineers require three things to end up with a successful creation, whether a wind turbine, a washing machine, or a weld. These three are: (i) predictability; (ii) consistency; and (iii) uniformity. Each is critical to success at each stage and in each aspect of the creation; from design conceptualization through fabrication to application.

Predictability, as a noun, is defined as "consistent repetition of a state, course of action, or behavior, making it possible to know in advance what to expect." A designer needs to know what to expect from every material, energy source (e.g. process), or bit of information used as an input to create a product. Most importantly, the designer must have predicable material properties and process attributes. Initially, material properties must be predictable (i.e. known with reasonable certainty) to allow their selection to meet design requirements for both manufacturability and service. Likewise, process attributes must be predictable (i.e. known with reasonable certainty) to assure chosen materials can be shaped, joined, and conditioned (e.g. heat treated, finished, etc.) to fabricate details and assemblies. But, in addition, how a material's initial properties may be changed by a process – to end up what they need to be or, at least, not to be degraded from what they need to be – must be predictable.[2] Material properties and process attributes can be obtained from a wide variety of sources, some reliable and some not. Hence, a designer needs to carefully and critically examine and assess (i.e. vet) a source. Without predictability, things are only a guess!

Consistency, a noun, is defined as "steadfast adherence to the same principles, course [of action or behavior], form, etc." A designer needs to know that how a material or process behaves the first time, it will be behave the next time as well, and that how it behaves today will be how it will behave tomorrow. Consistency is especially important when more than one of something is being created; whether one of a place-setting of a pattern of China for twelve, one of hundreds

1 An informal English word that appeared between 1870 and 1875, *druther* is a contraction of would rather, i.e. I'd rather. It means: One's own way, choice, or preference. (From www.dictionary.com, the source used for this chapter and most other chapters, unless noted otherwise.)
2 This harkens back to the inextricable interrelationship among a material's structure-property-processing for performance (Section 2.2, Figure 2.6).

A Practical Guide to Welding Solutions: Overcoming Technical and Material-Specific Issues,
First Edition. Robert W. Messler, Jr.

of jumbo jets, one of thousands of homes in a development, or one of millions of a modern make and model of an automobile. Consistency is sought by both the manufacturer for efficiency and by the end-user for confidence. It is achieved in production by quality assurance, with material and process control at its heart. Without consistency, one might be getting a pig in a poke.[3]

Uniformity, a noun, is defined as "the state or quality of being of overall sameness, homogeneity, or regularity." Uniformity of a material or product comes about from the proper and consistent application and execution of a process or processes in terms of its or their procedure and practice; which, in turn, comes about with governing specifications and standards. Nowhere may uniformity be more important than in the materials used in the creation of the product. In fact, while predictability and consistency are important, it could be argued that uniformity is the most important of all to the designer. Here's why …

A chain is only as good as its weakest link.[4] While applicable to all manner of non-physical events and physical entities, it has particular relevance in engineering, where a physical material or structure made therefrom is only as strong as the weakest "link" within it (Figure 20.1). Herein lies the particular need for

Figure 20.1 Image of the effect of the weakest link in a chain; literally or figuratively. Source: From the public domain.

3 The expression "a pig in a poke" means one doesn't know what one might get unless first checking or assessing it. The saying first appeared as "When ye proffer the pigge open the poke" in 1530 in an attribution to the English writer John Heywood (c. 1497–c. 1580), but appeared closer to its modern form in 1535 in Heywood's *Proverbs* (shortened title from a much longer title in the original Olde English, as "Though he love not to buy the pig in the poke."

4 The well-known phase "A chain is only as strong as its weakest link" – the premise for a short running game show entitled "Weakest Link" (adapted from the BBC) that appeared on NBC from 16 April 2001 through 14 July 2002 – first appeared in *Essays on the Intellectual Powers of Man* (1786) by Rev. Prof. Thomas Reid, a British philosopher, in the statement: "In every chain of reasoning the evidence of the last conclusion can be no greater than that of the weakest link of the chain, whatever may be the strength of the rest."

uniformity by design engineers and process engineers. Two examples will suffice to defend the claim.

If an actual chain was comprised of many identical-size links made from annealed AISI 1045 steel with a yield strength of 310 MPa/45 ksi and tensile strength (to fracture) of 565 MPa/82 ksi and a ductility of 16% elongation, but contained one link made from the pink synthetic rubber used to make erasers on common pencils, with a tensile strength of 18 MPa/2.6 ksi and a ductility of 380% elongation, everyone would know slow loading would concentrate stretching of the overall chain in the rubber link, as the rubber is much more elastic than steel, and the rubber link would be the first to fail as it is weakest in terms of strength. Under such slow loading, overall behavior is limited by strength. Under very fast, impact loading, the result would be the same, as overall behavior would still be limited by the strength of the weakest link.

If, on the other hand, the same annealed AISI 1045 steel chain contained one link made from fully hardened, water-quenched, untempered AISI 1045 steel with a tensile strength (to fracture) of over 2413 MPa/350 ksi and a ductility of less than 1% elongation, the results of slow loading and very fast, impact loading would be very different. Under slow loading, strength of the links would limit performance of the chain, so one of the annealed steel links would fail before the fully-hardened link failed. Which annealed steel link failed would generally be unpredictable if the chain were straight when loaded. However, if the chain ran over a capstan and was slow-loaded, the link right where the chain engaged the capstone would fail, as it would experience stress concentration. But, if the straight chain was subjected to an impact load, performance would be limited by impact toughness, and the hardened link would fail, as typical impact toughness (using the Izod versus Charpy method) values would be 54 J for the annealed link and well less than half that value for the hardened link, as the value for a water-quenched and tempered steel would be 30 J.

The point of these examples is: Performance is *always* limited by the weakest link in a physical entity; for whatever the *limiting property* is for the design and application.[5] The significance of this to welds and weldments is this: The designer can only rely on – and use as a limit for an allowable property – the *lowest value* of the key property anywhere in the weld or in the overall weldment. The weakest link dominates!

So, in a material, in general, and in a weld, in particular, *uniformity of properties* is often of greatest concern to the designer. The designer must know the worst properties across a weld zone (i.e. FZ + PMZ + HAZ); the lowest yield and tensile strength for load carrying, the lowest ductility for robustness, the lowest toughness for impact, the poorest resistance to corrosion for corrosion. It does a designer no good – and, in fact, it does the designer an injustice – for anyone involved to say a weld has an "average strength" of so-and-so or "overall the

5 Different design situations and requirements usually lead to a different property (or properties) that limit performance. For example, a vaulting pole must, obviously, be strong enough to sustain the weight of the vaulter, but performance of the pole is actually limited by its ability to absorb energy elastically as it bends, so as to return it as it straightens to lift the vaulter up and over the crossbar. The ability to absorb energy is controlled by a material's modulus of elasticity (E), so this is the limiting property.

strength is" so-and-so or "the filler metal should be at least as corrosion resistant as the base metal" if there is *any* location in the weld where this is not true. The majority of problems that arise from weld and/or welded structures (i.e. weldments) arise because properties were not what they were predicted to be, often because there was degradation in some location.

All said and done: If designers had their druthers, they'd pick uniformity every time!

Uniformity in welds manifests itself as chemical homogeneity and structure homogeneity. The former will be treated here. The latter has been treated throughout the rest of this book, as discontinuities in the form of either weld-induced or weld-related defects or weld-altered microstructure.

20.2 Chemical Inhomogeneity in Welds

A number of conflicting phenomena take place in the weld pool of fusion welds that tend to lead both the weld pool and, thus, the final weld to become both more heterogeneous[6] and more homogeneous at the same time. The net result is that actual fusion welds are never completely homogeneous and are usually severely heterogeneous. The obvious desire of designers and goal of manufacturing engineers and process engineers, despite everything acting to prevent it, is to obtain homogeneous welds. This means homogeneous in chemical composition, homogeneous in (micro)structure, and, as a consequence of both, uniform in properties. Only when this goal is achieved can a designer get the most from the weld and have confidence in his or her design.

The principal phenomenon responsible for heterogeneity in all fusion welds is segregation of solute rejected at the advancing solid–liquid interface as a molten alloy solidifies.[7] This is a natural consequence of solute partitioning between the liquid and solid phase of an alloyed material that would occur under equilibrium conditions[8] but is exacerbated and persists under always-prevailing non-equilibrium conditions associated with welding. Under true equilibrium, there would be infinite time for any tendency for the development of any solute concentration gradient in the molt pool to level out by diffusion. Hence, under true equilibrium, the final composition of the solidified weld would

---

6 *Inhomogeneous* and *heterogeneous,* as adjectives, are synonyms, with identical meanings as follows: Diverse in character or content. In chemistry, it means not having the same composition throughout.

7 Segregation from this cause is known as *microsegregation,* as it only spans microscopic dimensions of the order of 0.1–0.2 mm/0.004–0.008 in. More severe segregation can occur in large castings over larger dimension, but is especially prevalent when dissimilar metals or alloys are fusion welded. Known as *macrosegregation,* it is discussed in Chapter 21.

8 Partitioning of solute between the liquid and solid phases of an alloy is expressed by an equilibrium distribution coefficient $k_o = C_s/C_L$ that would exist at equilibrium at any particular temperature during the process of solidification. Depending on the phase diagram for a binary alloy, the slope of the lines of liquidus and solidus temperature can be downward (i.e. negative) or upward (i.e. positive), so the value of k_o can be <1 or >1, respectively.

be homogenous. With real-world welding, however, solidification takes place under conditions that deviate from equilibrium, often to a great degree. The result is incomplete removal of solute concentration gradients that build in the weld pool ahead of the advancing solid–liquid interface and persist in the final weld.

The principal phenomenon tending to cause homogeneity in a weld pool is convection. Convection usually arises from at least two and as many as four forces; i.e. (i) a buoyancy force from the effect of gravity (except in outer-space) on higher-density cooler liquid and lower-density hotter liquid; (ii) an ever-present surface tension gradient (or Marangoni) force that pulls cooler liquid with greater force than hotter liquid; (iii) a Lorentz electromotive force (emf) from interaction between the current in an arc, plasma, or electron beam and eddy currents induced in the circulating weld metal; and (iv) an impingement (or friction) force from circulating imparted by the momentum of incoming electrons and/or ions. The complex circulating flow patterns that result act to stir the weld pool, mixing the liquid to remove any composition gradients, regardless their source. However, as will be seen in Section 20.5, even the strongest convection, in the most turbulent weld pools, never results in perfect mixing.

Figure 20.2 schematically illustrates the flow patterns found in a molten weld pool due to various convection forces.

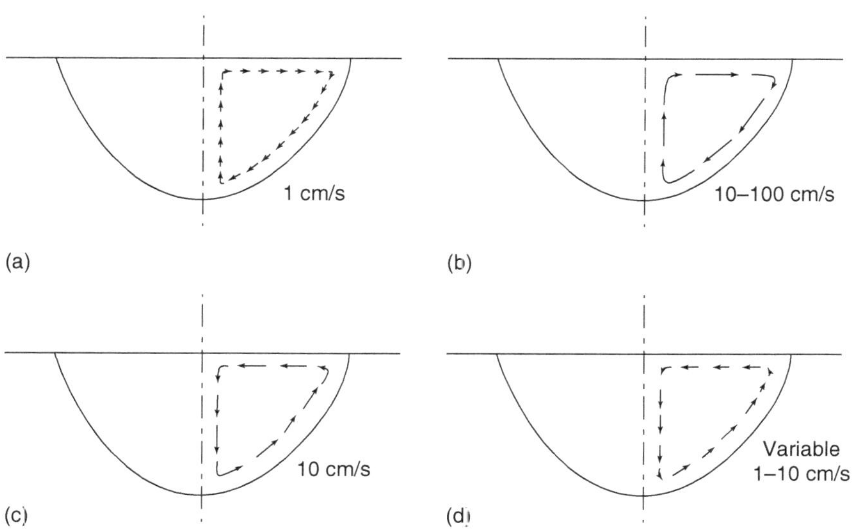

Figure 20.2 Schematic representation of the circulation patterns induced by each of the four forces that can give rise to convection in weld metal, i.e. (a) the buoyancy (or gravity) force, (b) the ever-present surface-tension-gradient (Marangoni) force, (c) the Lorentz emf force in electric arc, plasma or beam welding, and (d) the impingement (or friction) force. The arrows show direction, as well as relative velocity by the lengths. Source: Messler (1999). Used with permission of John Wiley & Sons.

20.3 Weld Dilution

Except in those unusual situations that joint elements of like metals or alloys can be tightly fitted to form straight- or square-butt joints, filler metal[9] must be added during fusion welding. For such unusual instances, so-called *autogenous welding* can be performed. Most situations call for the use of filler metal during either single-pass or multiple-pass welding for one or more of several purposes, including: (i) producing a weld of appropriate size/volume to safely carry design loads; (ii) properly filling a prepared joint configuration (e.g. single- or double-V; see Figure 13.7); (iii) obtaining chemical, mechanical, and thermo-mechanical (e.g. coefficient of thermal expansion, CTE) compatibility between joint element materials; (iv) obtaining special properties (e.g. under-matched strength to force ultimate failure to occur in the weld, as opposed to in the base metal joint elements); or (v) applying overlays for wear- or corrosion-resistance.

When a filler metal has exactly the same composition as base metals of the same composition as each other, the process is referred to as *homogeneous welding*, and the composition of the weld deposit is essentially the same as the two like base metals.[10] When the filler metal has a different composition than base metals of the same composition, and/or the base metals are not the same as one another, the process is known as *heterogeneous welding*, and the weld deposit becomes a mixture of the filler and like or dislike base metals.

In selecting a filler metal for a joint between like base metals, there are three generic options: (i) choose a *matching filler*; (ii) choose an *undermatching filler*; or (iii) choose an *overmatching filler*. The advantages of a matching filler are obvious, as follows: (i) absolutely certain chemical, thermophysical, and mechanical property compatibility; (ii) uniform chemical reactivity for equal corrosion resistance of base and weld metal; (iii) uniform mechanical properties in the weld and base metal, except for inherently inferior differences (especially in ductility, impact and fracture toughness, and fatigue strength) for cast versus wrought microstructures; and (iv) no concerns pertaining to mixing (i.e. inhomogeneity).

Despite the above-stated advantages of a homogeneous filler, a heterogeneous filler is often either required or preferred. A heterogeneous filler is obviously required when base metals are different than one another; with the choice of filler requiring compatibility with each and, inevitably, compromising some key property with one or the other. As general guidelines:

---

9 The term *filler metal* is used to denote the material added to a joint to produce a weld of proper size and shape or profile. The filler may be a pure metal to match identical pure-metal joint elements or, in some instances, to join like-alloy joint elements, dissimilar alloy joint elements, or pure metal and alloy joint elements. More often than not, the so-called filler metal is an alloy chosen to be compatible with the metals or alloys being joined.

10 "Essentially the same" as, in fact, minor losses of certain alloying elements in the filler occur largely by evaporation but also by oxidation if shielding is imperfect. Such losses can lead to mismatch, leaning out the filler in key elements and rendering it more susceptible to corrosion than the base metal.

- Favor undermatching the strength of the filler to at least the strongest (and least ductile) base metal; to allow the weld metal to accommodate minor overloads by plastically straining (like the synthetic rubber link in a chain of steel links) or fail before the base metal components to preserve them at the sacrifice of the weld
- Favor overmatching the corrosion resistance of the filler (choosing one with higher amounts of key solutes) compared to the base metal requiring the best resistance to corrosion; to offset inherent degradation of an alloy's corrosion resistance due to a cast versus wrought microstructure – all other things being equal.

Since, by definition, fusion welding requires that a portion of each base metal involved in a joint melt along with any filler, fillers are inevitably diluted by base metal(s) during heterogeneous welding. It is critical for many applications, and advisable with all applications, to known the *degree of dilution* that occurs, with key factors being: (i) the type of joint (see Figures 3.1 and 2.5) and type of joint edge or groove preparation (see Figures 3.2 and 3.3); (ii) the (a) welding process, (b) process parameters (including operating current mode, current, voltage, and, travel speed), and (c) technique (e.g. arc manipulation, greater or lesser bead overlap, use of buttering, etc.) used; and (iii) the mismatch between the filler and base metal(s). Figure 20.3 schematically illustrates the most general joint between two different base metals, A and B, and a backing C using a filler D. Also given in the lower portion of the figure are the formulas for calculating (i) *percent dilution* in

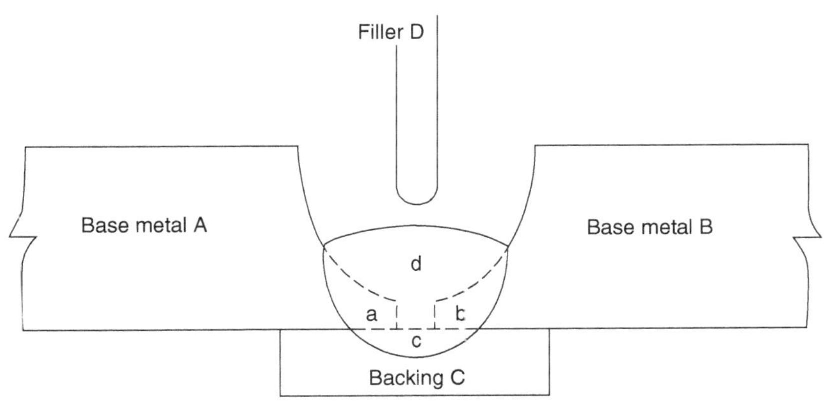

$$(1)\ \%\ \text{Dilution} = \frac{a+b+c}{a+b+c+d} \times 100$$

$$(2)\ \%\ \text{Element E in weld bead} = \frac{1}{a+b+c+d}\left[a(\%\ E\ \text{in}\ a) + b(\%\ E\ \text{in}\ b) + c(\%\ E\ \text{in}\ c) + d(\%\ E\ \text{in}\ d)\right]$$

Figure 20.3 At the top, a schematic illustration showing the region of a root pass being made between two different base metals A and B and a backing C using a filler D, and, at the bottom, formulas for calculating percent dilution and the percentage of each element in the final weld. Source: Estes and Turner (1964). Published by and used with permission of The American Welding Society.

the filler by the base metals and backing and (ii) the *percent of each element* (from all sources) *in the weld* (here the root pass of a multipass weld).

Weld metal dilution (or simply *weld dilution*) is particularly important in three situations, as follows:

1. Increased susceptibility for formation of martensite in the weld metal during cooling due to pick-up of elements from base metal(s) that increase hardenability (via C-equivalent, C_{eq}); with the risk of either quench cracking or hydrogen-induced cold cracking
2. Loss of corrosion resistance in the weld metal
3. Compromise of wear resistance in a wear-resistant weld overlay (i.e. hard-facing).

To prevent dilution in (1) and (2) above, a common technique is to apply a buttering layer or layers to the joint face of the offending base metal joint element, thick enough that when the joint proper (e.g. V-groove) is filled with either a single-pass or multiple passes, only material from the buttering layer, not the underlying base metal dilutes the weld metal. Typical buttering materials for preventing carbon pick-up are austenitic stainless steels (e.g. 304, 308, or 309) or pure nickel (e.g. Nickel 200). When buttering a corrosion-resistant base alloy, the best practice is to consult the alloy producer for recommendations. To prevent dilution in (3) above, the most common technique is to apply a cushion layer, typically consisting of not less than three to as many as five layers to reduce dilution from the base metal to an acceptably low level with three passes and to near zero with five passes. Typical cushion-layer materials for hard-facing alloys are Mn-austenitic, stainless or Hadfield steels or a low-carbon steel.

Knowing the degree of dilution is often critically important to the successful performance of a weld or a weldment. As always: Knowledge is strength.[11]

20.4 The Unmixed Zone in the Weld Metal

Like all fluids (i.e. gases and liquids), regardless of how fast that fluid moves over a stationary solid surface, every molten weld pool, irrespective of how violent the convection during fusion welding, has a stagnate layer at its boundary with the unmelted or partially melted surrounding region. Molten metal, like any fluid, must have continuity with the solid (metal) with which it comes in contact. At the interface between the two, the velocity of the fluid (molten metal) must be the same as the velocity of the solid. If the solid isn't moving, the fluid can be moving either, because of friction. In a weld pool, without movement, there is no stirring. Without stirring, there is no mixing of base metal and filler. The result is an *unmixed zone*.

Figure 20.4, schematically illustrates the unmixed zone in a single-pass fusion weld, along with other key zones and features of such welds. Figure 20.5 shows a micrograph of an unmixed zone in an actual fusion weld.

---

11 Check out all the proverbs and quotes relating to "knowledge is strength" on Google™, as there are many!

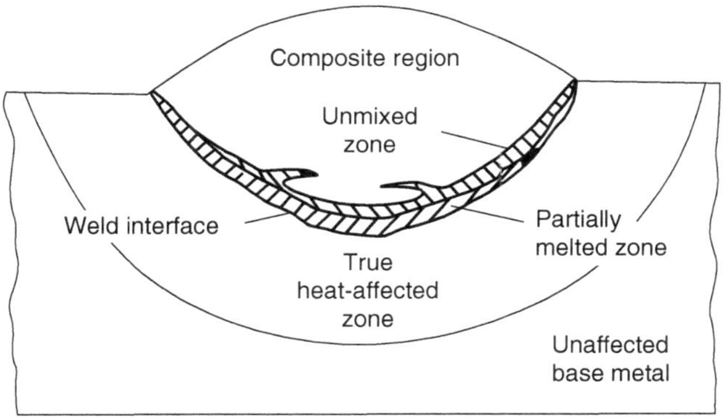

Figure 20.4 Schematic illustration showing the different discrete zones present in a single-pass fusion weld, including the unmixed zone. Source: ASM International (1993). Published by and used with permission of the ASM International.

While typically not very extensive, unmixed zones can still be problematic in welds in corrosion-resisting alloys, as they are often less resistant to corrosion than overmatching filler; thereby suffering preferential attack as shown in Figure 20.5b. The formation of undesirable phases, such as martensite, which can lead to hydrogen-induced cracking, has also been reported (e.g. when Cr–Ni–Mo low-alloy steels are welded with 20Cr–10Ni austenitic stainless steel filler).

One positive consequence is, the stagnate layer of liquid immediately adjacent to the boundary with the PMZ for an alloy (i.e. created by the liquidus temperature isotherm) allows a profile of solute rejected at the solid–liquid interface due to the distribution coefficient to build without disruption. The significance of this solute profile arises from the associated profile of effective liquidus temperature. Under non-equilibrium solidification, this latter profile leads to *constitutional supercooling* with the actual temperature gradient from the S–L interface into the liquid, which, in turn, gives rise to different sub-structure morphologies in the newly formed solid. The type and scale of this sub-structure influences properties and can either exacerbate or ameliorate hot crack formation.[12]

For better or worse, an unmixed zone is a natural consequence of fluid dynamics. The extend of an unmixed zone, on the other hand, can be minimized by maximizing convection. In fact, reports claim ultrasonic vibration or high-frequency electromagnetic vibration during welding can completely eliminate the unmixed zone in dissimilar metal fusion welds.[13]

---

12 Readers interested in details in this topic are encountered to seek information on "Case 3 solidification in alloys" in references on welding metallurgy (e.g. *Principles of Welding: Processes, Physics, Chemistry, and Metallurgy* by Robert W. Messler, Jr., 1993/2005, Wiley/Wiley-VCH, New York, NY and Berlin, Germany).

13 "Effect of ultrasonic vibration on mixed zone formation," by Y. Cui, C.L. Xu, and Q. Han, *Scripta Materialia*, 55(11), 975–978, 2006; "Effect of electromagnetic vibration on the unmixed zone formation in 25Cr–35Ni heat-resistant steel/Alloy 800 dissimilar metal welds," by R. Dehmolaei, M. Shamanian, and A. Kermanpur, *Materials Characterization*, 59(12), 1814–1817, 2008.

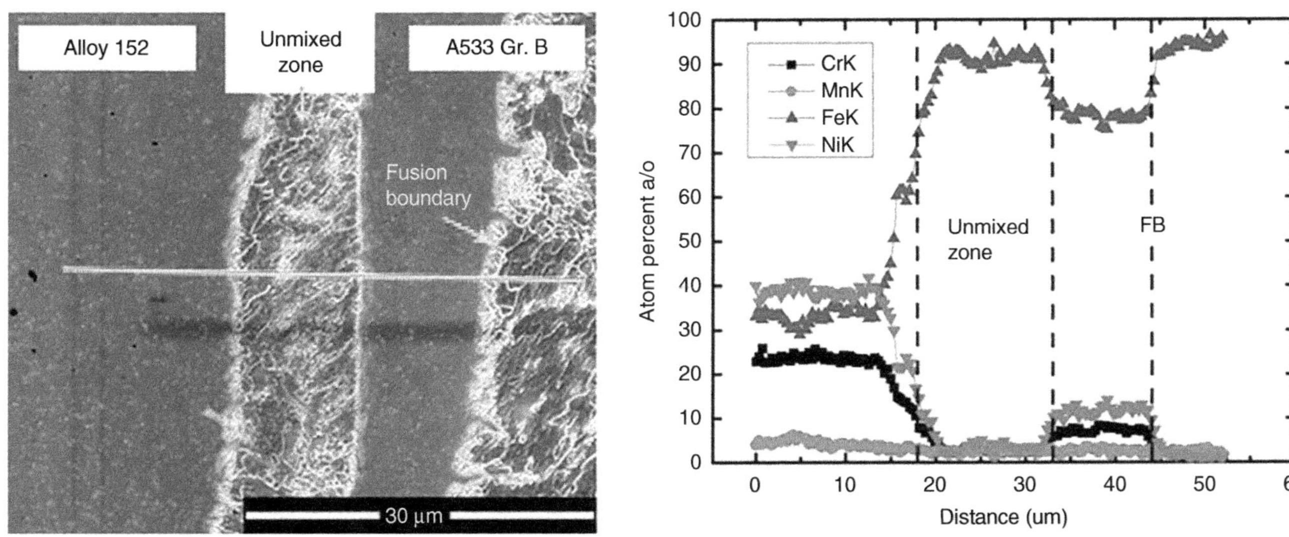

Figure 20.5 Image showing an elemental-analysis trace across the unmixed zone in a weld made using Alloy 152 to weld A533 Grade B. Source: Choi et al. (2012). Published by Elsevier Ltd. and used under mutual agreement between Wiley and Elsevier Ltd. for "fair use."

20.5 Impurities in the Weld Metal

Three general types of impurities can be found in weld metal: (i) gas impurities, usually in the form of gas porosity; (ii) nonmetallic impurities, usually in the form of slag inclusions from slag-protected welding processes and consumables; and (iii) foreign-metal inclusions, usually from transfer of tungsten from the non-consumable electrode used with GTA welding. These were adequately discussed in Chapter 5 (Porosity) and Chapter 7 (Inclusions), so will not be addressed further here.

The particular concern with each or any of these impurities is they lead to localized stress concentration, especially under fatigue or impact loading.[14]

A fourth source of impurities in weld metal is either base material or filler containing unacceptable levels of *impurity elements*. Such elements are commonly referred to as *residual elements* or *tramp elements*. The author, at least, tends to differentiate between these two terms depending on whether the particular impurity was left over from metallurgical refining (e.g. because of some technological or economic limitation) or brought in accidentally from some source (e.g. scrap added to a furnace-charge during remelting). In the author's mind, the former is residual, the latter is tramp. These impurities tend to be most detrimental in causing hot cracking (see Section 6.3).

20.6 Troubleshooting Problems from Weld Dilution and Chemical Inhomogeneity

Table 20.1 that follows lists the various problems that can arise from weld dilution and chemical inhomogeneity in fusion welds.

Table 20.1 List of the problems from weld dilution and chemical inhomogeneity in fusion welds.

Dilution of Weld Metal:	
Alteration of filler metal by base metal(s)	Consider alternate joint preparation
	Consider change of process or current operating mode
	Keep net linear heat input as low as feasible
	Employ thin, overlapping weld beads
	Butter joint face(s)
Unmixed Zone in Weld Pool and Fusion Zone:	
Stagnate layer near fusion line can lead to compromised corrosion resistance or martensite formation in steels	Consider aggressive pool stirring using either ultrasonic or electromagnetic vibration

(Continued)

14 The catastrophic failure of the tail-mounted GE CF6-6D high-by-pass-ratio turbofan engine on a DC-10 during United Airlines Flight 232 on 19 July 1989 led to the conclusion that impurities in cast Ti-alloy turbine components led to this and several other incidents with varying consequences. In each case, tiny impurities in cast structure, undetected during multiple inspections, led to catastrophic results.

Table 20.1 (Continued)

Impurities in Weld Metal:	
Lead to stress-concentrations under fatigue or impact loading	Avoid gas porosity per Section 5.5 Avoid slag inclusions per Section 7.2 Avoid foreign-metal inclusions per Section 7.3
Dissimilar Metal Welds:	
Macrosegregation	See Chapter 21

References

ASM International (1993). *ASM Handbook*, Volume 6: Welding, Brazing, and Soldering, 8e. Materials Park, OH: ASM International.

Choi, K.-J., Shin, S.H., Kim, J.-J., and Kim, J.H. (2012). Nano-structural and nano-chemical analysis of a Ni-base alloy/low-alloy steel dissimilar metal weld. *Nuclear Engineering and Technology* 44 (5): 481–500.

Estes, C.L. and Turner, P.W. (1964). Dilution in multipass welding of AISI 4130 to type 304 stainless steel. *Research Supplement of the Welding Journal* 43 (12): 541s–550s.

Messler, R.W. Jr. (1999/2005). *Principles of Welding: Processes, Physics, Chemistry, and Metallurgy*, figure 10.2, p. 293. New York, NY and Berlin, Germany: Wiley/Wiley-VCH.

Further Reading

Banovic, S.W., DuPont, J.N., and Marder, A.R. (2002). Dilution and microsegregation in dissimilar metal welds between super austenitic stainless steel and nickel-base alloy. *Journal of Science & Technology of Welding and Joining* 7 (6): 374–383.

Kou, S. (2003). *Welding Metallurgy*, 2e. Hoboken, NJ: Wiley Interscience, Wiley.

Messler, R.W. Jr. (2004/1999). *Principles of Welding: Processes, Physics, Chemistry, and Metallurgy*, Chapter 12. Weinheim, Germany/New York, NY: Wiley-VCH/Wiley.

21

Dissimilar Metal and Alloy Welding

21.1 Joining Dissimilar Materials

Our modern, increasingly technological world demands the growing use of dissimilar materials to enable the achievement of function where service and thus design requirements call for diverse and often divergent properties unobtainable in single materials, i.e. metals, ceramics, or polymers (Figure 21.1). So-called *hybrid structures* are frequently required (Figure 21.2).

Combining fundamentally dissimilar materials into a hybrid structure can offer one or more of several benefits, three of which are technical and one of which is economic. The first and usually principal technical benefit is that dissimilar materials optimize a design by matching the correct material to the needed property or behavior (e.g. refractoriness,[1] electrical or thermal conductivity, electric or thermal insulation, optical transparency or opacity, and corrosion or wear resistance) rather than compromising in some areas by settling for a less than optimal material to fabricate the entire structure. The second technical benefit is the minimization of structural weight by using the lowest density material with comparable strength- or stiffness-to-weight ratio in each area of the assembly or structure, as most appropriate. The third technical benefit is provision of enhanced damage tolerance to the overall structure by changing the material, and elastic properties, along a potential crack path, without having to rely on the use of discrete parts of like materials that are simply fastened together at physical interfaces.

In addition to these important property advantages, the use of dissimilar materials in hybrid assemblies or structures often allows the costs of fabrication and/or operation to be minimized by allowing optimum materials to be used in specific areas of the design. In this way, inherently expensive or difficult- and

1 *Refractoriness* refers to the property of a material (i.e. metal or ceramic) having resistance to very high temperatures, typically with a melting point over 2000 °C/3632 °F. Unfortunately, most refractory metals have densities (in g/cm^3) higher than lower melting point (°C/°F) Fe (1536/2797; 7.87), Co (1498/2728; 8.85), and Ni (1453/2647; 8.9) used in superalloys. The refractory metals always include, in descending order of MP (°C/°F), with density in g/cm^3: W (3422/6192; 19.25), Re (3186/5767; 21.02), Ta (3017/5463; 16.69), Mo (2623/4753; 10.23), and Nb (2477/4491; 8.57). The "wider list of refractory metals," in order with MP only, includes Os (2700/5491), Ru (2500/4233), Ir (2454/4435), Rh (1966/3567), V (1900/3470), Cr (1875/3465), Zr (1852/3371), and Ti (1668/3034).

A Practical Guide to Welding Solutions: Overcoming Technical and Material-Specific Issues,
First Edition. Robert W. Messler, Jr.

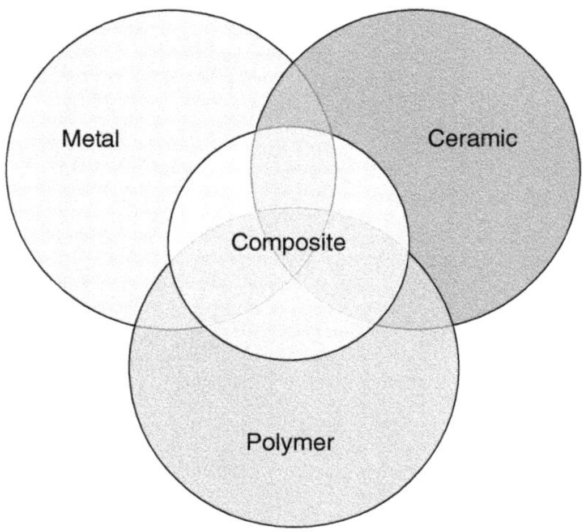

Figure 21.1 A general schematic showing the three fundamental classes or types of materials, metals, ceramics, and polymers, using a Venn diagram. Note that composite materials, being mixtures of two or more materials from like or dislike classes, are shown by the central circle of the Venn diagram. Source: A widely used untraceable generic image in the public domain.

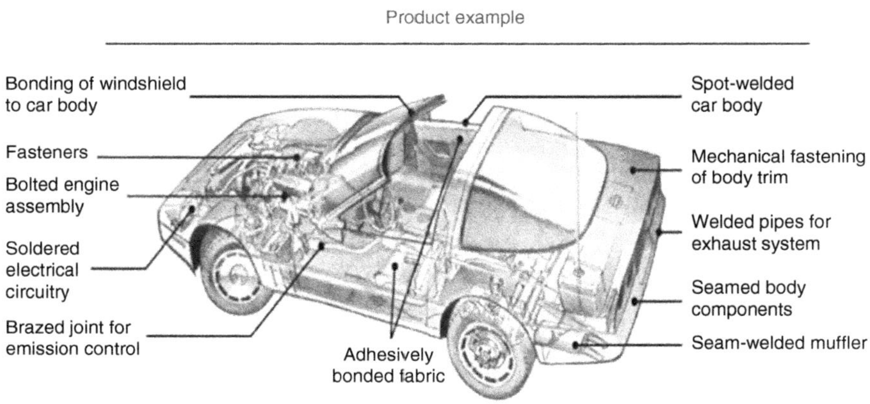

Figure 21.2 An illustration showing how various parts of a typical modern automobile are assembled using different joining processes to accommodate different and often dissimilar materials. Adhesive bonding, brazing, several types of welding, and mechanical fastening are shown. The example labeled "seamed body components" at the rear of the vehicle refers to mechanical interlocking using plastically formed-in features. Source: slideshare.net (2015). Used with permission of LinkedIn.

expensive-to-fabricate high-performance materials need to be used only where they are absolutely required and provide real benefit.

Among the three fundamental processes for joining materials (and parts thereof), mechanical joining (using either mechanical fasteners or interlocking design features[2]), chemical adhesive bonding, and welding (including brazing

2 Mechanical joining strictly employs mechanical forces to keep parts together, with one part physically interfering (as opposed to chemically or metallurgically bonding) with another to prevent movement in unwanted directions of translation or rotation (i.e. degrees-of-freedom). Supplemental devices (e.g. rivets or bolts) can be employed to accomplish *mechanical fastening*, or design features integral to mating parts can be employed (e.g. dovetails and grooves or mortise and tenon joints) to accomplish *integral mechanical attachment*.

and soldering[3]), mechanical joining and adhesive bonding are exclusively used to join the dissimilar types. Welding is restricted in its use to the joining of materials within a single class, e.g. one thermoplastic polymer to another thermoplastic polymer (in so-called *plastic welding*), perhaps one ceramic to another ceramic, and, most commonly, one metal to another metal. This said, welding dissimilar metals and alloys often poses special challenges and can lead to problems.

21.2 The Need for Welding Dissimilar Metals and Alloys

Dissimilar combinations of metals and/or alloys are often required to be joined, with welding, as usual for metal joining, being the preferred method in most cases. Six general applications requiring dissimilar metal and/or alloy welding are as follows:

1. Enhancing resistance to wet corrosion in one location versus another, e.g. on the corrosion environment side of a vessel, pipe, part, or structure
2. Enhancing resistance to dry corrosion by oxidation scaling or sulfurization in one location versus another, e.g. on the elevated temperature atmosphere or sulfur-rich side of a vessel, pipe, part, or structure
3. Enhancing resistance to adhesive or abrasive wear in one location versus another, e.g. conveyors, sluices, rock, or ore crushers.
4. Enhancing electrical or thermal conductivity or insulation/resistivity in one location versus another, e.g. at an external surface, interior surface, core, or in a laminate
5. Reducing weight locally by employing lower density, but less strong or stiff, metal or alloy
6. Enhancing resistance to ballistic penetration, e.g. in multilayer armor.

The choice of welding for joining seems logical, as obtaining material continuity should be easy given that all metals (and metallic alloys) are metallically bonded, with the atoms of one metal species readily bonding, in all cases, to the atoms of another metal species. Although this is true, there are, nevertheless, challenges. These challenges, with concomitant problems, stem from incompatibilities between materials, including (i) chemical incompatibility, (ii) mechanical incompatibility, or (iii) thermal incompatibility.

21.3 Chemical Incompatibility

Problems from chemical incompatibility during the fusion welding of metals or alloys manifest themselves in one of four ways, two of which are usually related.

3 Brazing and soldering are sub-types of welding in which a molten filler is used to join base material joint elements without causing any melting of those base materials. The low-melting filler is used between closely-spaced mating surfaces, with wetting of the base material surfaces and spreading by capillary flow. Brazing employs filler that melts above 450 °C/840 °F, while solders employ lower-melting fillers. Both brazing and soldering can be used to join metals to metals, some ceramics to ceramics, and some metals to ceramics, while some In-based solders can be used to join metals to glass and minerals (e.g. quartz) and glass-to-glass as well.

These four ways are (i) dilution effects, (ii) macrosegregation from incomplete mixing, (iii) immiscibility, and (iv) embrittlement by liquid metals (i.e. liquid metal embrittlement [LME]). Let us consider each.

Dilution occurs when a fusion weld is made between like or dislike base metals or alloys (and, perhaps, a backing strip) using a filler metal or an alloy with still a different composition, such that the final fusion zone (FZ) exhibits a composition that lies anywhere between these materials. The final composition and corresponding microstructure and properties of the fusion zone will be determined by the dilution level. The phenomenon is thoroughly discussed in Section 20.3, so will not be further discussed here.

Suffice to say that the problems that can arise from the dilution of a weld include, in an approximate descending order of frequency of occurrence, the following:

- Formation of brittle, as-quenched martensite in the FZ because of increased hardenability of the FZ by dilution by base material(s), with possible quench cracking or delayed hydrogen-induced cold cracking
- Loss of corrosion resistance in the weld fusion zone of a joint or in a clad layer or overlay
- Loss of wear resistance in a hard-facing overlay
- Hot cracking in the weld FZ because of pickup from a "dirty" base material.[4]

Methods for avoiding problems from dilution are given in Section 20.3.

Macrosegregation during fusion welding of dissimilar metals or alloys is the result of incomplete mixing (Figure 21.3). Potential problems are obvious, i.e. nonuniform, unpredictable, and unrepeatable composition, microstructure, and properties across the weld. Mechanical properties could be degraded, and corrosion resistance would almost certainly be degraded, both in accordance

Figure 21.3 A transverse macrograph showing a Cu-to-steel arc weld showing three features associated with the resulting macrosegregation between these dissimilar metals; i.e. a steel peninsula (at the right in the image), an irregular-shaped Cu-beach (at the left in the image), and a layered weld metal (in the bulk of the fusion zone). Source: Soysal et al. (2016). Used under reciprocal agreement between Wiley and Elsevier for fair use.

4 The risk of picking up tramp elements that can promote hot cracking (e.g. S, P, As, and Se) is the greatest from base materials of unknown composition and or pedigree, as they are old or have been in service for a long time and are being repair welded.

with "the weakest-link" factor (Section 20.1). The only conceivable remedies involve aggressive stirring during fusion welding using (i) ultrasonic vibration, (ii) high-frequency electromagnetic vibration, or, to a lesser extent, (iii) current pulsing for arc or plasma-arc welding processes.

Related to macrosegregation as both a possible cause and effect is *immiscibility*. Miscibility is defined as the property of substances to mix in all proportions (i.e. fully dissolve) forming a homogeneous solution. Immiscibility is the opposite. The term is most often applied to liquids but is equally applicable to gases and solids. In fact, immiscibility of liquids almost inevitably results in immiscible solids following solidification.

Known binary systems of metals that exhibit immiscible liquids include, but are not limited to (as there are others among far more infrequently used metals), in alphabetical order, Al–Pb, Cu–Co, Cu–Pb, Pb–Ag, and Pb–Zn. Fortunately, these are rarely, if ever, encountered in welding, as there are no remedies. Immiscibility is mentioned here for completeness.[5]

The final form of chemical incompatibility that can have profound consequences when fusion welding is *LME*, also known as *liquid-metal-induced embrittlement*. What makes LME especially troubling is that it most often leads to a drastic loss of tensile ductility, often leading to brittle fracture, in certain highly ductile metals. Although LME most precisely involves the loss of ductility in normally ductile metals or alloys when stressed in contact with liquid metal, similar metal embrittlement effects can be observed even in the solid state, when one of the metals is brought close to its melting point. This phenomenon is properly known as *solid metal embrittlement*.

Although several theories have been proposed for LME (as listed below), the general rules for the possible occurrence of the phenomenon are (i) low mutual solubility between the liquid and the solid metals and (ii) absence of intermetallic compound formation between the solid–liquid couple. An intriguing aspect of LME is its apparent specificity; it is often being the case that certain liquid metals embrittle certain solid metals, but not others.[6] Table 21.1 provides a fairly comprehensive matrix of the occurrence of LME.

The major theories proposed for LME are as follows:

- o Absorption of the liquid metal on the solid metal induces dissolution and inward diffusion, which can lead to crack nucleation and propagation under tensile stress (Google search "dissolution-diffusion model of LME" by Robertson or by Glicksman)
- o Adsorption of the liquid metal atoms at a crack tip weakens interatomic bonds and causes the crack to propagate (Google search "brittle fracture theory of LME" by Stoloff and Johnson or by Wetswood and Kamdar).

---

5 One use of immiscibility with industrial importance is that liquid zinc and liquid silver are immiscible in liquid lead, whereas silver is miscible in zinc. This leads to the Parkes process, an example of liquid–liquid extraction, whereby lead containing any amount of silver is melted with zinc. The silver migrates to the zinc, which is skimmed off the top of the two-phase liquid, and the zinc is boiled away, leaving nearly pure silver.

6 As two examples, liquid gallium embrittles solid aluminum but not magnesium, and liquid mercury embrittles solid zinc but not cadmium.

Table 21.1 Metal combinations prone to liquid metal embrittlement (liquid at across the top; solid metal/alloy down the side).

	Hg	Ga	Cd	Zn	Sn	Pb	Bi	Li	Na	Cs	In	Cu	Fe
Aluminum	X	X		X	X				X		X		
Bismuth	X												
Cadmium		X			X					X	X		
Cobalt												X	
Copper	X						X	X	X		X		
Iron	X	X	X	X				X			X	X	
Magnesium				X					X				
Silver	X	X											
Tin	X	X											
Titanium												X	X
Zinc	X	X			X	X							

Source: Rostoker et al. (1965) and Carlston (1965), with additions by R.W. Messler, Jr. Used with permission of Naval Research Laboratory (as Reinhold Publishing Corporation ceased publishing in 1995).

- o Diffusion–penetration of liquid metal atoms to nucleate cracks that grow under tensile stress to cause fracture (Google search "diffusion-penetration model of LME" by Gordon).
- o Adsorption of the liquid metal leads to weakening of atomic bonds in the solid metal, nucleating dislocations that move under stress, pile up, and work-harden the solid metal (Google search "ductile failure model of LME" by Lynch or by Popovich).

During fusion welding, one metal in a pair of metals prone to LME is liquefied by melting, whereas the other is solid. Some especially common situations prone to LME during fusion welding are, in an approximate descending order of frequency of encounter, as follows:

- o Steels (especially austenitic stainless steels) embrittled by zinc (usually from Zn galvanized coatings) at temperatures above 400 °C/~700 °F, greatly exacerbated by stresses.
- o Mild, carbon steels embrittled by mercury (Hg) in stressed or unstressed situations at low temperatures in the presence of local stress risers. (Solute additions of Ni, Si, Co, or Al increase susceptibility to Hg embrittlement.)
- o Ni alloys can suffer LME by Zn or Hg, but to a lesser degree than carbon and austenitic stainless steels, with most to least susceptible being Monel 400, Monel K500, Alloy 624, Alloy X750, Alloy 718, Alloy 600, pure Ni (Nickel 200), Alloy 825, and Alloy 800
- o Embrittlement of Co alloys and Fe alloys by liquid Cu (known as *copper contamination cracking*), with austenitic stainless steels being especially prone, even when a weld is made through a scuff from copper (e.g. a chill block)
- o Ti and Ti alloys by Fe, including heavy scuffs from an iron-containing (e.g. steel) tool

- Hg embrittlement of Al alloys, usually from mercury spills attacking welded joints with unrelieved residual stresses.

Prevention is the only solution to LME during fusion welding.

21.4 Mechanical Incompatibility

Problems from mechanical incompatibility tend to arise, if they arise, in one of the following situations: (i) lack of weld metal strain accommodation under high thermal shrinkage and/or contraction stresses or (ii) lack of a shock-absorbing buffer layer under a wear-resistant overlay (i.e. hard facing).

Situations can arise during the solidification of a fusion weld and, especially, during post-weld cooling in the heat-affected zone (HAZ) when the tensile stresses resulting from volumetric shrinkage in the FZ (see Section 4.3) and thermal contraction in the HAZ (see Section 4.4) exert high enough stresses on weld metal to cause cracking or fracture (see *transverse cracking* in Section 6.3). Other types of cracking can also occur from thermal contraction stresses or residual stresses in the HAZ during post-weld heat treatment (PWHT; see Chapter 17). In all of these situations, were the weld metal able to accommodate the strain induced by the operative stresses, cracking and possible fracture could be avoided. From this standpoint, a problem arose from the mechanical incompatibility of the weld metal for the situation that could arise from shrinkage and contraction stresses in thick sections, say, or in high-restrained joints *or* during expected PWHT. In each case, the use of an undermatched filler material in terms of yield strength compared to the base material(s) might have helped.

During the application of a hard, wear-resisting material on wear-prone areas of a part's surface (e.g. railroad switch points, bulldozer blades, rock crusher hammers or cones, and sugar cane crusher rolls), it is often desirable to first apply *buffer layers* to the surface (see Section 13.1). Buffer layers provide one or both of the two benefits: (i) they help prevent dilution of the hard-facing alloy by base material, which may compromise the hard-facing material's hardness (if base material pick-up would soften the hard-facing alloy, e.g. Cr carbide on a low-C mild steel snowplow blade) or which might render the hard-facing alloy unacceptably brittle, e.g. Cr carbide over a high-C gray cast iron cane crusher roll) and/or (ii) they may provide protection of a brittle overlay to shock loads (e.g. rock or ore crusher hammers or cones). From this standpoint, problems can arise from the mechanical incompatibility of a brittle overlay on a base material that is either significantly softer or itself unacceptably brittle.

21.5 Thermal Incompatibility

A too-often overlooked or underappreciated problem that can occur when dissimilar metals or alloys are welded is thermal incompatibility in the form of *thermal expansion mismatch-induced stress.* Although this problem is of far greater concern when inherently brittle materials, such as ceramics or glasses, are joined

Table 21.2 Coefficient of thermal expansion (α_l) and modulus of elasticity (E) for common metals and engineering alloys.

	α_l		E	
	ppm/°C	ppm/°F	10^6 psi	GPa
Al, pure	23.6	13.1	10.0	69
2xxx Al alloys	22.3–23.0	12.1–12.8	10.2	70.3
5xxx Al alloys	23.4–23.8	13.0–13.3	10.2	70.3
6xxx Al alloys	23.0–23.4	12.8–13.0	10.2	70.3
Cast alloys	19.0–20.4	10.0–11.4	10.5–10.9	73–75
Be, pure	21.5	13.0	42	290
Be–Cu alloy	17.8	10.0	18.1–18.9	125–130
Co, pure	13.8	7.7	30.3	209
Co alloys	9.8–13.0	5.1–6.9	30.4–31.9	210–220
Cu, pure	16.5	9.2	16.0	110
Brass, red	18.7–19.1	10.4–11.1	17.0	117
Brass, yellow	19.9–20.3	11.1–11.3	16.0	110
Phosphor bronze	18.2–18.4	10.1–10.3	16.8	116
Mn bronze	21.2	11.8	16.0	110
Cupro-nickel	16.2	9.0	17.4–21.7	120–150
Fe, pure	11.7	6.5	28.5	197
0.1–0.6 C-steels	12.0–11.0	6.5–6.1	29.3–29.5	202–203
Low-alloy steels	11–14	5.9–7.5	29.0–29.6	200–204
Cast iron, gray	10.5	5.7	18.9	130
Austenitic SS	16.7–17.3	9.3–9.6	28.3	195
Ferritic SS	9.3–11.0	5.2–6.1	28.3–29.0	195–200
Martensitic SS	10.1–12.0	5.6–6.6	27.6–30.4	190–210
Mg, pure	25.2	14.1	6.6	45.5
Mg alloys	25.2–26.0	14.1–14.6	5.8–6.4	40–44
Mo, pure	5.0	3.0	40	276
Ni, pure (nickel 200)	13.3	7.4	31.0	214
Ni–Cu Monels	14.0	7.8	26.0	179
Inconels	11.5	6.4	31.0	214
Ti, pure (CP-Ti)	8.4	4.7	15.5	103
Ti alloys	7.7–10.0	4.3–5.7	15.2–17.4	105–120
W, pure	4.6	2.7	58.0–59.5	400–410

to one another or to a dissimilar material, most often a metal, the potential for a problem cannot be ignored when dissimilar metals are welded.

The underlying issue is mismatched coefficients of thermal expansion (CTE), as these can lead to (i) warpage or distortion in a weld or weldment that is free to move, (ii) residual stresses in a weld or weldment that is restrained, or, in the worst case, and (iii) cracking and, ultimately, fracture in a restrained weld or weldment. Without going into details of the required analysis, the level of thermally induced stress depends directly on the difference in CTE ($\Delta\alpha$) between materials in contact and the modulus (E) of the stiffer material, and the temperature change (ΔT) or gradient over a length $\Delta T/L_0$). Distortion in free-to-move structures and residual stresses in constrained structures is more likely for metals or alloys with low stiffness E, as these can yield. Cracking and fracture are more likely in metals and alloys with a high stiffness E, as these tend to be more brittle. Although no hard-and-fast rule seems to exist as to how great a mismatch is acceptable, consensus among designers seems to be not more than 25% difference are shown by the regions of intersection between or among circles of the Venn diagram.

Table 21.2 lists the linear CTE (α_l) and modulus of elasticity (E) for the most commonly welded metals and alloys.

Methods for avoiding problems from CTE mismatch between joint materials include

- preheating and controlled slow cooling
- use of one or more transition pieces welded between joint materials having radically different values of CTE, with each transition piece (or joint) stepping CTE by about 25%.

An excellent example of the use of a transition piece to overcome mismatched CTEs is shown in Figure 21.4, in which pure aluminum was fusion welded to a steel "besi" (in the native language of the source) using a transition piece of Al–bronze.

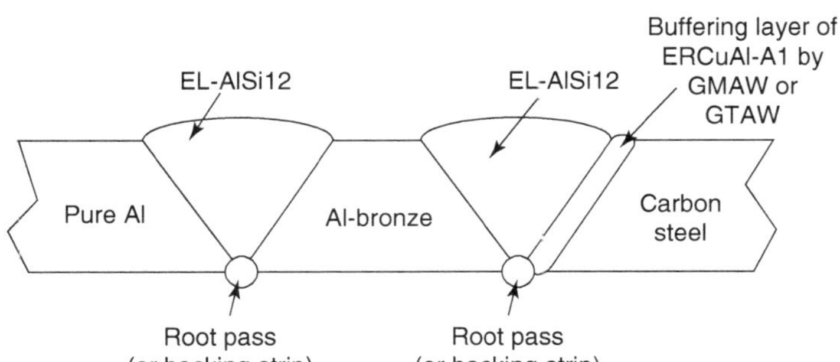

Figure 21.4 Schematic for a dissimilar metal joint made between chemically and thermally incompatible pure Al (with a CTE = 23.6 ppm/°C, 13.1 ppm/°F) and steel, here labeled as "BESI" (with a CTE = 11.5 ppm/°C, 6.3 ppm/°F), using two fusion welds to insert a transition piece of Al–bronze (with CTE = 16.2 ppm/°C, 9.0 ppm/°F). The transition piece reduced a CTE mismatch between Al and steel of over 50% to around half that value using the Al–bronze inset and aided by two soft fusion zones of Al—Si filler. Note dilution of the EL-AlSi12 filler by steel was prevented by buttering with ERCuAl-A1.

Table 21.3 List of problems from dissimilar metal and alloy welding.

Chemical Incompatibility:	
Dilution of WM by BM to adversely affect strength, toughness, or corrosion resistance	Alter joint preparation Butter joint faces
Macrosegregation in WM adversely affecting property uniformity	Employ aggressive U/S or high-frequency EM stirring Consider pulsed current operation
Liquid metal embrittlement and cracking in HAZ	Avoid known LME-prone combinations
Mechanical Incompatibility:	
Transverse cracking in FZ or IG cracking in HAZ during PWHT	Employ ductile strain-accommodating filler Minimize restraint during welding and PWHT
Cracking or spalling of hard-facing overlays under impact or high-contact stress	Employ buffer layers beneath the hard-facing alloy
Thermal Incompatibility:	
Mismatched CTEs lead to distortion in free-to-move welds/weldments and cracking in restrained welds/weldments	Reduce CTE mismatch to under 25% using stepped-CTE transition pieces or inserts

21.6 Troubleshooting Problems with Dissimilar Metal and Alloy Welding

Table 21.3 lists the various problems that can arise during dissimilar metal or alloy welding.

References

Carlston, R.C. (1965). Liquid metal embrittlement. *Naval Research Reviews.*

Rostoker, W., McCaughey, J.M., and Markus, H. (1965). *Embrittlement by Liquid Metals.* New York, NY: Reinhold.

slideshare.net (2015). Joining processes. Topic 7 – joining process welding brazing soldering fastening. www.slideshare.net. Created by Huai123 and posted 05 November 2015. A cropped image of Slide #4/144 found in a presentation on LinkedIn's.

Soysal, T., Kou, S., Tat, D., and Pasang, T. (2016). Macrosegregation in dissimilar-metal fusion welding. *Acta Materialia* 110: 149–160.

Further Reading

Messler, R.W. Jr. (2004a). *Joining of Materials and Structures: From Pragmatic Process to Enabling Technology* (Chapter 15), 1e. Burlington, MA: Elsevier Butterworth-Heinemann.

Messler, R.W. Jr. (2004b). *Principles of Welding: Processes, Physics, Chemistry, and Metallurgy* (Chapter 12). Weinheim/New York, NY: Wiley-VCH Verlag/Wiley.

Closing Thoughts

Welding is extremely important to our modern world and all its conveniences. It helps create buildings in which we work and highways and bridges over which the cars, trucks, and buses it helps assemble move. It allows the practical and sound assembly of the engines that propel the vehicles we drive, the airplanes that whisk us from place to place, the merchant and cruise ships that ply the seas, and the passenger and freight trains that crisscross our land. It helps construct the rigs that locate and extract crude oil and natural gas on land and offshore under the sea, the pipelines that enable the easy transport of those raw materials to the plants that process them into useful fuels and essential chemicals, and the equipment that enables that processing.

When done properly, welding produces sound joints, restores metal lost to wear or corrosion, and overlays surfaces of parts and equipment to extend their life in harsh service environments, and it does so quickly, efficiently, and safely. When done wrong, welding produces problems that cause delays, erode productivity, produce worthless scrap, and cost money. Worse, done improperly, welding can cost lives.

If one performs welding and produces welds and weldments, one will inevitably encounter problems. The demands for sound understanding of the materials being welded and for stringent control of the process almost make the occurrence of problems inevitable. However, as the American televangelist, pastor, motivational speaker, and author Robert H. Schuller (1929–2015) once said: "Problems are not stop signs, they are guidelines." This said, everyone, in general, and engineers, in particular, must heed what Albert Einstein (1879–1955) once said about problems: "We cannot solve our problems with the same thinking we used when we created them."

It is the goal of this book to provide you with guidelines for solving problems with welding, welds, and weldments, perhaps by changing your thinking. It was my goal as its author to hopefully entertain you while I informed you.

I will end with a quote from Edward Kennedy "Duke" Ellington (1899–1974), the American composer, pianist, and bandleader of a jazz orchestra from 1923 until his passing: "A problem is a chance for you to do your best."

In writing this book, I tried to do my best. I hope I succeeded.

A Practical Guide to Welding Solutions: Overcoming Technical and Material-Specific Issues,
First Edition. Robert W. Messler, Jr.

Index

A Practical Guide to Welding Solutions: Overcoming Technical and Material-Specific Issues,
First Edition. Robert W. Messler, Jr.

www.ingramcontent.com/pod-product-compliance
Ingram Content Group UK Ltd.
Pitfield, Milton Keynes, MK11 3LW, UK
UKHW050943290926
15470UKWH00005B/1616

9 783527 345434